Urban and Periurban Forest Diversity and Ecosystem Services

Special Issue Editors

Francisco Escobedo
Stephen John Livesley
Justin Morgenroth

MDPI

Special Issue Editors

Francisco Escobedo
Universidad del Rosario
Colombia

Stephen John Livesley
The University of Melbourne
Australia

Justin Morgenroth
University of Canterbury
New Zealand

Editorial Office
MDPI AG
St. Alban-Anlage 66
Basel, Switzerland

This edition is a reprint of the Special Issue published online in the open access journal *Forests* (ISSN 1999-4907) in 2016 (available at: http://www.mdpi.com/journal/forests/special_issues/urban_forests).

For citation purposes, cite each article independently as indicated on the article page online and as indicated below:

Author 1; Author 2; Author 3 etc. Article title. *Journal Name*. **Year**. Article number/page range.

ISBN 978-3-03842-410-9 (Pbk)
ISBN 978-3-03842-411-6 (PDF)

Table of Contents

Section 1: Urban Forest Structure and Biodiversity

Section 2: Socioecological Systems; Non-Market Valuation: Perceptions and Attitudes

Section 3: Ecosystem Service Tradeoffs; Climate Change

Section 4: Management and Planning

About the Guest Editors

Francisco J Escobedo is a Professor at the Universidad del Rosario in Bogotá, Colombia. Prior to this he was an Associate Professor and Extension Specialist at the University of Florida in the United States and a Research Forester with the USDA Forest Service. His research and interests include understanding the dynamics between ecosystems and people, especially how humans directly and indirectly influence forests in both urban and rural settings, as well as nature's benefit and costs to society. His socio-ecological research uses multidisciplinary methods and approaches such as: ecosystem process modelling, policy analysis, forest inventories, urban and forest ecology, geospatial analyses, and socioeconomic valuation.

Stephen Livesley is currently an Associate Professor in the School of Ecosystem and Forest Sciences at The University of Melbourne. He has been involved in natural and managed forest environmental research for over 15 years. Since 2011 Stephen's research and teaching has focused on the role of urban trees and green spaces to provide environmental, social and biodiversity benefits. Stephen is a member of the Green Infrastructure Research Group and the newly formed Australian 'Clean Air and Urban Landscapes' (CAUL) hub.

Justin Morgenroth is a researcher and Senior Lecturer at the New Zealand School of Forestry, University of Canterbury, located in Christchurch, New Zealand. He received a Bachelor of Science degree in Computer Science (University of Western Ontario, Canada), a Master of Forest Conservation (University of Toronto, Canada) and a PhD in Forestry (University of Canterbury, New Zealand). His research interests are focused on the interactions between trees and their surroundings, with a particular interest in urban environments. He has authored dozens of scientific and general interest articles across a range of urban forestry subjects, including urban forest planning and management; measurement and monitoring; and green infrastructure. He is an associate editor for *Urban Forestry & Urban Greening* and also *Arboriculture & Urban Forestry* and is the current chair of the International Society of Arboriculture's Science and Research Committee.

Preface to "Urban and Periurban Forest Diversity and Ecosystem Services"

1. Introduction

The term "ecosystem service" is used prodigiously with respect to current urban ecosystems and in urban forest research (Haase et al., 2014). These studies often regard ecosystem functions and benefits in an abstract fashion, without connecting with, or considering fully, the human–environment interactions that pervade our urban landscapes. Furthermore, many of these studies frequently refer to ecosystem services without attempting to quantify those services or qualify what enhances—or detracts—from the level of benefits to society. These two issues may be holding back the use of urban trees as a strategy that can help tackle many of the "wicked problems" that urban society faces, such as climate change, community welfare and wellbeing, and biodiversity conservation and management. The aim of this Special Issue is to help fill this void in the current research by focusing on the diversity of urban forests and the benefits (e.g., cultural, environmental, personal, and economic) that different societies across the world gain through the biodiversity and ecological functions that urban and peri-urban forests provide.

This Special Issue was conceptualised by Professor Francisco Escobedo as a means to progress our global discussion of urban forest function using a more social-ecological approach. His proposal for this Special Issue coincided with preparations for the 2nd International Conference on Urban Tree Diversity, held in Melbourne in February 2016, and co-organised by Stephen Livesley and Justin Morgenroth amongst others. Despite the simple title, this conference aimed to provide a research and management platform to discuss the many "diverse" services and functions that urban trees provide us and our urban landscape. As such, this Special Issue was promoted at that conference, through our combined research networks and by MDPI itself. We have been able to bring together a large number of international studies covering a wide spectrum of ecosystem services, ecological functions that urban trees and urban forests can provide—from supporting faunal biodiversity to the diversity of urban forests themselves; from urban forests for carbon sequestration to air quality improvements through particulate deposition; from indicators of resilience and health in urban forest planning to socio-economic drivers and inequity in urban forest cover. This Special Issue includes research performed in every continent except Antarctica. These studies originate from the USA, Germany, Canada, Colombia, Sweden, South Korea, Chile, South Africa, Mexico, Italy, and China. We could not have hoped to create a more internationally inclusive and relevant Special Issue, and are very proud to present, as Guest Editors, this collection of urban forest studies.

2. Human–Tree Interactions in an Urbanised Society

Östberg and Kleinschmit (Östberg et al., 2016) describe the role of the media in reporting and maintaining local and national interest in the removal of a significant urban tree in Stockholm, Sweden. This case study highlights the important role of "champions", be they from the media or private individuals making shrewd use of the media. Changes to urban forests can lead to passionate protest and demand, and this can come from any one of the many stakeholders concerned with the vegetation environment of their local street, neighbourhood, or city. A good way to minimise confusion, anger, and protest is to provide information in advance of tree-related changes, to educate communities and stakeholders on the issues at hand, and to consult and involve them in the high-level planning and decision-making for that change. This has been successfully demonstrated by several cities across the world, including Melbourne, through their exhaustive round of neighbourhood workshops to communicate, educate, consult, and co-plan the future of their urban forests.

The changing face of urban forest management and consultation is further investigated by Barron et al. (2016) in their contribution to the Special Issue that looks at the disconnect between what we as urban forest managers or researchers measure and monitor and what we actually expect or want urban forests to deliver. With greater management and public demands from our urban forests, comes the need for clear indicators of performance that can track progress and the success or failure of initiatives and interventions. Barron et al. (2016) tackle this issue using the Delphi method to rank issues and indicators that international urban forest managers or researchers regarded as important, followed by targeted interviews with Canadian urban forest professionals. The study noted that many indicators regarded as being of "high importance" are not being measured in many municipal urban forestry programs, particularly social indicators of human health and well-being. This is a real concern for managers seeking (or being asked) to track the efficacy of funded urban forest programs to deliver the socio-ecological and ecosystem service benefits they claim and expect.

3. Urban Tree Inequity

It is now widely recognised that there can be considerable inequity within urban society as to access to green space or urban forests and tree cover itself (Schwarz et al., 2015). In this Special Issue, Nesbitt and Meitner (2016) assess the correlation between urban vegetation cover in Portland, Oregon, and socio-economic variables collected in the census of the United States. Neighbourhoods of higher population density, lower average household income, and fewer residents identifying as white or Asian had less vegetation cover. This study provides more evidence that green inequity is a very real phenomenon, and future research needs to tackle what impacts this inequity may be having upon physical and mental health and well-being. The approach presented in this study provides a guide that can be used to identify and target areas that need urban forest intervention to address the stark inequity in urban greening. Escobedo et al. (2016) add a temporal level of understanding to this issue of green inequity through a study of spatial and temporal dynamics in Santiago's urban forest over 12 years. Average tree mortality and overall tree basal area remained stable across the city, whereas tree canopy cover and basal area increased in the more affluent suburbs, whilst decreasing in the intermediate- and low-income suburbs. The study further reinforces the observation that green inequity is a universal issue, and a contemporary issue as tree canopy cover is changing now and progressing towards increasingly negative outcomes.

Tackling green inequity will mean conserving the trees and green space that exist, whilst adding new tree plantings and hopefully new green spaces in the areas of our cities that need it most. Widney et al. (2016) examine the growth and survival of urban tree planting initiatives in three U.S. cities (Detroit, Indianapolis, and Philadelphia) to model the expected ecosystem service benefits 5 and 10 years in the future. The news is not good, because the current (and accepted) levels of planted tree mortality in these three cities means that these new tree planting initiatives cannot keep up with concurrent mortality and the loss of the larger "legacy" trees already in the urban landscape. Widney et al. [8] make a plea for improved and early intervention measures to raise tree survival rates in those crucial establishment years, so that the social, ecological, and ecosystem service benefits these trees were planted to maintain, if not increase, can be realised.

4. Climate Change Mitigation through Carbon Sequestration in Our Own Neighbourhoods

Mitigation of global climate change may not be the most recognised function that urban forests can provide for society at a global scale, but there is great regional and local interest in the carbon sequestration potential of urban vegetation systems both above- and below-ground. This is probably because society needs more information so as to become more pro-active and empowered as to how green space and vegetation in "their" landscape can help in some way. In Colombia, Clerici et al. (2016) developed a cost-effective method combining high-resolution, remotely sensed imagery classification with ground-truthed plot data to estimate and monitor the above-ground tree biomass and carbon stocks in peri-urban Andean forests. In China, Lv et al. (2016) studied above- and below-ground carbon stocks in more than 200 plots and surmised that soil carbon increases in urban green spaces have sequestered an additional 25% on top of that stored above-ground in the existing or planted urban forests of the Harbon City region. In a similar study of South Korean cities that have developed rapidly in recent decades, Yoon et al. (2016) were able to estimate soil carbon density in a range of urban green space and forest types and then scale up to make whole-of-city estimates for Seoul, Daegu, and Daejeon.

5. Urban Biodiversity: The Trees Themselves and the Fauna Habitat They Provide

Interest in maintaining and even enhancing biodiversity within urban landscapes is increasing, not only for the inherent value of biodiversity conservation itself, but also because of the tangible societal benefits (e.g., environmental awareness, and the mental health and well-being) realised from viewing and interacting with biodiversity. MacGregor-Fors et al. (2016) report on an extensive city-wide study of fauna and flora biodiversity in Mexico covering ten taxonomic groups in a very little studied region of the world. They are able to relate species richness to key size and location traits of the urban green space and forests that they measured. A common and passionate debate that runs throughout urban biodiversity research relates to the use of exotic, native, or indigenous plant species as the cornerstone of faunal biodiversity habitats (Sjöman et al., 2016). In this Special Issue, Shackleton (2016) contributes to this debate with a simple but intriguing study of over 1200 street trees in Grahamstown, Eastern Cape, South Africa. Shackleton (2016) is able to demonstrate the importance of native trees for bird species richness and abundance. However, at the same time, exotic trees are important for supporting parasitic mistletoes that provide interesting habitats for invertebrates in their own right, and as such, foraging resources to insectivorous animals. This study adds a layer of complexity to the debate of urban forests being "novel ecosystems" and reiterates that the native-and-exotic tree debate in novel urban landscapes is far from black and white. Nitoslawski and Duinker (2016) look at the diversity of urban forests themselves, and again through a native and exotic lens. By assessing the impacts of sub-division development on the tree species composition of urban forests in Halifax and London in Canada, they are able to determine whether the pre-urban landscape (woodlands or agriculture) lead to differences in urban tree diversity following urbanization. In both cities, regardless of the previous landscape, the newer neighbourhoods had greater tree species richness and evenness and are characterised by substantially more native tree species. This study provides hope that these newer suburbs will provide high quality, native tree habitats to support faunal biodiversity, albeit highly fragmented habitats on small building lots, interspersed with non-native trees.

6. Summary and Future Directions

We are pleased to present this Special Issue and believe that many of the studies from across the world will make a lasting contribution to raising the recognition of the ecological, environmental and socio-economic value of trees to our towns and cities. Urban trees play a vital role in maintaining and enhancing the resilience and integrity of many social, cultural, and ecosystem functions. There is a real need to recognise and tackle the issues of "green inequity" or "tree inequity" in the towns and cities of all countries. Without accepting the need for action, we cannot effectively use urban trees as a single mechanism to provide greater social, ecological, and ecosystem service benefits in urban landscapes while minimising the ecosystem disservices. Urban tree planting initiatives should not contribute to the growing divide between the haves and have-nots in modern urban society. For tree planting to provide the greatest and most cost-effective ecosystem service benefit, the first areas to be planted should be those with the least green space or tree canopy cover (Norton et al., 2015). If this can be done, it will provide a great opportunity for urban forest researchers to concurrently monitor and measure the gradual, long-term delivery of benefits that increased urban tree cover, tree diversity, and tree health can provide. Measuring the relevant social and ecosystem indicators will of course be essential to evaluating this and the success of separate municipal urban forest programs.

Engaging the urban population with greenery and nature is a must and can indeed improve our awareness, appreciation, and willingness to tackle all our pressing environmental issues, be they urban, rural, local, national, or global. Soon, the majority of the world's population will be urban residents and they will have a profound relationship with the trees around them, providing cues for memories and a "sense of place", and invoking emotions that these trees are perceived as active participants in urban life (Pearce et al., 2015). As such, the role of a vibrant, diverse, and healthy urban forest cannot be underestimated. Several studies point to the important role of diversity in tree populations, as well as the positive role that urban forests can have for maintaining fauna biodiversity, creating opportunities for local communities to make a greater connection with nature. There is a real need for greater research on the human health and wellness benefits of urban biodiversity and urban forests themselves. These urban forests may not provide critical habitats for threatened or endangered animals in the same way that more remote or larger nature reserves might. Similarly, these urban forests may sequester only a small fraction of the carbon sequestered by managed plantations and natural forest systems. However, an increasing number of us live an "urban life", so it is this urban forest that provides the best, or most frequent, opportunity for society to interact with nature, to be environmentally aware in the truest sense, to directly observe the impacts of climate change, and to feel empowered that your urban landscape contributes in some small way to a better world.

Francisco Escobedo, Stephen John Livesley and Justin Morgenroth
Guest Editors

Section 1:
Urban Forest Structure and Biodiversity

forests

MDPI

Article

The Urban Environment Can Modify Drought Stress of Small-Leaved Lime (*Tilia cordata* Mill.) and Black Locust (*Robinia pseudoacacia* L.)

Astrid Moser [1],*, Thomas Rötzer [1], Stephan Pauleit [2] and Hans Pretzsch [1]

[1] Forest Growth and Yield Science, School of Life Sciences Weihenstephan, Technische Universität München, Hans-Carl-von-Carlowitz-Platz 2, Freising 85354, Germany; thomas.roetzer@tum.de (T.R.); Hans.Pretzsch@lrz.tum.de (H.P.)

[2] Strategic Landscape Planning and Management, School of Life Sciences Weihenstephan, Technische Universität München, Emil-Ramann-Str. 6, Freising 85354, Germany; pauleit@wzw.tum.de

* Correspondence: astrid.moser@lrz.tum.de; Tel.: +49-8161-71-4719

Academic Editors: Francisco Escobedo, Stephen John Livesley and Justin Morgenroth
Received: 8 January 2016; Accepted: 11 March 2016; Published: 17 March 2016

Abstract: The urban environment characterized by various stresses poses challenges to trees. In particular, water deficits and high temperatures can cause immense drought stress to urban trees, resulting in reduced growth and die-off. Drought-tolerant species are expected to be resilient to these conditions and are therefore advantageous over other, more susceptible species. However, the drought tolerance of urban trees in relation to the specific growth conditions in urban areas remains poorly researched. This study aimed to analyze the annual growth and drought tolerance of two common urban tree species, namely small-leaved lime (*Tilia cordata* Mill. (*T. cordata*)) and black locust (*Robinia pseudoacacia* L. (*R. pseudoacacia*)), in two cities in southern Germany in relation to their urban growing conditions. Marked growth reductions during drought periods and subsequent fast recovery were found for *R. pseudoacacia*, whereas *T. cordata* exhibited continued reduced growth after a drought event, although these results were highly specific to the analyzed city. We further show that individual tree characteristics and environmental conditions significantly influence the growth of urban trees. Canopy openness and other aspects of the surrounding environment (water supply and open surface area of the tree pit), tree size, and tree species significantly affect urban tree growth and can modify the ability of trees to tolerate the drought stress in urban areas. Sustainable tree planting of well adapted tree species to their urban environment ensures healthy trees providing ecosystem services for a high quality of life in cities.

Keywords: drought tolerance; mixed models; standardized precipitation-evapotranspiration index; superposed epoch analysis; urban trees

1. Introduction

Urban trees are of great value to a city: their performance and esthetics are beneficial to the climate and human population. By providing ecosystem services such as evaporative cooling [1] and shading [2], trees in an urban landscape are able to ameliorate negative effects of urban climates and climate change by reducing irradiances, and surface and air temperatures [3–5]. Furthermore, urban trees store carbon [6], reduce rainwater runoff [7], and filter pollutants [8]. The moderation of microclimates and improvement of environmental conditions by urban trees has been analyzed by several studies [9–13], with results highlighting the importance of trees for cities and the city climate. Therefore, urban forests can also become a key component to the adaptation of cities to climate change [14]. However, healthy and well growing trees provide the greatest benefits.

In addition to the functions and effects of trees on the climate of a city, tree growth and site conditions are closely related to the services that trees can provide. Cities are characterized by varying growth conditions that highly influence the growth, resilience, and mortality of trees [15–17]. Urban environments can be stressful habitats for trees [18], with hindering growth conditions such as restricted water availability [19], restricted soil volume [20], de-icing salt in winter [21], mechanical injury [22], and insect infestation [23]. Climate change with an associated higher frequency of years of extreme weather events [24] will expose urban trees to even more restrictive growth conditions. Drought years can cause high stress levels for trees with excessive evaporative demands [25] and less photosynthesis. Frequent drought years with short time periods in between may result in less time for tree recovery. High water stress can induce a change in tree growth with modified tree allometry [26,27], overall reduced growth [28], and tree die-off [29,30]. Tree species with a higher drought tolerance are expected to be more resistant and are therefore advantageous to altered climates than those with less tolerance. Moreover, the individual tree structure can additionally influence the drought tolerance of a tree, with higher crown volume increasing drought resistance [31] and higher age decreasing growth during drought episodes [32].

The annual growth patterns of urban trees can indicate their growth conditions and provide information regarding the climate influencing individual tree growth [33] and drought tolerance. Dendrochronology describes the study of tree rings with the aim of examining events through time recorded by tree-ring widths [34]; it can be used to interpret urban ecosystem dynamics and the impact of land use on trees as well as to analyze the climate-growth relationships of trees [33]. Tree-ring analysis of forest trees has been conducted in many studies [29,31,35–37], whereas studies on urban trees are relatively scarce [3,15,33]. Gillner *et al.* [38] and He *et al.* [39] pointed out that the effects of past climates on growth can provide valuable information on tree performance, including performance under climate change, using the growth patterns of urban trees based on dendrochronology. Knowledge regarding the drought tolerance and sensitivity of common urban tree species could assist urban space planners in selecting long-living and healthy trees, thereby providing sustainable ecosystem services, such as evaporative cooling, to mitigate the effects of a changing climate [38]. Dendrochronology allows a retrospective study of how tree species perform within urban climates and provides important information regarding the choice of optimal species for planting at a certain location.

The aims of this study were to analyze the annual growth rates and drought response of two common urban tree species, namely the small-leaved lime (*Tilia cordata* Mill. (*T. cordata*)) and the black locust (*Robinia pseudoacacia* L. (*R. pseudoacacia*)), in two cities in southern Germany with differing climates. Furthermore, the influence of individual tree structure and site conditions (canopy openness, distance to neighboring trees and buildings, and open surface area (OSA)) on annual tree-ring development was assessed. In more specific detail, we focused on the following questions:

- Can the average growth rates of the analyzed tree species be quantified in respect to their growing sites?
- How stable and sensitive is the growth of trees in urban environments?
- How does the urban climate (temperature and precipitation) and environment (light, open surface, neighboring trees, and close buildings) influence tree growth?
- What are the responses of urban tree species with varying drought tolerances to drought years?
- Do the urban environment and individual tree structure modify the drought stress of trees?

2. Materials and Methods

2.1. Site Description and Data Collection

Increment cores were collected in München (48°09′ N, 11°35′ E, 519 m above sea level (a.s.l.)) and Würzburg (49°48′ N, 9°56′ E, 177 m a.s.l.), two major cities in southern Germany. As shown in Figure 1, the climatic characteristics of the two cities differ. The long-term annual precipitation values of München and Würzburg are 959 and 596 mm, respectively [40], whereas the mean annual temperature (1961–1990) of both cities is 9.4 °C [40].

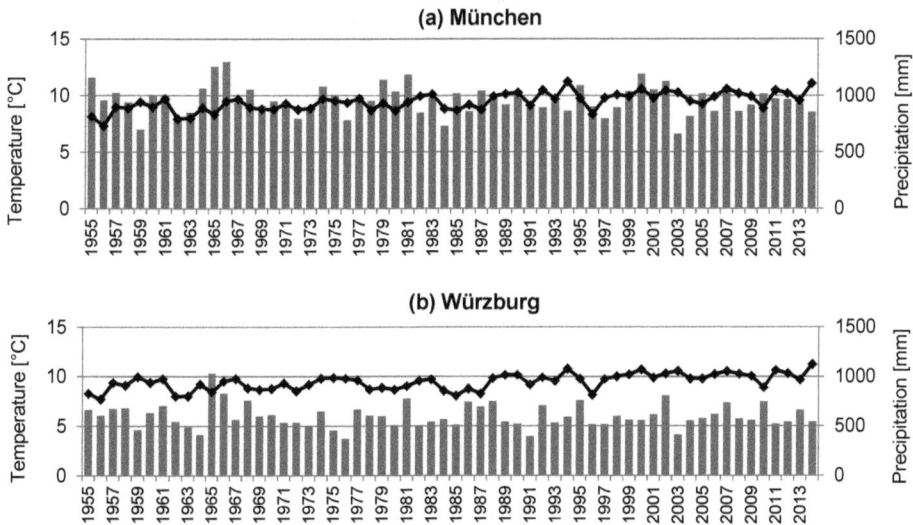

Figure 1. Climate graphs of München (a) and Würzburg (b) from 1955 to 2014 (data source: DWD [40]). Black dots and line represent the mean annual temperature in °C and the gray bars represent the annual precipitation in mm.

For this study, two common urban tree species were selected: *T. cordata* and *R. pseudoacacia*. These two species were selected as they are most relevant common urban tree species in Germany and markedly differ in their ecological features [41]. Whereas *T. cordata* is shade tolerant and moderately drought tolerant [42,43], *R. pseudoacacia* requires a certain amount of light (shade intolerant) and is very tolerant to drought as well as fast growing [43–46].

Tree selection was based on visual impression, that is, damaged, pruned, or low-forking trees were excluded. Increment cores were collected from June 2014 to September 2014. Altogether, increment cores were taken from 68 individual *T. cordata* trees and 62 individual *R. pseudoacacia* trees. All trees were located in street canyons, parks, and public squares and were randomly distributed in both cities. Two cores perpendicular to each other were extracted per tree at a height of 1.3 m, in a northern and eastern direction. Due to the main wind direction (southwest), coring from north and east yields more representative tree ring widths [29].

Moreover, stem diameter (dbh) at a height of 1.3 m, tree height, open surface area of the tree pit (OSA) of all trees, and distance to neighboring trees as well as distance to adjacent buildings were recorded [47]. Hemispheric photographs of the tree crowns were taken using a Nikon Coolpix P5100 camera with a fisheye lens and Mid-OMount. The resulting hemispherical photos were analyzed using WinSCANOPY (Régent Instruments Inc., Ville de Québec, Canada) to derive the canopy openness (percentage of open sky visible, degree of development, equivalent to the sky view factor SVF) of every individual tree [48,49].

2.2. Quantification of Urban Tree Growth in Relation to Growing Site Based on Tree Ring Analysis

All cores were mounted on grooved boards with glue and sanded using progressively finer sand papers. The first sanding was applied to flatten the cores, whereas the subsequent sanding episodes polished the cores for better visualization of the cross-sectional area [34]. The annual tree-ring widths of the cores were measured using a Lintab digital positioning table with a resolution of 1/100 mm [50]. For cross-dating of the time-series, the software package TSAP-Win [51] was used.

Further analyses were carried out in R [52] using package dplR [53]. With dplR, all tree-ring series were indexed using a double detrending process: first, modified negative exponential curves were applied followed by cubic smoothing splines (20 years rigidity, 50% wavelength cutoff). The detrending was conducted to remove low frequency trends, which are age associated [38,54]. The resulting detrended series were averaged using Tukey's biweight robust mean to build chronologies for both species and each city. As a result of detrending, standardized chronologies with a yearly ring width index (RWI) averaging around 1 were obtained. Values smaller than 1 indicate growth below normal, while values greater than 1 indicate growth higher than normal. Further, for chronology building, the autocorrelation of every individual series was removed using autoregressive models with a maximum order of three. This procedure of detrending ensured a removal of all long-term growth trends, thereby obtaining a chronology containing only tree ring variability with climate fluctuations [3,55]. The statistical validity of the chronologies was assessed using the expressed population signal (EPS) for the common period of the time series of all analyzed tree individuals.

2.3. Investigated Variables

Linear regressions and linear mixed models were used to assess the influence of structural and environmental variables on the annual growth rates (response variable). Tested explanatory variables were the analyzed species (*R. pseudoacacia* and *T. cordata*), the sampling city (München and Würzburg), the growing site (park, public square, or street canyon), the stability of growth (Equation (1)), the sensitivity of growth in relation to environment (see below), the dbh, the tree height, the estimated age (Equations (2) and (3)), estimation of vitality by Roloff [56], the OSA (Equation (4)), the distance to neighboring trees (Equation (5)), the distance to adjacent buildings (Equation (6)), and the canopy openness. According to Roloff [56] the vitality was rated from very good (0) to very poor (3) conditions regarding the branching structure of the crown. Stability of growth was computed following Jucker *et al.* [57] by

$$\text{Stability} = \text{Average growth rate per tree/standard deviation of growth} \qquad (1)$$

with values ranging from 0 to 10, whereby low values indicate less stability and high values high stability.

The average growth rates and sensitivity of each tree were derived using the R package dplR. The mean sensitivity describes the year-to-year variability of tree ring data in relation to the previous year ring width [58]. The ages of *T. cordata* trees were calculated by the formula of Lukaszkiewicz and Kosmala [59]:

$$\text{age} = a + \exp^{(b + c \times \frac{\text{dbh}}{100} + d \times \text{tree height})} \qquad (2)$$

where $a = 264.073$, $b = 5.5834$, $c = 0.3397$, and $d = 0.0026$; dbh in cm and tree height in m ([47]).

To estimate the age of *R. pseudoacacia* we applied a species-dependent age factor of 0.996, which was computed by the measurements of Dwyer [60] for honey locust (*Gleditsia triacanthos*):

$$\text{age} = 0.996 \times \text{dbh} \qquad (3)$$

The open surface area (OSA) of the tree pit is calculated by

$$\text{OSA (m}^2) = (\sqrt{(r_N^2 + r_{NE}^2 + \ldots + r_{NW}^2)/8}) \qquad (4)$$

where r_N is the length of the visible open surface in the northern direction and r_{NW} the length of the visible open surface in the northwest direction.

Mean distances to neighboring trees d_t and adjacent buildings d_b were computed as follows:

$$d_t(m) = \sqrt{(((t_{d_N} - r_N)^2 + (t_{d_{NW}} - r_{NW})^2)/8)} \qquad (5)$$

$$d_b(m) = \sqrt{(((b_{d_N} - r_N)^2 + (b_{d_{NW}} - r_{NW})^2)/8)} \qquad (6)$$

where t_{d_N} is the distance to neighboring trees in the northern direction, $t_{d_{NW}}$ is the distance to neighboring trees in the northwest direction, b_{d_N} is the distance to adjacent buildings in the northern direction, $b_{d_{NW}}$ is the distance to adjacent buildings in the northwest direction, r_N is the maximum crown extension in the northern direction, and r_{NW} the maximum crown extension in the northwest direction.

To analyze the effects of climate in terms of the water supply on tree growth, we calculated a monthly, multiscalar climatic drought index, the SPEI [61]. The SPEI uses precipitation and potential evapotranspiration (PET) as input data, whereby PET was calculated according to the Hargreaves approach [62,63]. The time scale was set to 4 months, with a Gaussian kernel to consider the water supply, and a log-logistic distribution was applied [61]. Using the calculated SPEI, the influence of the identified drought years on the growth patterns of urban trees was investigated. A yearly SPEI for 1955–2013 was computed by averaging the monthly index values of the growing season from April to September. Years with an SPEI smaller than -1 were classified as moderate drought years, whereas years with an SPEI smaller than -2 were interpreted as years with severe droughts [61,64].

2.4. Statistical Analyses

The stability of growth was assessed with linear regressions following Pretzsch *et al.* [26]:

$$y = b \times x^a \text{ or } = \ln(a) + b \times \ln(x) \qquad (7)$$

where y is the response variable, x is the explanatory variable, a is the intercept, and b is the slope.

Further, using the R package nlme [65], linear mixed models of the following form were developed to estimate the influence of climate, environment, and tree structure (explanatory variables) on the annual tree growth (response variable) derived by increment cores:

$$\text{Growth rate}_{ij} = \beta_1 \times x_{1ij} + \ldots + \beta_n \times x_{nij} + b_{i1} \times z_{1ij} + \ldots + b_{in} \times z_{nij} + \varepsilon_{ij} \qquad (8)$$

where the growth rate is the response variable for the jth of n_i observations in the ith of M groups or clusters, $\beta_1, \ldots, \beta_n$ are the fixed-effect coefficients, which are identical for all groups, $x_{1ij}, \ldots, x_{nij}$ are the fixed-effect regressors for observation j in group i; the first regressor is usually for the constant, $x_{1ij-1}, b_{i1}, \ldots, b_{in}$ are the random-effect coefficients for group i, $z_{1ij}, \ldots, z_{nij}$ are the random-effect regressors, and ε_{ij} is the error for observation j in group i.

The derived annual growth rates over the past twenty years (1994–2013) for each tree were used as input for the response variable. The individual tree number and the species were set as random effects. The back-calculated dbh, the water supply (measured with SPEI), the distance to neighboring trees, the distance to adjacent buildings, the OSA, the vitality, and the canopy openness were used as fixed effects along with the species (*R. pseudoacacia*, *T. cordata*), city (Würzburg, München), and growing site (park, public square, and street canyon). In addition, all fixed effects were tested with interactions. Nonsignificant terms were gradually removed from the models. The models with significant *p*-values for the fixed effects and the overall lowest Akaike's information criterion (AIC) were chosen as final models.

To investigate the influence of the urban environment and tree structure on growth during low growth episodes and episodes with extraordinarily high growth, another two linear mixed models were calculated. First, we investigated all years with a ring width index of <1 (detrended values averaged around 1). Second, all years with a ring width index of >1 were analyzed in relation to the tree structure and environment as explained earlier.

Extreme drought years, such as 2003, can have a great impact on the growth of trees [31]. The effects of extreme droughts on tree growth patterns may persist several years after the drought. Therefore, we analyzed the influence of single years on the tree growth using superposed epoch analysis (SEA). SEA investigates the significance of a mean tree growth response to certain events (such as droughts) to pre- and post-drought growth periods [53]. According to Lough and Fritts [66], Orwig and Abrams [36], and Gillner *et al.* [38], deviations from the mean ring width index of each core were calculated for the following three periods. The growth of the 5 years prior to the analyzed drought year (pre-drought), of the drought year and of the 5 years after the drought year (post-drought) was averaged to detect significant departures between those superposed epochs. The SEA was computed using the R package dplR [53], using random sets of 11 years from 1000 bootstrapped sets [38,53].

3. Results

3.1. Quantification of Urban Tree Growth in Relation to Growing Site Based on Tree Ring Analysis

The average dbh of the analyzed trees was 44.5 cm in München and 44.3 cm in Würzburg for *R. pseudoacacia* and 34.2 cm in München and 33.1 cm in Würzburg for *T. cordata* with a maximum range between 11 cm and 102.2 cm (Table 1). The tree individuals chosen for coring therefore represented a broad size spectrum of urban trees. The annual growth rates of both species were similar in München and Würzburg, with *R. pseudoacacia* displaying a higher overall growth rate compared with *T. cordata* (3.9 and 4.0 mm·$year^{-1}$ *versus* 3.2 and 2.9 mm·$year^{-1}$, respectively). After crossdating and detrending, four tree ring chronologies were derived. The EPS values of all series varied between 0.85 and 0.90 for the common period of all tree ring series, thereby exceeding the required threshold of 0.85 [67].

Table 1. Statistical characteristics of the tree ring series of *Tilia cordata* and *Robinia pseudoacacia* in München and Würzburg.

	n	dbh min [cm]	dbh Avg [cm]	dbh Max [cm]	Avg Tree Height [m]	Avg OSA [m²]	Avg Age [a]	Growth Rate ± SD [mm·$Year^{-1}$]	Mean Sensitivity	EPS [1]
					Robinia pseudoacacia					
München	30	14.0	44.5	101.9	15.7	102.39	44.3	3.9 ± 1.7	0.33	0.90
Würzburg	32	11.0	44.3	102.2	15.1	109.55	44.1	4.0 ± 1.9	0.34	0.85
					Tilia cordata					
München	37	12.0	34.2	86.7	13.1	146.48	42.6	3.2 ± 1.6	0.38	0.89
Würzburg	30	14.0	33.1	71.5	12.5	123.34	44.0	2.9 ± 1.6	0.39	0.89

[1] Based on the ring width index of tree ring chronologies for the common period, obtained by double detrending of the tree ring series. dbh = stem diameter; OSA = open surface area; EPS = expressed population signal; SD = standard deviation.

All tree ring series showed a similar mean sensitivity of 0.33 to 0.39 and were within the range given by Speer [34] (Table 1). The slightly higher sensitivity of *T. cordata* indicated a higher susceptibility to climatic variables and poor growth conditions [3,38,68,69].

In total, the chronologies of *R. pseudoacacia* covered the period from 1954 to 2013 in München and 1960 to 2013 in Würzburg, whereas the tree ring data of *T. cordata* ranged from 1941 to 2013 in München and 1910 to 2013 in Würzburg (Figure 2). Of the analyzed species and cities, *T. cordata* in Würzburg showed the longest tree ring series covering 103 years. The chronologies of *R. pseudoacacia* of München and Würzburg were both shorter than that of *T. cordata*, covering 59 years in München and 53 years in Würzburg.

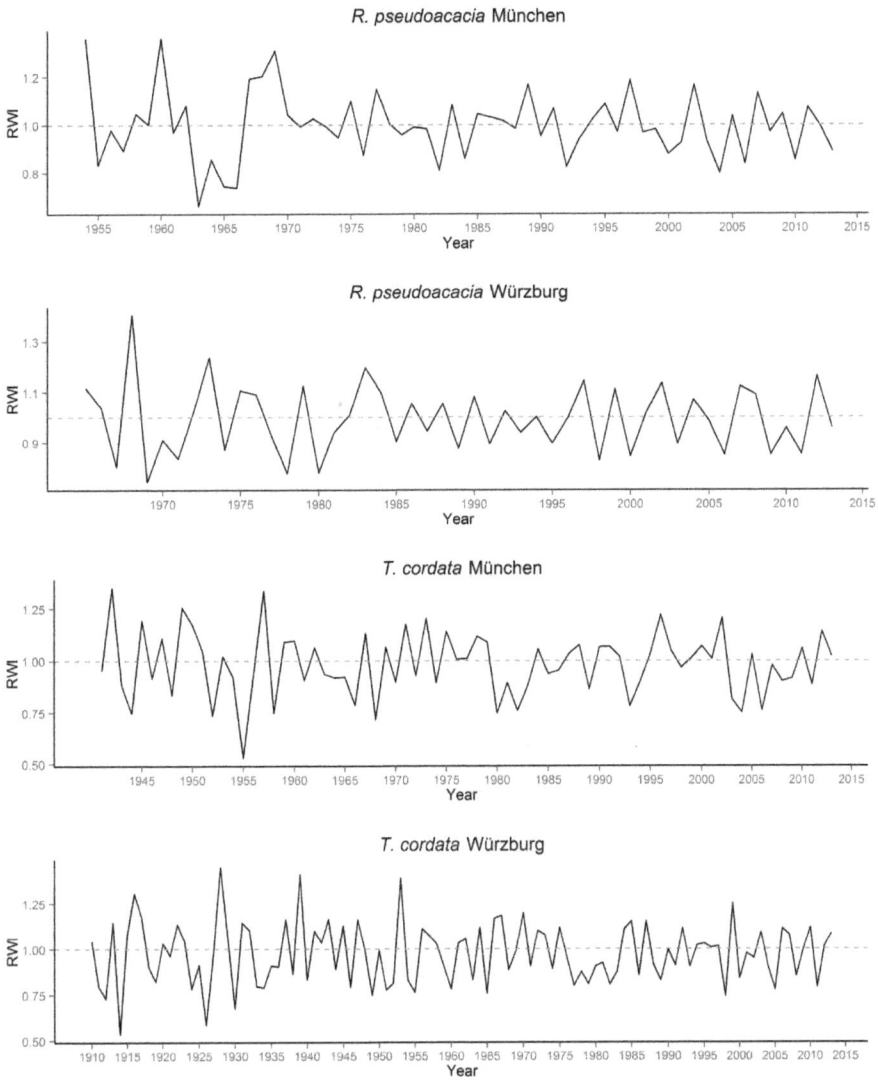

Figure 2. Ring width index of *Robinia pseudoacacia* and *Tilia cordata* in München and Würzburg after double detrending (negative exponential function and 2/3 cubic smoothing spline).

3.2. Stability, Sensitivity, and Modeling of Tree Growth in Relation to Their Environment

To reveal any species-specific reaction patterns and the effect of the sampled cities on the stability of growth, simple linear regressions were applied (Figure 3). With a *p*-value of 0.08, the relationship of stability with city was not significant, whereas there was a significant relationship of species with stability ($p = 0.001$, $r^2 = 0.1$); however, the low coefficient of determination indicated a very weak fit. The stability value between the species was significantly different, with *R. pseudoacacia* showing a higher stability on average as compared to *T. cordata*.

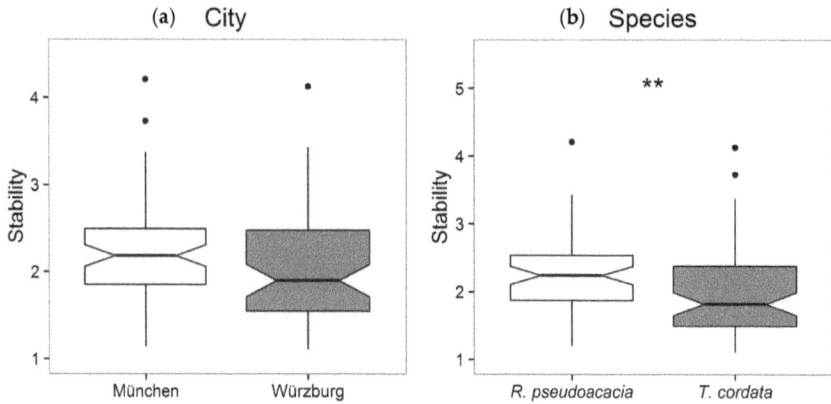

Figure 3. Boxplots showing the relationships of growth stability with city (**a**) and species (**b**). Significances are marked by asterisks.

The regressions of the further analyzed variables tree height, vitality, OSA, distance to adjacent buildings, and trees as well as the canopy openness resulted in non-significant *p*-values. Significant relationships were only found for sensitivity, dbh, and age (Table 2).

Table 2. Results of linear regressions with stability as response variable and the variables sensitivity, dbh, and age as individually tested explanatory variables for the stability of growth of *Tilia cordata* and *Robinia pseudoacacia* in München and Würzburg (equation: response variable = $a + b \times$ explanatory variable). The table lists the regression coefficients (a, b), coefficients of determination (r^2), RSE for bias correction, *F*-values, *p*-values, and SEs of regression coefficients as well as the sample size (n).

Response Variable	Explanatory Variable	n	$a \pm$ SE	$b \pm$ SE	r^2	RSE	F	p
ln(Stability)	ln(Sensitivity)	128	0.14 ± 0.09	-0.54 ± 0.08	0.25	0.28	42.83	<0.001
Stability	Dbh	128	1.68 ± 0.15	0.01 ± 0.003	0.09	0.77	12.85	<0.001
Stability	Age	128	1.74 ± 0.16	0.01 ± 0.003	0.06	0.78	7.77	0.006

Dbh = stem diameter; RSE = residual standard error; SE = standard error.

For dbh and age, the r^2 values obtained were <0.1 with small *F*-values and therefore not suitable for explaining the variance in the stability of growth. In contrast, sensitivity displayed a higher r^2 and *F*-value. The r^2 of stability in relation to sensitivity and age strongly increased when adding the terms "species" and "city" as dummy variables to the regression (Figure 4).

As shown in Table 2 and Figure 4, stability decreased with increasing sensitivity. In contrast, stability increased with increasing age. To derive a final model best fitting the data, "city" and "species" were tested as explanatory variables. Therefore, a regression with both "species" and "city" added to explain stability on the basis of sensitivity was obtained. With an r^2 of 0.29 ($F = 17.37$, $p < 0.01$), the model resulted in the highest r^2 values and could explain most variance in the data of all computed models. The form of the final model was:

$$\ln(\text{Stability}) = 0.18 - 0.48 \times \ln(\text{Sensitivity}) + 0.11 \times \text{"Species"} - 0.09 \times \text{"City"} + \varepsilon \qquad (9)$$

The derived regression illustrates, again, that stability and sensitivity were opposing variables. Further, the analyzed individuals of *T. cordata* showed a higher sensitivity than those of *R. pseudoacacia* and the trees in München had a slightly higher stability than the trees in Würzburg.

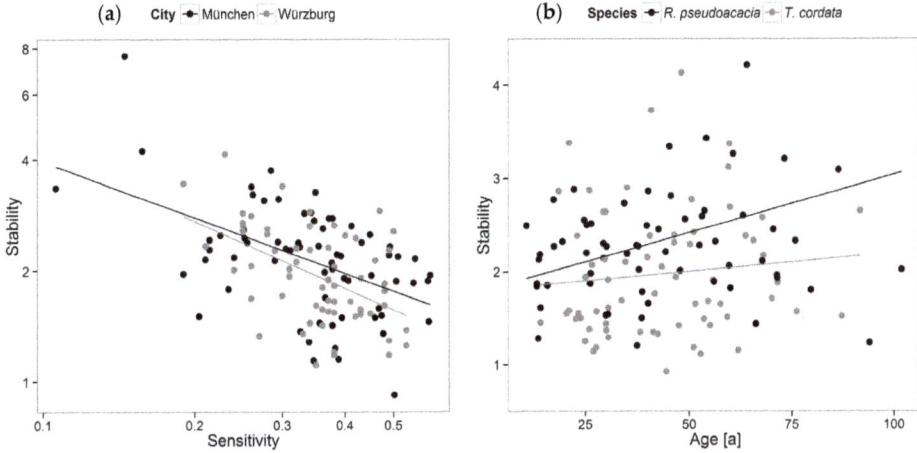

Figure 4. Regressions of stability with sensitivity (**a**) and stability with age (**b**). Highlighted are the regression lines of trees in München and Würzburg (a, ln(Stability) = 0.18 − 0.53 × ln(Sensitivity) − 0.08 × "City" + ε), and *Tilia cordata* as well as *Robinia pseudoacacia* (b, Sensitivity = 1.57 + 0.01 × age + 0.36 × "Species" + ε). For the regression of stability with sensitivity (b), both variables were log transformed.

Using the water supply (measured with SPEI), the growth of *R. pseudoacacia* and *T. cordata* before, during, and after certain drought years was investigated (Figure 5). In München, the SPEI of 2003 was −2.0, whereas the indices for 1998 and 1992 were −1.3 and −1.1, respectively. Several years showed an SPEI close to −1: 2004 (−0.96) and 1976 (−0.93). The drought years of Würzburg as indicated by the SPEI were 2003 (−1.6), 1976 (−1.5), and 1947 (−1.2). Moreover, 2012 (−1.1), 1993 (−1.1), and 1964 (−1.1) showed SPEI values smaller than −1. The SPEI additionally identified years with a high positive index, therefore revealing years with extraordinarily positive growth conditions. In München, those being 1979 (1.2), 1966 (1.5), 1965 (2.0), and 1955 (1.5). Positive SPEI values in Würzburg were found for 1968 (1.4), 1966 (1.2), and 1965 (1.65).

In the following step, linear mixed models of tree growth over the past 20 years in relation to site conditions were developed. Table 3 presents the statistical results of the final model including significant fixed effects and lowest AIC.

(a)

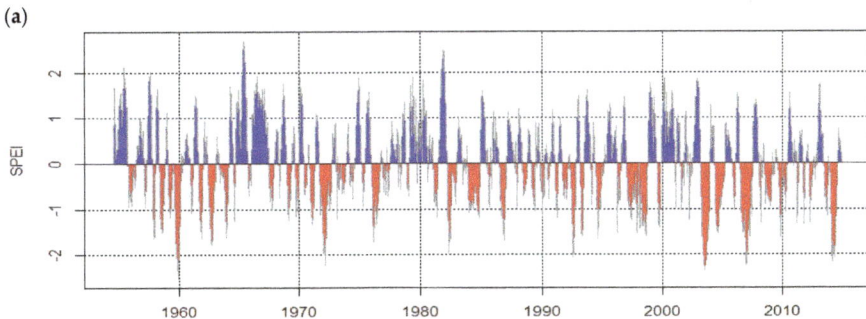

Figure 5. *Cont.*

11

(b)

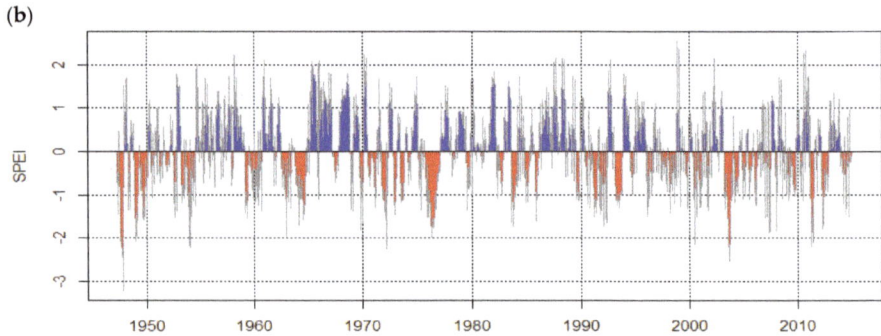

Figure 5. Calculated standard precipitation-evapotranspiration index (SPEI) for München (**a**) from 1955 to 2014 and Würzburg (**b**) from 1947 to 2014 with a time scale of four months. Blue-colored series represents positive SPEI values (>0) of years with a positive climatic water balance, and red-colored series represents negative SPEI (<0) values of a negative climatic water balance.

Table 3. Linear mixed model on the annual growth rate (mm·year^{-1}) during the last 20 years (1994–2013) of all analyzed trees (response variable) with the individual tree code as random effect and a random slope added for the effect of dbh, and fixed effects dbh, canopy openness, OSA, species, and water supply (measured with SPEI) of the form: Growth rate$_{ij}$ = $\beta_1 \times x_{1ij}$ + ... + $\beta_n \times x_{nij} + b_{i1} \times z_{1ij}$ + ... + $b_{in} \times z_{nij} + \varepsilon_{ij}$ where the growth rate is the response variable for the *j*th of n_i observations in the *i*th of *M* groups or clusters, $\beta_1, \ldots, \beta_n$ are the fixed-effect coefficients, which are identical for all groups, $x_{1ij}, \ldots, x_{nij}$ are the fixed-effect regressors for observation *j* in group *i*; the first regressor is usually for the constant, $x_{1ij-1}, b_{i1}, \ldots b_{in}$ are the random-effect coefficients for group *i*, $z_{1ij}, \ldots, z_{nij}$ are the random-effect regressors, and ε_{ij} is the error for observation *j* in group *i*.

	Value ± SE	*p*
Intercept	8.46 ± 1.74	<0.001
Dbh	−0.27 ± 0.03	<0.001
Canopy openness	0.10 ± 0.03	<0.001
OSA	−0.02 ± 0.01	0.03
Species	−0.33 ± 1.90	0.86
Water supply (SPEI)	0.18 ± 0.04	<0.001
Dbh:OSA	0.001 ± 2.1 × 10^{-4}	<0.001
Canopy openness:Species	0.09 ± 0.04	0.03
SD Intercept	5.72	-
SD Dbh	0.19	-
ε	1.18	-

Dbh = stem diameter; OSA = open surface of the ground; SPEI = standardized precipitation-evapotranspiration index; SD = standard deviation; SE = standard error. Levels of species: 2 (*Robinia pseudoacacia* and *Tilia cordata*).

Of all tested variables, the dbh, the canopy openness, the OSA, and the water supply (measured with SPEI) proved to be highly significant variables, whereas the vitality, the city, the distance to neighboring trees and adjacent buildings, and the growing site (parks, public squares, and street canyons) had no marked effect and were therefore removed from the final model (Table 3). The variable species was significantly influencing tree growth especially in interaction with the canopy openness. The model showed that an increasing dbh had a strong negative influence on the growth rate, highlighting the age trend of tree growth. A higher canopy openness, in contrast, was beneficial for the growth of the analyzed urban trees; however a smaller OSA was more positive for tree growth. Furthermore, a higher water supply (measured with SPEI) increased the annual growth rate of the analyzed trees. The variable species was non-significant in the final model; however, this variable displayed a strong influence on tree growth in interactions with the canopy openness. According to the model, *R. pseudoacacia* displayed a higher growth with a higher canopy openness and consequently higher light availability. *T. cordata* was less influenced in its growth by the available light. Those results reflect the individual shade tolerance of both species, since *R. pseudoacacia* is characterized as a light-demanding species, whereas *T. cordata* is very shade-tolerant.

Another significant interaction was dbh with OSA; younger trees with a smaller OSA displayed growth reductions, while bigger, older trees with a larger OSA showed increased growth.

The model highlighted how the growth of the analyzed urban tree species was influenced by the local climate (in terms of the water supply (SPEI) calculated with the potential evapotranspiration), the individual tree structure (dbh), and the surrounding environment (OSA), whereas the variable species was highly significant in relation with another explanatory variable (the canopy openness, representing the light availability). Other variables such as the distance to adjacent buildings and the city were of less importance for growth.

3.3. Growth of Urban Trees Under Drought Stress in Relation to the Tree Structure and Environment

With the obtained SPEI of the drought years, a drought year analysis (SEA) was performed to investigate the different drought strategies of both tree species (Figures 6 and 7).

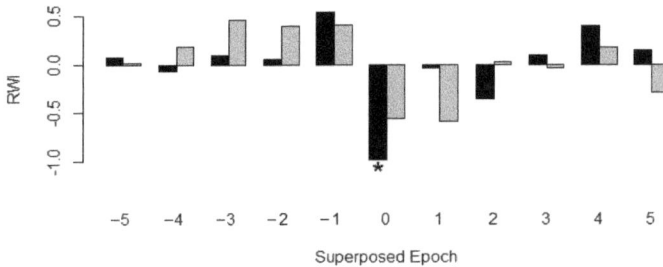

Figure 6. Drought year analysis (superposed epoch analysis) of the ring width index (RWI) during drought years (0), pre-drought (−5 to −1) and after drought (1–5) for *Robinia pseudoacacia* (black) and *Tilia cordata* (gray) in München. Input drought years are 2004, 2003, 1998, 1992, 1982, and 1976 for *R. pseudoacacia* and 2004, 2003, 1998, 1992, 1982, and 1976 for *T. cordata*. Marked columns (asterisk) represent a departure that is greater than would have occurred randomly as determined from 1000 bootstrap simulations at $p < 0.05$.

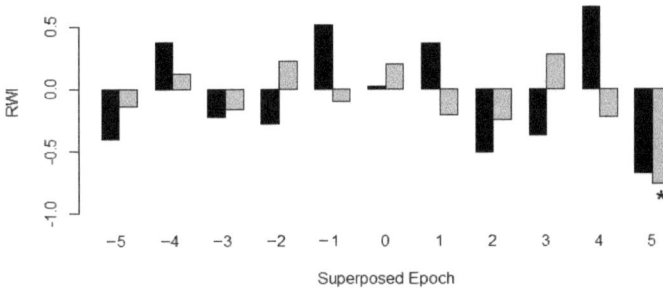

Figure 7. Drought year analysis (superposed epoch analysis) of the ring width index (RWI) during drought years (0), pre-drought (−5 to −1) and after drought (1–5) for *Robinia pseudoacacia* (black) and *Tilia cordata* (gray) in Würzburg. Input drought years are 2012, 2003, 1993, 1976, and 1964 for *R. pseudoacacia* and 2012, 2003, 1976, and 1947 for *T. cordata*. Marked columns (asterisk) represent a departure that is greater than would have occurred randomly as determined from 1000 bootstrap simulations at $p < 0.05$.

Both species showed positive growth rates for most of the five pre-drought years in München and a decrease in growth during the drought year. The decline in growth was significant (asterisk) for *R. pseudoacacia* but not for *T. cordata*, which exhibited the highest growth decrease in the first year after drought. *R. pseudoacacia* could recover very quickly from the drought and reached nearly the former growth during the first year after the drought event. In the third year after a drought, *R. pseudoacacia* again had a positive average ring width index. *T. cordata* had recovered from the drought during the second post-drought year.

The growth patterns of both species in Würzburg were completely different compared with München (Figure 7). Although there were several growth reductions and positive ring width index values during the pre-drought years, during the drought years no decline in growth was observable for both species. *T. cordata* showed a reduction in the average ring width index value during the first and second years following the drought. However, a significant growth decline occurred five years after the analyzed drought years. *R. pseudoacacia* displayed no growth decrease during the first year after the drought but showed a very quick recovery during the fourth year. During the fifth year, *R. pseudoacacia* additionally exhibited a severe growth reduction.

To analyze the influence of the urban environment and tree structure on growth during drought years, we investigated all years with an RWI of <1 (Table 4).

Table 4. Linear mixed model on ring width index (RWI) development of all years with an index < 1 (growth lower than normal, drought years) of all analyzed trees as the response variable with the individual tree code as a random effect and a random slope added for the effect of dbh, and fixed effects water supply (measured with SPEI), canopy openness, analyzed species, growing city, and dbh of the form: $RWI_{ij} = \beta_1 \times x_{1ij} + \ldots + \beta_n \times x_{nij} + b_{i1} \times z_{1ij} + \ldots + b_{in} \times z_{nij} + \varepsilon_{ij}$ where RWI is the response variable for the *j*th of n_i observations in the *i*th of *M* groups or clusters, $\beta_1, \ldots, \beta_n$ are the fixed-effect coefficients, which are identical for all groups, $x_{1ij}, \ldots, x_{nij}$ are the fixed-effect regressors for observation *j* in group *i*; the first regressor is usually for the constant, $x_{1ij-1}, b_{i1}, \ldots b_{in}$ are the random-effect coefficients for group *i*, $z_{1ij}, \ldots, z_{nij}$ are the random-effect regressors, and ε_{ij} is the error for observation *j* in group *i*.

	Value ± SE	*p*
Intercept	0.82 ± 0.04	<0.001
Water supply (SPEI)	0.03 ± 0.01	<0.001
City	-0.01 ± 0.01	0.67
Species	-0.05 ± 0.01	<0.001
Dbh	$-0.001 \pm 7.4 \times 10^{-4}$	0.27
Canopy openness	$-0.002 \pm 7.4 \times 10^{-4}$	0.03
Water supply (SPEI): City	-0.03 ± 0.01	0.009
Canopy openness: Dbh	$0.0001 \pm 1.7 \times 10^{-5}$	0.002
SD Intercept	0.09	-
SD Dbh	0.001	-
ε	0.17	-

RWI = ring width index, dbh = stem diameter; SPEI = standardized precipitation-evapotranspiration index; SD = standard deviation; SE = standard error. Levels of species: 2 (*Robinia pseudoacacia* and *Tilia cordata*), levels of city: 2 (Würzburg and München).

The factors driving growth during drought years (Table 4) were partially altered compared with the model for the overall growth (Table 3). In the previous model on the annual growth rates over the past 20 years, dbh, the canopy openness, the OSA, and the water supply (measured with SPEI) as single terms as well as the interaction terms canopy openness with the species and the OSA with dbh were significantly affecting tree growth. During drought years (Table 4), the surrounding climate in terms of the water supply (SPEI) and the species proved to be highly influential for tree growth. Contrary to the overall growth, the dbh was not significantly affecting the growth patterns. The stand climate, especially in interaction with the city, was of most influence. Würzburg, with far less precipitation than

München resulted in more frequent drought years, and was therefore more inhibiting for tree growth than München. This result was compliant with the overall study design, as the trees in Würzburg exhibit an even higher effect of drought years on growth. Further, the model highlighted the better growth of *R. pseudoacacia* during drought, displaying less growth reductions than *T. cordata* in years with poor growth conditions. The canopy openness, representing the available light conditions, once again had a pronounced effect on growth, but reversed to the previous model. The influence of light availability on growth was additionally dependent on size and age; less canopy openness inhibited the growth of younger trees to a stronger degree than those of older, bigger trees.

Since we used detrended values with a removed age trend for the drought model, the variable dbh alone did not significantly affect growth in contrast to the previous overall growth model we ran. Further, the OSA did not affect growth during drought years. The growth in drought years appeared to be influenced by overall environmental factors including the water supply (SPEI), the species, the growing city, and the available light, and was less significantly influenced by the tree structure (age and dbh) and surroundings such as the OSA, neighboring trees, and buildings.

To reveal the differences in the growth patterns during drought years (low growth episodes) and favorable years (high growth periods), we ran an additional model using a ring width index greater than 1 (Table 5). In contrast to the model with a ring width index of smaller than 1 (Table 4), the canopy openness positively affected growth, which was in accordance with the overall linear mixed model on tree growth (Table 3). Moreover, during years with extraordinarily good growing conditions, *T. cordata* could display higher growth than *R. pseudoacacia*. Converse to the model of years with poor growth, younger trees with a higher light availability obtained higher growth rates. Other factors such as the distance to neighboring trees as well as the OSA were of less importance for growth.

Table 5. Linear mixed model on ring width index (RWI) development of all years with an index > 1 (growth higher than normal, favorable years) of all analyzed trees as the response variable with the individual tree code as a random effect and a random slope added for the effect of dbh, and fixed effects canopy openness, analyzed species, and dbh of the form: $RWI_{ij} = \beta_1 \times x_{1ij} + \ldots + \beta_n \times x_{nij} + b_{i1} \times z_{1ij} + \ldots + b_{in} \times z_{nij} + \varepsilon_{ij}$ where RWI is the response variable for the *j*th of n_i observations in the *i*th of *M* groups or clusters, $\beta_1, \ldots, \beta_n$ are the fixed-effect coefficients, which are identical for all groups, $x_{1ij}, \ldots, x_{nij}$ are the fixed-effect regressors for observation *j* in group *i*; the first regressor is usually for the constant, $x_{1ij-1}, b_{i1}, \ldots b_{in}$ are the random-effect coefficients for group *i*, $z_{1ij}, \ldots, z_{nij}$ are the random-effect regressors, and ε_{ij} is the error for observation *j* in group *i*.

	Value ± SE	*p*
Intercept	1.24 ± 0.04	<0.001
Canopy openness	0.002 ± 8.9 × 10⁻⁴	0.01
Species	0.05 ± 0.02	0.001
Dbh	0.001 ± 7.5 × 10⁻⁴	0.30
Canopy openness: Dbh	−0.0001 ± 1.8 × 10⁻⁵	0.003
SD Intercept	0.11	-
SD Dbh	0.001	-
ε	0.22	-

RWI = ring width index; dbh = stem diameter; SD = standard deviation; SE = standard error. Levels of species: 2 (*Robinia pseudoacacia* and *Tilia cordata*).

4. Discussion

4.1. Tree Ring Analysis: Quantification of Growth of Urban Trees in Relation to the Growing Site

Although numerous studies regarding the drought tolerance and resilience of forest trees in relation to their site conditions exist [29,31,37,70–72], the behavior of urban trees to drought is rarely researched [33,38,39]. However, urban trees are of great benefit to the climate of a city and the well-being of its citizens. The current study points out a high variability in the growth of urban trees. Since urban trees experience increased stress situations at their growing sites, for example a stronger limitation of water availability, higher temperatures, limited space, and a high variation in their growing conditions, the variability in growth indicates a high adaptability of urban trees to their environment. As expected from the ecological characteristics of both species, *R. pseudoacacia* showed a higher annual growth rate than *T. cordata* [44,47].

München and Würzburg were chosen as study sites since they differ in regard to climate characteristics; with 300 mm· year^{-1} less precipitation Würzburg is considerably drier than München. Minor growth differences in relation to the differing climates of München and Würzburg were observable for *T. cordata*; this species showed a slightly higher growth in München than in Würzburg. This is not surprising regarding the low drought tolerance of the species, whereas *R. pseudoacacia* had similar growth rates in both cities. *R. pseudoacacia* is considered as a drought tolerant species [43]; therefore, the individuals in Würzburg are expected to show improved adaptation to a drier climate compared with the less drought-tolerant species *T. cordata*. The measured growth based on tree ring data of both analyzed species was similar to those reported in other dendrochronological studies with regard to urban trees. *T. cordata* had an average growth rate of 3.05 mm· year^{-1}, which is higher than the growth found by Gillner [73] in Dresden (1.6 and 2.43 mm· year^{-1} for two different streets). Iakovoglou *et al.* [18] found that for the honey locust (*G. triacanthos*), growth rates were approximately 4.8 mm· year^{-1} in the United States Midwest, whereas *R. pseudoacacia* in München and Würzburg exhibited a growth of 3.95 mm· year^{-1}.

4.2. Stability, Sensitivity, and Modeling of Tree Growth in Relation to Their Environment

The stability of growth was similar in both cities, but a difference regarding species was found. *R. pseudoacacia* had a significantly higher stability than *T. cordata*. This is consistent with the slightly higher sensitivity of *T. cordata* than that of *R. pseudoacacia*, which can be explained by a higher vulnerability to climate and poor sites [3,38,68,69] Therefore, the sensitivity proved to be inversely related to the stability of growth; however, stability increased with age. With higher age an equilibrium between growth and the environmental influences is achieved, which leads to a stabilization of the growth rates over time, reducing the variation in growth [57]. This pattern is similar to the age trend of trees.

Trees growing in an urban landscape are affected by various environmental factors and site conditions, which influence the annual growth rate aside from their structure and species characteristics [18–20]. We could prove using a linear mixed model approach that in particular the available light, dbh, OSA, and water supply (measured with SPEI) influenced the growth of the analyzed tree species. Light is one of the driving forces of growth [74,75], not only in forests but also in urban areas. Trees growing solitarily without neighboring higher trees and shading buildings can reach their growth maximum [76,77]. These findings are strongly dependent on the species, with *T. cordata* being less affected by shade as a shade-tolerant species in comparison with *R. pseudoacacia*, a light-demanding species experiencing growth deficits when growing in shade. Furthermore, the analyzed growth over the last 20 years was not significantly affected by the city although Würzburg is characterized by 300 mm· year^{-1} less precipitation than München. The growing site with its local climate and individual tree characteristics are more important for tree growth than the overall growing city. A negative or positive city effect was therefore not observable.

4.3. Growth of Urban Trees under Drought Stress in Relation to the Tree Structure and Environment

The results of the drought year analysis (SEA) for München could not confirm the high drought resistance of *R. pseudoacacia*, which exhibited a significant growth reduction during the drought year, but recovered very quickly from the drought, proving a high drought resilience. When examining the original distribution of *R. pseudoacacia*, those results are not surprising. The original distribution of this species is the Atlantic North America at sites with yearly precipitation rates of 1000–1500 mm [44]. Roloff *et al.* [43] classified *R. pseudoacacia* as well suited for dry sites but not fit for more humid sites.

The former distribution of this species suggests that *R. pseudoacacia* can additionally grow in regions of very high precipitation rates and is not as drought tolerant as expected. However, a fast recovery from the drought events highlights a high resilience to drought. *T. cordata* had the highest growth decrease during the first year after the drought event, followed by a quick recovery. Those results are similar to the results of Roloff *et al.* [43] and Gillner *et al.* [38] for two *Acer* species and confirm the classification of *T. cordata* as a moderate drought tolerant-species.

The drought year analysis (SEA) could prove that the studied species react to drought events with different water management strategies. Anisohydric species such as *Quercus rubra* close their stomata very slowly after a drought event; they show high fluctuations in their water management [78,79]. In contrast, isohydric species such as *Picea abies* react immediately to drought by closing their stomata; therefore, they exhibit fewer disruptions in their water management. *Fagus sylvatica* shows an intermediate water management type, showing similar patterns to *T. cordata* [78,79]. Although *T. cordata* reacted slowly, with growth reductions visible 1–2 years after the drought, *R. pseudoacacia* showed an immediate growth decrease during drought, with a fast recovery in München. Those species traits indicate that *T. cordata* is a more anisohydric species, whereas *R. pseudoacacia* displays characteristics more indicative of an isohydric species. This is in line with the findings of Peters *et al.* [80], who characterized ring-porous species such as *R. pseudoacacia* as having a higher capacity for regulation of their stomata as protection against drought, whereas diffuse-porous species such as *T. cordata* experience higher rates of water loss during summer [81].

Trees sampled in Würzburg showed no clear patterns regarding growth during and after drought. During the drought year, both species displayed no growth decline, whereas *R. pseudoacacia* showed a distinct growth reduction two years after the drought and *T. cordata* during the first- and second-year post-drought year. In particular, *R. pseudoacacia* displayed rapid recovery after the growth reduction; however, both species exhibited the highest drop in growth five years after the analyzed drought events (significant for *T. cordata*), which was presumably caused by an additional drought year. The SPEI of Würzburg revealed more frequent drought years than in München. Furthermore, the frequency of drought years in Würzburg is very high; after approximately five years, an additional year with an SPEI of <-0.5 occurs, adding further stress to the trees, which have only a very short time to recover from the previous drought. Moreover, Würzburg is characterized by extreme growth conditions with frequent late frosts and an overall low precipitation rate. Altogether, the factors driving the growth of urban trees in Würzburg are more complex compared to München, and tree growth can be influenced by other factors besides the extreme climate, including different planting methods or soil material and different fertilization methods. In particular, the method and soil used for planting, along with soil compaction, may have a strong effect on water management and growth of urban trees [82,83].

4.4. Effect of Urban Environment and Individual Tree Structure on the Drought Stress of Trees

A linear mixed model describing the growth behavior of the analyzed trees during low-growth episodes revealed that under stress, light availability, species traits, and the water supply (measured with SPEI) of the city are the driving forces of growth. The found results are mainly related to the analyzed species characteristics (higher shade tolerance of *T. cordata*), and contrary to the model for the overall growth, the variable "city" (München and Würzburg) in relation to the water supply (SPEI) proved significant, showing that specific city effects like, e.g., less precipitation and smaller planting pit in Würzburg [84,85] can modify the growth and drought stress of urban trees. During years with high growth episodes, the water supply (measured with SPEI) and the city had no influence on growth. All other variables (species, dbh, and canopy openness) significantly affected growth, but with reversed effects. While during drought years, a smaller canopy openness diminished tree growth, during favorable years the effect of higher canopy openness was positive. Both models of growth during drought years and during favorable years revealed the driving forces of urban tree growth reacting opposing on growth dependent on the surrounding environmental conditions.

5. Conclusions

The study highlights the growth patterns of two common European urban tree species (*R. pseudoacacia* and *T. cordata*) and how their growth changed with age. Different strategies were found depending on the analyzed city and species, with the ring-porous *R. pseudoacacia* exhibiting immediate growth reductions followed by fast recovery, while the diffuse-porous *T. cordata* showed delayed growth reductions. Further, the influence of the surrounding environment on tree growth was analyzed. Hereby, light conditions, water supply, and the species were most important for tree growth, which proved to be detrimental for tree growth during drought. However, the effects of the urban environment such as higher temperatures, highly sealed sites, late frosts, and de-icing salting in winter on tree growth, in particular within the context of drought years and climate change, have rarely been analyzed up to date. Further studies regarding the driving forces of urban tree growth during drought in relation to their environment, structure, and ecophysiology, as well as in view of climate change are necessary. These further studies would help to gain valuable information on sustainable tree planting and tree performance for healthy, long-living trees, thereby providing ecosystem services such as evaporative cooling, shading, carbon sequestration, and pollutant filtering for increasing the quality of life of human populations in cites.

Acknowledgments: This research was funded by the Bavarian State Ministry of the Environment and Consumer Protection (project TUF01UF-64971 "Urban trees under climate change: their growth, environmental performance, and perspectives"). Further, this work was supported by the German Research Foundation (DFG) and the Technical University of Munich (TUM) in the framework of the Open Access Publishing Program. We thank the departments for the municipal green areas of München and Würzburg for their support and encouragement. The authors also thank Chan Ka Nok, Alexander Hellwig, and Karin Beer for their assistance in field data collection as well as Peter Biber for statistical support.

Author Contributions: Thomas Rötzer, Stephan Pauleit, and Hans Pretzsch initiated the project; Astrid Moser performed the experiments and analyzed the data; Astrid Moser and Thomas Rötzer wrote the manuscript; and Stephan Pauleit and Hans Pretzsch revised the manuscript.

Conflicts of Interest: The authors declare no conflict of interest.

Abbreviations

The following abbreviations are used in this manuscript:

AIC	Akaike's information criterion
Dbh	diameter at breast height
EPS	expressed population signal
OSA	open surface area of the tree pit
PET	potential evapotranspiration
RSE	residual standard error
RWI	ring width index
SD	standard deviation
SE	standard error
SEA	superposed epoch analysis
SPEI	standardized precipitation-evapotranspiration index
SVF	sky view factor

References

1. Shashua-Bar, L.; Hoffman, M.E. Geometry and orientation aspects in passive cooling of canyon streets with trees. *Energy Build.* **2003**, *35*, 61–68. [CrossRef]
2. Akbari, H.; Pomerantz, M.; Taha, H. Cool surfaces and shade trees to reduce energy use and improve air quality in urban areas. *Sol. Energy* **2001**, *70*, 295310. [CrossRef]
3. Gillner, S.; Vogt, J.; Roloff, A. Climatic response and impacts of drought on oaks at urban and forest sites. *Urban For. Urban Green.* **2013**, *12*, 597–605. [CrossRef]
4. Leuzinger, S.; Vogt, R.; Körner, C. Tree surface temperature in an urban environment. *Agr. For. Meteorol.* **2010**, *150*, 56–62. [CrossRef]

5.   Mueller, E.C.; Day, T.A. The effect of urban ground cover on microclimate, growth and leaf gas exchange of oleander in Phoenix, Arizona. *Int. J. Biometeorol.* **2005**, *49*, 244–255. [CrossRef] [PubMed]

6.   Nowak, D.J.; Crane, D.E. Carbon storage and sequestration by urban trees in the USA. *Environ. Pollut.* **2002**, *116*, 381–389. [CrossRef]

7.   Bolund, P.; Hunhammar, S. Ecosystem services in urban areas. *Ecol. Econ.* **1999**, *29*, 293–301. [CrossRef]

8.   Nowak, D.J.; Hirabayashi, S.; Bodine, A.; Hoehn, R. Modeled PM2.5 removal by trees in ten US cities and associated health effects. *Environ. Pollut.* **2013**, *176*, 395–402. [CrossRef] [PubMed]

9.   Dimoudi, A.; Nikolopoulou, M. Vegetation in the urban environment: Microclimatic analysis and benefits. *Energy Build.* **2003**, *35*, 69–76. [CrossRef]

10. Dobbs, C.; Kendal, D.; Nitschke, C.R. Multiple ecosystem services and disservices of the urban forest establishing their connections with landscape structure and sociodemographics. *Ecol. Indic.* **2014**, *43*, 44–55. [CrossRef]

11. Dobbs, C.; Escobedo, F.J.; Zipperer, W.C. A framework for developing urban forest ecosystem services and goods indicators. *Landsc. Urban Plan.* **2011**, *99*, 196–206. [CrossRef]

12. Xiao, Q.; McPherson, E.G.; Ustin, S.L.; Grismer, M.E.; Simpson, J.R. A new approach to modeling tree rainfall interception. *J. Geophys. Res.* **2000**, *105*, 29–173. [CrossRef]

13. Xiao, Q.; McPherson, E.G.; Ustin, S.L.; Grismer, M.E.; Simpson, J.R. Winter rainfall interception by two mature open-grown trees in Davis, California. *Hydrol. Process.* **2000**, *14*, 763–784. [CrossRef]

14. Tyrväinen, L.; Pauleit, S.; Seeland, K.; Vries, S. Benefits and Uses of Urban Forests and Trees. In *Urban Forests and Trees: A Reference Book*; Konijnendijk, C., Nilsson, K., Randrup, T., Schipperijn, J., Eds.; Springer Berlin Heidelberg: Berlin, Germany, 2005; pp. 81–114.

15. Fahey, R.T.; Bialecki, M.B.; Carter, D.R. Tree Growth and Resilience to Extreme Drought Across an Urban Land-use Gradient. *Arboric. Urban For.* **2013**, *39*, 279–285.

16. Iakovoglou, V.; Thompson, J.; Burras, L.; Kipper, R. Factors related to tree growth across urban-rural gradients in the Midwest, USA. *Urban Ecosyst.* **2001**, *5*, 71–85. [CrossRef]

17. Nowak, D.J.; Kuroda, M.; Crane, D.E. Tree mortality rates and tree population projections in Baltimore, Maryland, USA. *Urban For. Urban Green.* **2004**, *2*, 139–147. [CrossRef]

18. Iakovoglou, V.; Thompson, J.; Burras, L. Characteristics of trees according to community population level and by land use in the U.S. Midwest. *J. Arboric.* **2002**, *28*, 59–69.

19. Whitlow, T.H.; Bassuk, N.L. Trees in difficult sites. *J. Arboric.* **1986**, *13*, 10–17.

20. Day, S.D.; Bassuk, N.L.; van Es, H. Effects of four compaction remediation methods for landscape trees on soil aeration, mechanical impedance and tree establishment. *J. Arboric.* **1995**, *22*, 144–150.

21. Petersen, A.; Eckstein, D. Roadside trees in Hamburg-their present situation of environmental stress and their future chance for recovery. *Arboric. J.* **1988**, *12*, 109–117. [CrossRef]

22. Beatty, R.A.; Heckman, C.T. Survey of urban tree programs in the United States. *Urban Ecol.* **1981**, *5*, 81–102. [CrossRef]

23. Rhoades, R.W.; Stipes, R.J. Growth of trees on Virgina Tech campus in response to various factors. *J. Arboric.* **1999**, *25*, 211–217.

24. IPCC. Climate Change 2007: The Physical Science Basis. In *Contribution of Working Group I to the Fourth Assessment Report of the Intergovernmental Panel on Climate Change*; Cambridge University Press: Cambridge, UK, 2007.

25. Kjelgren, R.K.; Clark, J.R. Microclimates and Tree Growth in Tree Urban Spaces. *J. Environ. Hortic.* **1992**, *10*, 139–145.

26. Pretzsch, H.; Matthew, C.; Dieler, J. Allometry of Tree Crown Structure. Relevance for Space Occupation at the Individual Plant Level. In *Growth and Defence in Plants*; Matyssek, R.E.A., Ed.; Springer Verlag: Berlin, Germany, 2012.

27. Rötzer, T.; Seifert, T.; Gayler, S.; Priesack, E.; Pretzsch, H. Effects of Stress and Defence Allocationon Tree Growth: Simulation Results at the Individual and Stand Level. In *Growth and Defence in Plants, Ecological Studies*; Al, R.M.E., Ed.; Springer-Verlag: Berlin, Germany, 2012; Volume 220, pp. 401–432.

28. Pretzsch, H.; Dieler, J. The dependency of the size-growth relationship of Norway spruce (*Picea abies* [L.] Karst.) and European beech (*Fagus sylvatica* [L.]) in forest stands on long-term size conditions, drought events, and ozone stress. *Trees* **2011**, *25*, 355–369. [CrossRef]

29. Pretzsch, H.; Schütze, G.; Uhl, E. Resistance of European tree species to drought stress in mixed *versus* pure forests: Evidence of stress release by inter-specific faciliation. *Plant Biol.* **2013**, *15*, 483–495. [CrossRef] [PubMed]

30. McDowell, N.; Pockman, W.T.; Allen, C.D.; Breshears, D.D.; Cobb, N.; Kolb, T.; Plaut, J.; Sperry, J.; West, A.; Williams, D.G.; *et al.* Mechanisms of plant survival and mortality during drought: Why do some plants survive while others succumb to drought? *New Phytol.* **2008**, *178*, 719–739. [CrossRef] [PubMed]

31. Rais, A.; van de Kuilen, J.-W.G.; Pretzsch, H. Growth reaction patterns of tree height, diameter, and volume of Douglas-fir (*Pseudotsuga menziesii* [Mirb.] Franco) under acute drought stress in Southern Germany. *Eur. J. For. Res.* **2014**, *133*, 1043–1056. [CrossRef]

32. Ryan, M.G.; Binkley, D.; Fownes, J.H. Age-Related Decline in Forest Productivity: Pattern and Process. *Adv. Ecol. Res.* **1997**, *27*, 213–262.

33. Bartens, J.; Grissino-Mayer, H.D.; Day, S.D.; Wiseman, P.E. Evaluating the potential for dendrochronological analysis of live oak (*Quercus virginiana* Mill.) from the urban and rural environment—An explorative study. *Dendrochronologia* **2012**, *30*, 15–21. [CrossRef]

34. Speer, J.H. *Fundamentals of Tree-Ring Research*; The University of Arizona Press: Tucson, AZ, USA, 2012.

35. Cavin, L.; Mountford, E.P.; Peterken, G.F.; Jump, A.S. Extreme drought alters competitive dominance within and between tree species in a mixed forest stand. *Funct. Ecol.* **2013**, *27*, 1424–1435. [CrossRef]

36. Orwig, D.A.; Abrams, D.A. Variation in radial growth responses to drought among species, site, and canopy strata. *Trees* **1997**, *11*, 474–484. [CrossRef]

37. Rybníček, M.; Čermák, P.; Kolář, T.; Žid, T. Growth responses of Norway spruce (*Picea abies* (L.) Karst.) to the climate in the south-eastern part of the Českomoravská Upland (Czech Republic). *Geochronometria* **2012**, *39*, 149–157.

38. Gillner, S.; Bräuning, A.; Roloff, A. Dendrochronological analysis of urban trees: Climatic response and impact of drought on frequently used tree species. *Trees* **2014**, *28*, 1079–1093. [CrossRef]

39. He, X.; Chen, Z.; Chen, W.; Shao, X.; He, H.; Sun, Y. Solar activity, global surface air temperature anomaly and pacific decadal oscillation recorded in urban tree rings. *Ann. For. Sci.* **2007**, *64*, 743–756. [CrossRef]

40. DWD. Deutscher Wetterdienst. Available online: http://www.dwd.de.

41. Pauleit, S.; Jones, N.; Garcia-Marin, G.; Garcia-Valdecantos, J.-L.; Rivière, L.M.; Vidal-Beaudet, L.; Bodson, M.; Randrup, T.B. Tree establishment practice in towns and cities—Results from a European survey. *Urban For. Urban Green.* **2002**, *1*, 83–96. [CrossRef]

42. Radoglou, K.; Dobrowolska, D.; Spyroglou, G.; Nicolescu, V.-N. A review on the ecology and silviculture of limes (*Tilia cordata* Mill., *Tilia platyphyllos* Scop. and *Tilia tormentosa* Moench.) in Europe. *Die Bodenkult.* **2009**, *60*, 9–19.

43. Roloff, A.; Bonn, S.; Gillner, S. Klimawandel und Baumartenwahl in der Stadt—Entscheidungsfindung mit der Klima-Arten-Matrix (KLAM). Available online: http://www.frankfurt.de/sixcms/media.php/738/klam_stadt.pdf (accessed on 16 March 2016).

44. Keresztesi, B. *The Black Locust*; Akadémiai Kiadó: Budapest, Hungary, 1988.

45. Baker, F.S. A revised tolerance table. *J. For.* **1949**, *47*, 179–181.

46. Niinemets, Ü.; Valladares, F. Tolerance to shade, drought, and waterlogging of temperate northern hemisphere trees and shrubs. *Ecol. Monogr.* **2006**, *76*, 521–547. [CrossRef]

47. Moser, A.; Rötzer, T.; Pauleit, S.; Pretzsch, H. Structure and ecosystem services of small-leaved lime (*Tilia cordata* Mill.) and black locust (*Robinia pseudoacacia* L.) in urban environments. *Urban For. Urban Green.* **2015**, *14*, 1110–1121. [CrossRef]

48. Matzarakis, A.; Rutz, F.; Mayer, H. Modelling radiation fluxes in simple and complex environments—Application of the RayMan model. *Int. J. Biometeorol.* **2007**, *51*, 323–334. [CrossRef] [PubMed]

49. Yamashita, S.; Sekine, K.; Shoda, M.; Yamashita, K.; Hara, Y. On relationships between heat island and sky view factor in the cities of Tama River Basin, Japan. *Atmos. Environ.* **1986**, *20*, 681–686. [CrossRef]

50. Rinn, F. TSAP Reference Manual. Available online: http://www.rinntech.com/.

51. TSAP-Win: Time Series Analysis and Presentation for Dendrochronology and Related Applications, version 0.55. Rinn Tech. Heidelberg, Germany, 2010.

52. R, A language and environment for statistical computing. R Core Team, Vienna, Austria, 2014.

53. Bunn, A.; Korpela, M.; Biondi, F.; Campelo, F.; Mérian, P.; Qeadan, F.; Zang, C.; Buras, A.; Cecile, J.; Mudelsee, M.; *et al.* (Eds.) Package "dplR", version 1.6.3; Dendrochronology Program Library in R. 2015.

54. COFECHA and ARSTAN, Chronology Series VI. In *Tree-Ring Chronologies of Western North America: California, Eastern Oregon and Northern Great Basin with Procedures Used in the Chronology Development Work Including Users Manuals*; Laboratory of Tree-Ring Research: University of Arizona, Tuscon, AZ, USA, 1986.

55. Cook, E.R.; Holmes, R.L. User manual for computer program ARSTAN. In *Tree-Ring Chronologies of Western North America: California, Eastern Oregon and Northern Great Basi*; Holmes, R.L., Adams, R.K., Fritts, H.C., Eds.; University of Arizona: Tucson, AZ, USA, 1986; Volume 6, pp. 50–57.

56. Roloff, A. *Baumkronen. Verständnis Und Praktische Bedeutung Eines Komplexen Naturphänomens*; Ulmer: Stuttgart, Germany, 2001.

57. Jucker, T.; Bouriaud, O.; Avacaritei, D.; Coomes, D.A. Stabilizing effects of diversity on aboveground wood production in forest ecosystems: linking patterns and processes. *Ecol. Lett.* **2014**, *17*, 1560–1569. [CrossRef] [PubMed]

58. Fritts, H.C. *Tree Rings and Climate*; Academic Press: London, UK, 1976.

59. Lukaszkiewicz, J.; Kosmala, M. Determining the Age of Streetside Trees with Diameter at Breast Height-based Multifactorial Model. *Arboric. Urban For.* **2008**, *34*, 137–143.

60. Dwyer, J.F. How old is that tree? *Illinois Trees* **2009**, *24*, 13.

61. Vicente-Serrano, S.M.; Beguería, S.; López-Moreno, J.I. A Multiscalar Drought Index Sensitive to Global Warming: The Standardized Precipitation Evapotranspiration Index. *J. Clim.* **2010**, *23*, 1696–1718. [CrossRef]

62. Droogers, P.; Allen, R.G. Estimating reference evapotranspiration under inaccurate data conditions. *Irrig. Drain. Syst.* **2002**, *16*, 33–45. [CrossRef]

63. Hargreaves, G.H. Defining and using reference evapotranspiration. *J. Irrig. Drain. Eng.* **1994**, *120*, 1132–1139. [CrossRef]

64. *Quinoa: Improvement and Sustainable Production*; Wiley-Blackwell: Oxford, UK, 2015.

65. Pinheiro, J.; Bates, D.; DebRoy, S.; Sarkar, D.; R Core Team (Eds.) Nlme: Linear and Nonlinear Mixed Effects Models, 2015.

66. Lough, J.M.; Fritts, H.C. An assessment of the possible effects of volcanic eruptions on North American climate using tree-ring data, 1602 to 1900 A.D. *Clim. Chang.* **1987**, *10*, 219–239. [CrossRef]

67. Wigley, T.M.L.; Briffa, K.R.; Jones, P.D. On the Average Value of Correlated Time Series, with Applications in Dendroclimatology and Hydrometeorology. *J. Clim. Appl. Meteor.* **1984**, *23*, 201–213. [CrossRef]

68. Schweingruber, F.H. *Tree Rings and Environment*; Swiss Federal Institute for Forest, Snow and Landscape Research: Birmensdorf, Switzerland, 1996.

69. Beck, W.; Müller, J. Impact of heat and drought on tree and stand vitality–Dendroecological methods and first results from level II-plots in southern Germany. *Schr. Forstl. Fak. Univ. Gött. Niedersächs. Forstl. Versanst.* **2006**, *142*, 120–127.

70. George, J.P.; Schueler, S.; Karanitsch-Ackerl, S.; Mayer, K.; Klumpp, R.T.; Grabner, M. Inter- and intra-specific variation in drought sensitivity in Abies spec. and its relation to wood density and growth traits. *Agr. For. Meteorol.* **2015**, *214–215*, 430–443. [CrossRef]

71. Juday, G.P.; Alix, C. Consistent negative temperature sensitivity and positive influence of precipitation on growth of floodplain Picea glauca in Interior Alaska. *Can. J. For. Res.* **2012**, *42*, 561–573. [CrossRef]

72. Pretzsch, H.; Dieler, J.; Seifert, T.; Rötzer, T. Climate effects on productivity and resource-use efficiency of Norway spruce (*Picea abies* [L.] Karst.) and European beech (*Fagus sylvatica* [L.]) in stands with different spatial mixing patterns. *Trees* **2012**, *26*, 1343–1360. [CrossRef]

73. Gillner, S. Stadtbäume im Klimawandel—Dendrochronologische und physiologische Untersuchungen zur Identifikation der Trockenstressempfindlichkeit häufig verwendeter Stadtbaumarten in Dresden. Ph.D. Thesis, Universität Dresden, Dresden, Germany, 2012.

74. Monteith, J.L. Climate and the efficiency of crop production in Britain. *Philos. Trans. R. Soc. B* **1977**, *281*, 277–294. [CrossRef]

75. Walters, M.B.; Reich, P.B. Are shade tolerance, survival, and growth linked? Low light and nitrogen effects on hardwood seedlings. *Ecology* **1996**, *77*, 841–853. [CrossRef]

76. Hasenauer, H. Dimensional relationship of open-grown trees in Austria. *For. Ecol. Manag.* **1997**, *96*, 197–206. [CrossRef]

77. Uhl, E.; Metzger, H.G.; Seifert, T. *Dimension und Wachstum von Solitären Buchen und Eichen, Sektion Ertragskunde, Jahrestagung, Staufen*; Deutscher Verband Forstlicher Forschungsanstalten: Staufen, Germany, 2006.

78. Río, M.D.; Schütze, G.; Pretzsch, H. Temporal variation of competition and facilitation in mixed species forests in Central Europe. *Plant Biol.* **2014**, *16*, 166–176. [PubMed]

79. Pretzsch, H.; Rötzer, T.; Matyssek, R.; Grams, T.E.E.; Häberle, K.-H.; Pritsch, K.; Kerner, R.; Munch, J.-C. Mixed Norway spruce (*Picea abies* [L.] Karst) and European beech (*Fagus sylvatica* [L.]) stands under drought: From reaction pattern to mechanism. *Trees* **2014**, *28*, 1305–1321. [CrossRef]

80. Peters, E.B.; McFadden, J.P.; Montgomery, R.A. Biological and environmental controls on tree transpiration in a suburban landscape. *J. Geophys. Res.* **2010**. [CrossRef]

81. Rahman, M.A.; Armson, D.; Ennos, A.R. A comparison of the growth and cooling effectiveness of five commonly planted urban tree species. *Urban Ecosyst.* **2015**, *18*, 371–389. [CrossRef]

82. Rahman, M.A.; Smith, J.G.; Stringer, P.; Ennos, A.R. Effect of rooting conditions on the growth and cooling ability of Pyrus calleryana. *Urban For. Urban Green.* **2011**, *10*, 185–192. [CrossRef]

83. Zaharah, S.S.; Razi, I.M. Growth, stomata aperture, biochemical changes and branch anatomy in mango (*Mangifera indica*) cv. Chokanan in response to root restriction and water stress. *Sci. Hortic.* **2009**, *123*, 58–67. [CrossRef]

84. Prügl, B.J. *Vegetationstechniken für Innerörtliche Baumstandorte*; Bodeninstitut Johannes Prügl: Hallertau, Germany, 2016.

85. Stadt Würzburg. STELLPLATZSATZUNG der Stadt Würzburg (SPS). Stellplatzsatzung vom 25. März 2014 (MP und VBl Nr. 73/2014 vom 28. März 2014)

forests

Article

Tree Species and Their Space Requirements in Six Urban Environments Worldwide

Jens Dahlhausen *, Peter Biber, Thomas Rötzer, Enno Uhl and Hans Pretzsch

Chair for Forest Growth and Yield Science, Center of Life and Food Sciences Weihenstephan, Technical University of Munich, Hans-Carl-von-Carlowitz-Platz 2, Freising 85354, Germany; peter.biber@lrz.tum.de (P.B.); thomas.roetzer@lrz.tum.de (T.R.); enno.uhl@lrz.tum.de (E.U.); hans.pretzsch@lrz.tum.de (H.P.)
* Correspondence: jens.dahlhausen@lrz.tum.de; Tel.: +49-8161-71-4713; Fax: +49-8161-71-4721

Academic Editors: Francisco Escobedo, Stephen John Livesley and Justin Morgenroth
Received: 23 March 2016; Accepted: 19 May 2016; Published: 25 May 2016

Abstract: Urban trees have gained in importance during recent decades, but little is known about the temporal dynamic of tree growth in urban areas. The present study investigated the allometric relationships of stem diameter, tree height, and crown radius for six different tree species in six metropolises worldwide. Increment cores of the trees were used for identifying the relationship of basal area and basal area increment and for extrapolating the temporal dynamics for each species in relation to the allometric parameters and growth extensions. Space limitation and its direct influence on growth were quantified, as well as the aboveground woody biomass and the carbon storage capacity. The results show that, among the investigated species, *Quercus nigra* and *Khaya senegalensis* have the highest growth rates for stem diameter and crown radius, whereas *Tilia cordata* and *Aesculus hippocastanum* remain on a lower level. A significant reduction of tree growth due to restricted non-paved area was found for *Aesculus hippocastanum* and *Khaya senegalensis*. Estimations of aboveground biomass were highest for *Quercus nigra* and lowest for *Tilia cordata*. These results show the species-specific allometries of urban trees over a projected time period. Thus, the data set is highly relevant for planners and urban green managers.

Keywords: growing space requirements; metropolises; tree allometry; tree growth dynamics; urban trees; biomass

1. Introduction

Along with urban greening the interest in tree growth rapidly increased during recent decades. On the one hand, it can be explained by the fact that trees provide many ecosystem services and functions, like cooling the environment, filtering harmful particulate matter from the air, and having aesthetical and structural functions within build-up areas [1,2]. On the other hand, several disservices of urban trees exist, like litter fall, damage to foot paths, and fallen branches which cause additional costs [1,3,4] and can endanger the public's safety.

Due to the benefits and disadvantages, the growth of urban trees and its space requirements are of high interest for urban planners and managers, as well as for arborculturists. First, it is important to know the future space requirements for the existing trees at a certain age. Second, for future planning the growing space requirements can be crucial for the choice of tree species before planting at a certain place [5]. Lastly, information on allometric relationships of urban trees is essential for the treatment measures of arborculturists for ensuring public safety. While there are several studies on forest trees, the space development of urban trees (solitary and open grown trees) is rarely known. Only a few studies [6,7] have investigated the differences in allometry and biomass between forest and urban trees. Our study aims at generating knowledge for urban trees by analyzing the space requirements, growth dynamics, and allometry of urban trees.

Taking a deeper look at the benefits, one major benefit of urban trees is their carbon sequestration. The carbon storage capacity in turn is species-specific as Moser *et al.* [8], for example, showed for *Robinia pseudoacacia* and *Tilia cordata* in two German cities. They found a storage capacity of 461 kg C per tree for *Robinia pseudoacacia* (mean age: 39 years) and with 196 kg C a remarkable lower storage capacity for *Tilia cordata* (mean age: 41 years). Yoon *et al.* [9] estimated 24.9 t of total average C storage for the tree species *Acer buergerianum*, *Ginkgo biloba*, *Prunus yedoensis*, and *Zelkova serrata* in Daegu, Korea, and 69.7 t C ha^{-1} for *Platanus orientalis*.

A key variable for space occupation is the crown size. Species-specific crown projection areas were found by Pretzsch *et al.* [10] while analyzing the growing space requirements of 22 tree species worldwide. They focused on crown size parameters and determined five different crown extension types. Large crown expansions, crown sizes, and tree heights result in higher service functions such as shading and filtering particulates. The dimensions of the crown are responsible for shading and cooling, two of the most important ecosystem services of trees in urban areas. They play an important role not only on a tree scale, cooling down the microclimate, but also on a street scale. This means that through shading, trees can increase human thermal comfort [11]. Further, on the city scale this service can reduce the urban heat island effect [4]. Additionally, the shading of street trees can reduce the maintenance costs for streets by protecting the asphalt from solar radiation. McPherson and Muchnick [3] report an extended lifetime of 10 to 25 years for road and pavement surfaces in California that are highly shaded in comparison to those that are not shaded.

A further benefit at all three aforementioned scales is the reduction of water runoff. Besides tree number, increasing tree and crown dimension also reduces water runoff due to higher evapotranspiration and interception rates. This reduction in the amount of runoff is species-specific [4,12,13] and might also reduce stormwater runoff [14,15].

While the studies mentioned above show that trees have many benefits, the related space requirements of urban trees are a potential conflict. On the one hand, space in urban areas is limited and therefore expensive. On the other hand, open space is essential for housing prices, especially close to the city center [16]. Thus, knowing the space requirements of certain tree species is of high interest for urban planning. As tree growth is species-specific, the growth rate and the spatial dimensions of different tree species in dependence on tree age could be essential information for urban planning. While the growth behavior (*i.e.*, the crown radius, diameter at breast height or tree height) is well known for forest stands, little is known for solitary urban trees [17]. Studies have shown that information about growth dynamics of trees from forest stands cannot be directly transferred to urban trees [6]. This is due to the fact that in urban areas mainly mechanical stresses outweigh effects in forest stands, such as competition between trees or forest management measures [18].

The discussion about space requirements of trees in urban areas is not only relevant for the aboveground tree structure, but also for the rooting zone. The root growth is influenced by the available soil fraction, soil compaction, and the volume of root-penetrable soil [2]. Further, the effects of soil compaction, reduced soil volume, and paved surface are investigated in several studies [18,19]. If the soil volume is becoming too shallow, with time the roots start uplifting and cracking the pavement [20]. This in turn causes high costs for road or pavement maintenance, which is reported to be one of the first reasons for removing trees in urban areas [21].

In summary, while there are several studies dealing with space requirements of trees [10,22,23], the present study differs from the existing studies as it (1) is specific to urban trees; and (2) investigates the space requirements for six different tree species worldwide; and (3) provides a data set for each tree species representing the temporal dynamic of the tree structure. It is important to state that we do not see this study as a species comparison in the usual sense, which would not be possible with our data. Rather, we intend to compare different urban tree species, each under conditions where it can be considered to be typical and highly adapted. Based on the analyses, conclusions for space requirements of existing and planned tress, as well as conclusions on the potential benefits of urban trees, can be drawn. This leads to the following research questions:

(1) Is there a relationship between space limitations of typical urban tree species (in the sense mentioned above) and allometric relationships?
(2) Does soil sealing limit the growth of urban trees?
(3) How do typical urban tree species (in the sense mentioned above) differ in terms of aboveground biomass and carbon storage?

2. Materials and Methods

2.1. Data Acquisition

For this study we used data from urban trees that were collected in six metropolises worldwide. The data are a subset from samples acquired in the overarching project Response of Urban Trees on Climate Change funded by the AUDI Environmental Foundation. The field campaigns took place between October 2010 and June 2014 in the cities of Berlin, Brisbane, Hanoi, Houston, Munich, and Paris. Altogether, our data comprise 1097 trees, with an average of 183 trees per city, with a minimum of 126 (Brisbane) and a maximum of 240 trees (Berlin). In each city the most common local urban tree species was chosen. While the overarching project included trees in the city centers, suburban zones, and in rural outskirts, only the former were selected for the study at hand.

Selecting criteria for sample trees were: (1) health; and (2) that the trees were not pruned. An overview of the selected cities and species is given in Table 1.

Table 1. Sample description of tree species, within the cities, geography (latitude, longitude), sampling size, and time. Only trees in an urban environment are considered.

Species	Common Name	City	Country	Location	Number of Trees	Sampling Time
Aesculus hippocastanum L.	Horse-chestnut	Munich	Germany	48°8′ N 11°35′ E	231	2013
Araucaria cunninghamii AITON ex. D.DON	Hoop pine	Brisbane	Australia	27°28′ S 153°1′ E	126	2013
Khaya senegalensis (DESR.) A.JUSS.	African mahogany	Hanoi	Vietnam	21°2′ N 105°51′ E	163	2012
Platanus x hispanica MÜNCHH.	London plane	Paris	France	48°51′ N 2°21′ E	171	2013
Quercus nigra L.	Water oak	Houston	USA	29°45′ N 95°23′ W	166	2014
Tilia cordata MILL.	Small-leaved lime	Berlin	Germany	52°31′ N 13°24′ E	240	2010, 2012, 2013

For each tree, diameter increment time series are available from two increment cores taken from orthogonal directions at breast height. As the tree diameter d was measured, the diameter time series could be reconstructed backwards. Additional information about the sampling is described in Pretzsch *et al.* [10]. In the study at hand we transformed the diameter time series into basal area (ba) and basal area increment (iba) time series, because in contrast to diameter increment, basal area increment is directly related to a tree's biological production [24] and thus more appropriate for the allometric analyses shown below. Total tree height (h) and height to crown base (hcb) are also available, together with crown radii (r) measured in the eight cardinal and sub-cardinal directions, as suggested by

Preuhsler [25]. The mean crown radius (*cr*) was calculated as the quadratic mean of these eight crown radii in Equation (1).

$$cr = \sqrt{\left(r_N^2 + r_{NW}^2 + \ldots + r_{NE}^2\right)/8} \tag{1}$$

Based on the mean crown radius, the crown projection area is then given by the following Equation (2), ensuring a bias-free transition between crown radius to crown projection area.

$$cpa = cr^2 \times \pi \tag{2}$$

Moreover, tree height and height to crown base (*cb*) were recorded by using a Vertex IV ultrasonic hypsometer.

In order to quantify a tree's spatial confinement by soil sealing like asphalted pavement or concrete surfaces, the unsealed area around the stem base was measured for each tree in four cardinal directions with a measuring tape. We used this information for calculating the variable *SCON*, which expresses the degree of confinement due to soil sealing. It is calculated as follows in Equation (3):

$$SCON = 1 - \frac{npa}{cpa} \tag{3}$$

where *cpa* is the crown projection area (m^2) and *npa* is the measured non-paved area (m^2). Table 2 gives an overview of the tree characteristics covered by our data. *A. cunninghamii* covers the broadest diameter range (15.7–129.5 cm) and the largest diameter was measured for *P. x hispanica* with 144 cm. The largest maximum *cr* with 14.8 m was also measured for *P. x hispanica*. The tallest tree was of the species *K. senegalensis* with 36 m, and this species also showed the highest range for ring widths. *A. hippocastanum* (232 years) and *P. x hispanica* (234 years) covered the largest age range within our sampled trees. *Npa* was largest for *A. cunninghamii* with 410 m^2, and smallest for *T. cordata*, *A. hippocastanum*, and *K. senegalensis* with 0.5 to 100 m^2.

2.2. Allometric Analyses

Plant allometry is a concept that is used in this study to relate tree dimensions to each other as well as to relate plant size and increment. It has a strong theoretical foundation [26–28] which, together with its mathematical compactness, makes it a powerful tool for applications in science and practice [22]. It is very common to use the double-logarithmic form of the allometric equation, especially in empirical study, because it makes the relation between two plant variables *x* and *y*, say *x* = stem diameter, *y* = height, accessible to linear regression models (Equation (4)):

$$\ln y = k + c \cdot \ln x \tag{4}$$

Here, the constants c and e^k are the allometric scaling coefficient and the allometric factor, respectively. Embedded in this context, the backbone of our analyses is a linear mixed effects model (Equation (5)), which describes the allometric relationship between the annual basal area increment *iba* (cm^2/y) and the corresponding basal area *ba* (cm^2) at the beginning of the corresponding year.

$$\ln iba_{ij} = \beta_0 + \beta_1 \cdot \ln ba_{ij} + b_i + \varepsilon_{ij} \tag{5}$$

This model was fitted to the retrospective tree growth data of each city separately. The indexes *i* and *j* represent the jth observation of the ith tree, β_0 and β_1 are fixed effect parameters, b_i is a random effect on tree level with $b_i \sim N\left(0, \tau^2\right)$, and ε_{ij} are i.i.d. errors.

In order to estimate the influence of spatial restriction on basal area increment we extended this model to Equation (6) by including the variable *SCON* which describes the confinement of a tree's growing area by soil sealing as defined in Equation (3) above.

$$\ln iba_{ij} = \beta_0 + \beta_1 \cdot \ln ba_{ij} + \beta_2 \cdot SCON_i + b_i + \varepsilon_{ij} \tag{6}$$

When fitting this model, we restricted the tree growth data to the years after 1980, because *npa* was only measured once at survey time, and we refrained from assuming it was in effect for four decades or more. With preliminary versions of Equation (6) we also tested for the presence of interactions between *SCON* and *ba*, which, however, were not significant. Due to the comparably short time series from the restricted data set, we did not obtain plausible estimates for the allometric slope β_1 when fitting Equation (6). We therefore decided to keep the allometric slopes β_1 from the fitted Equation (6), and to use the following model (Equation (7)) for assessing the influence of *SCON*:

$$\ln iba_{ij} - \beta_1 \cdot \ln ba_{ij} = \alpha_0 + \alpha_1 \cdot SCON_i + b_i + \varepsilon_{ij} \tag{7}$$

Here, α_0 and α_1 are the fixed effects parameters to be estimated with the model. Thus, for estimating a tree's basal area increment *iba* from its basal area *ba* and a given spatial confinement *SCON*, without considering tree-specific random effects, the Equation (8) can be used:

$$\widehat{\ln iba} = \alpha_0 + \beta_1 \cdot \ln ba + \alpha_1 \cdot SCON \tag{8}$$

Analogously, if the spatial confinement effect shall not be considered explicitly in the estimation of basal area increment, the fit results of Equation (6) can be applied with Equation (9):

$$\widehat{\ln iba} = \beta_0 + \beta_1 \cdot \ln ba \tag{9}$$

Both equations allow for estimating a tree's temporal basal area development by using them in the following alternative serial calculations (Equations (10) and (11)):

$$ba_{t+1} = ba_t + e^{\alpha_0 + \beta_1 \cdot \ln ba_t + \alpha_1 \cdot SCON} \tag{10}$$

$$ba_{t+1} = ba_t + e^{\beta_0 + \beta_1 \cdot \ln ba_t} \tag{11}$$

The index *t* represents one point of time, $t+1$ is the point of time one year later. Starting with a given initial value for basal area ba_{t_0}, basal area development for any reasonable time span can be projected. As a common initial value we chose $ba_{t_0} = 78.54$ cm^2, which corresponds to a diameter at breast height of 10 cm, a size at which urban trees are often planted. We projected basal area development for 200 years, in case of Equation (10) with three different values for *SCON*, representing total absence of non-paved area limitation, medium, and maximum limitation as found in the data for each tree species.

In order to link basal area with vertical and horizontal space requirements, we fitted two additional allometric models (Equations (12) and (13)):

$$\ln h_i = \beta_0 + \beta_1 \cdot \ln d_i + \varepsilon_i \tag{12}$$

$$\ln cr_i = \beta_0 + \beta_1 \cdot \ln d_i + \varepsilon_i \tag{13}$$

where *h* is tree height, *cr* crown radius, and d is a tree's diameter at breast height. The latter can be easily obtained if basal area is known by $d = \sqrt{ba \cdot 4/\pi}$. Both models were fitted city-wise by ordinary least squares (OLS) regression. As only one height and crown radius per tree were available from the field survey, including tree-level random effects was not necessary.

Table 2. Tree characteristics for each species.

Species	d (cm)			cr (m)			h (m)			rw (mm)[1]			age (year)			npa (m²)[2]			SCON		
	Min	Mean	Max	Min	Mean	Max	Min	Mean	Max	Min	Mean	Max	Min	Mean	Max	Min	Mean	Max	Min	Mean	Max
Aesculus hippocastanum	27.2	63.6	116.9	2.9	5.5	9.0	7.4	16.1	27.2	0.32	1.61	7.91	17	118	249	1.00	15.08	100.00	0.00	0.83	0.99
Araucaria cunninghamii	15.7	40.7	129.5	1.6	3.5	11.6	6.1	17.3	33.5	0.75	3.63	6.50	19	52	182	3.90	79.85	410.00	0.00	0.36	0.96
Khaya senegalensis	44.1	73.4	123.1	3.1	6.5	11.7	14.1	22.6	36	0.63	5.30	18.55	13	53	142	1.00	11.18	100.00	0.07	0.91	1.00
Platanus x hispanica	40.3	64.8	144.0	2.7	6.6	14.8	6.84	18.8	34.5	0.41	2.54	10.93	20	104	254	2.00	33.59	140.00	0.20	0.78	0.99
Quercus nigra	34.2	61.5	98.0	3.4	7.1	11.9	10	15.9	22.8	0.82	3.96	12.55	16	52	101	1.50	33.82	150.00	0.00	0.66	0.98
Tilia cordata	25.2	45.5	81.1	2.5	5.1	9.5	8.1	16.3	29.1	0.22	1.63	5.38	34	85	194	0.50	25.31	100.00	0.00	0.81	1.00

[1] n rw *A. hippocastanum* = 193, *A. cunninghamii* = 62, *K. senegalensis* = 144, *P. x hispanica* = 144, *Q. nigra* = 133, *T. cordata* = 133, *T. cordata* = 144. *P. x hispanica* = 183, *Q. nigra* = 183, *T. cordata* = 133, *T. cordata* = 144; [2] n npa *A. hippocastanum* = 231, *A. cunninghamii* = 126, *K. senegalensis* = 163, *P. x hispanica* = 171, *Q. nigra* = 166, *T. cordata* = 144. *d*: diameter at breast height (1.3 m), *cr*: crown radius, *h*: tree height, *rw*: ring width, tree age, *npa*: non-paved area, and SCON: degree of a trees spatial confinement.2.2. Allometric Analyses

2.3. Tree Biomass Estimates

Based on the statistical models shown above, temporal tree development in terms of basal area (diameter), height, and crown radius can be estimated. In order to link those projected tree dynamics to biomass and carbon storage, the tree biomass has to be estimated. We focus on total aboveground woody biomass (w), which can usually be estimated as a function of stem diameter at breast height (d) and tree height (h).

For all species covered by this study, such equations for estimating w were available as by Jenkins *et al.* [29] for *A. hippocastanum*, Eamus *et al.* [30] for *A. cunninghamii*, Clément [31] for *K. senegalensis*, Yoon *et al.* [9] for *P. x hispanica*, Clark *et al.* [32] for *Q. nigra*, and Čihák *et al.* [33] for *T. cordata*.

The biomass equations are based on forest tree data, and thus on trees under competition by other trees, excepting the equation for *P. x hispanica*, which is based on urban tree data. Trees in urban areas are mostly open-grown without competing trees, which calls the applicability of the above-mentioned equations into question. To account for this we apply a factor of 0.8 to the traditional biomass equations.

Based on the aboveground biomass we determine the related carbon storage by multiplying the biomass with 0.5 [34].

3. Results

3.1. Growth and Space Requirements of Urban Trees Based on Allometric Relationships

In Table 3 the results of the linear regression models, one per species, for the 'basal area-basal area increment' allometric relationship Equation (5) are shown for all tree species. The allometric exponent β_1 was in all cases different from zero on a significance level $p < 0.0001$ and varied from $\beta_1 = 0.4747$ for *A. cunninghamii* in Brisbane to $\beta_1 = -0.0057$ for *A. hippocastanum* in Munich. The scaling parameter β_0 was in a range between $\beta_0 = -4.5288$ for *A. cunninghamii* and $\beta_0 = -5.8743$ for *T. cordata*. τ^2 describes the variance of the random effects on tree level.

Table 3. Linear regression models for the basal area to basal area increment allometry (Equation (5)) for all six species (ordered according to latitude).

Species	β_0	SE (β_0)	p (β_0)	β_1	SE (β_1)	p (β_1)	τ^2
Araucaria cunninghamii	−4.5288	0.0533	<0.0001	0.4747	0.0127	<0.0001	0.0637
Khaya senegalensis	−4.9173	0.0523	<0.0001	0.0028	0.0260	0.9145	0.1711
Quercus nigra	−4.8455	0.0416	<0.0001	0.1077	0.0097	<0.0001	0.2201
Platanus x hispanica	−5.5322	0.0570	<0.0001	0.0325	0.0138	0.0185	0.3447
Aesculus hippocastanum	−5.8001	0.0322	<0.0001	−0.0057	0.0084	0.4924	0.1535
Tilia cordata	−5.8743	0.0302	<0.0001	0.1425	0.0061	<0.0001	0.1453

Parameter estimates (β_0, β_1) with standard errors (SE), significances (p), and the variance of the random effect (τ^2).

Further model results of the allometric relationships between "tree height-stem diameter" and "crown radius-stem diameter" are shown in Tables 4 and 5.

Table 4. Linear regression models for stem diameter (*d*) to tree height (*h*) Equation (12) for all six species (ordered according to latitude).

Species	β_0	SE	β_1	SE	r^2	RSE	F	p
Araucaria cunninghamii	0.6599	0.1417	0.5945	0.0391	0.6510	0.1955	231.72	<0.0001
Khaya senegalensis	−0.1319	0.2416	0.7560	0.0566	0.5550	0.1510	178.51	<0.0001
Quercus nigra	1.8093	0.2149	0.2312	0.0524	0.1060	0.1521	19.47	<0.0001
Platanus x hispanica	0.9630	0.2343	0.4708	0.0566	0.2900	0.2135	69.07	<0.0001
Aesculus hippocastanum	0.1092	0.1986	0.6410	0.0480	0.4370	0.1651	178.03	<0.0001
Tilia cordata	1.6825	0.2246	0.2868	0.0591	0.0900	0.1976	23.56	<0.0001

Parameter estimates (β_0, β_1) with standard errors (SE), coefficients of determination (r^2), residual standard errors (RSE), F-values, and significances (p).

Table 5. Linear regression models for stem diameter (*d*) to crown radius (*cr*) Equation (13) for all six species (ordered according to latitude).

Species	β_0	SE	β_1	SE	r^2	RSE	F	p
Araucaria cunninghamii	−1.3881	0.1209	0.7147	0.0333	0.7880	0.1668	460.15	<0.0001
Khaya senegalensis	−0.7786	0.3394	0.6134	0.0793	0.2710	0.2157	59.79	<0.0001
Quercus nigra	−0.7962	0.2130	0.6691	0.0519	0.5030	0.1508	165.96	<0.0001
Platanus x hispanica	−0.5391	0.2298	0.5775	0.0556	0.3900	0.2094	108.06	<0.0001
Aesculus hippocastanum	−0.9663	0.1695	0.6431	0.0410	0.5180	0.1409	245.77	<0.0001
Tilia cordata	−0.0336	0.1929	0.4342	0.0507	0.2360	0.1675	73.27	<0.0001

Parameter estimates (β_0, β_1) with standard errors (SE), coefficients of determination (r^2), residual standard errors (RSE), F-values, and significances (p).

The relationship of stem diameter to tree height shows the highest scaling parameter for *K. senegalensis* (β_1 = 0.76) in Hanoi and lowest for *Q. nigra* (β_1 = 0.23) in Houston. For β_0 only *K. senegalensis* (β_0 = −0.13) shows a negative value, all other species have positive values with the highest for *Q. nigra* (β_0 = 1.81). In comparison to this, allometric exponents for the relationship of stem diameter to crown radius are in a smaller range. The scaling factors are positive for all species and in a very narrow range (β_1 = 0.43 for *T. cordata*, β_0 = 0.71 for *A. cunninghamii*). The lowest allometric exponent is β_0 = −1.39 for *A. cunninghamii* and the highest is for *T. cordata* with β_0 = −0.03.

Based on the allometric relationship of 'basal area-basal area increment' and the estimated coefficients (Table 3) the diameter growth of a distinct tree species at a specific age can be projected. It has to be mentioned that the term age concisely means the age at breast height of the tree. Further, by taking the results of the other allometric relationships 'tree height-stem diameter' and 'crown radius-stem diameter' (Tables 4 and 5) into account this kind of time-specific extrapolation can be extended to the tree height and the crown diameter growth.

A time-series for 200 years was calculated on the base of a stem diameter of 10 cm starting in year 0. The results for the three estimated parameters *d*, *cr*, *h*, and for all six tree species are shown in Figure 1. For all three allometric relationships a visualization of the standard errors of β_1 is added in the appendix (Figure A1).

For the stem diameter growth (Figure 1a) *K. senegalensis* in Hanoi shows the highest growth over time, while the lowest was calculated for *T. cordata* in Munich. At an age of 100 years, for example, *T. cordata* has a diameter at breast height of 42 cm, whereas *K. senegalensis* is about 92 cm. Similar findings can be seen for the temporal dynamics of the tree height-diameter relationship (Figure 1b). *K. senegalensis* marks the upper level, in contrast the lowest level, which is obvious for *A. hippocastanum* until an age of 130 years. Between this range the species *Q. nigra*, *T. cordata*, and *P. x hispanica* remain at the lower border, unlike *A. cunninghamii*, which is close to the level of *K. senegalensis*. A different temporal distribution was found for the crown radius (Figure 1c). The upper limit is represented by *Q. nigra* in Houston, having a *cr* of about 9 m at an age of 100 years. In contrast, at the same time the species *A. cunninghamii* in Brisbane, *T. cordata* in Berlin, and *A. hippocastanum* in Munich have crown radii of about 5 m. The study is a comparison of well-adapted trees for specific regions, but not a mere species comparison.

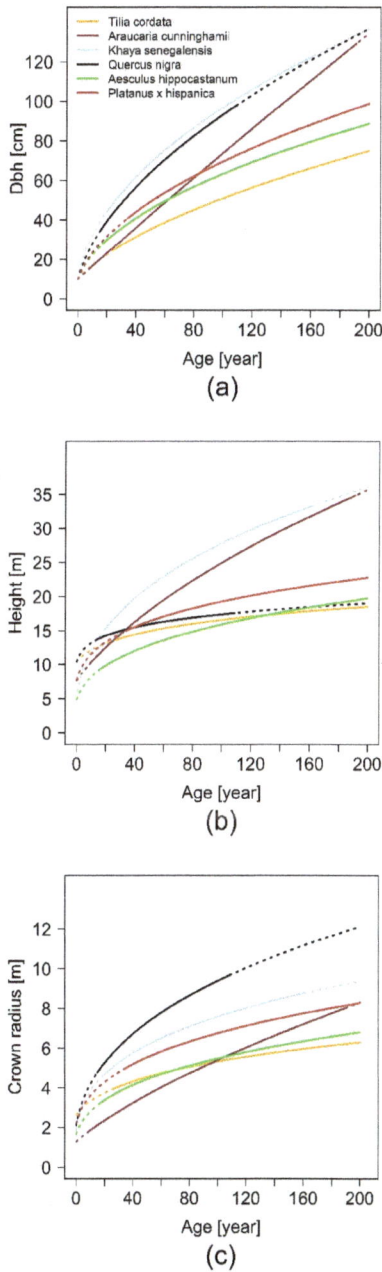

Figure 1. Allometric relationships for (**a**) diameter at breast height; (**b**) tree height; and (**c**) crown radius based on allometric model parametrizations for all six tree species (based on values of Tables 3–5). Dotted lines show projections which are not covered by measured data.

3.2. Impact of Paved Surface Area on Urban Tree Growth

As mentioned above, urban trees are often limited by non-paved area, especially in city centers. During our measurements the "non-paved surface area" (*npa*) was also recorded for tree species, if possible. This parameter, in the form of SCON Equation (3), is included in linear mixed effect models, one per species, for the ba-iba allometry Equation (7). The output of the models can be seen in Table 6, and the significances of α_1 indicate that the additional factor SCON is only significant for two species, *K. senegalensis* and *A. hippocastanum*. *Q. nigra* was also significant, but the results are not further shown, due to a not-plausible positive coefficient α_1, which can be explained by a very narrow range of the recorded values for *npa*.

Table 6. Linear regression results for the basal area—basal area increment allometry and the influence of SCON Equation (7) for all six species (ordered according to latitude).

Species	α_0	SE (α_0)	p (α_0)	α_1	SE (α_1)	p (α_1)	τ^2
A. cunninghamii	−4.6109	0.0769	0.0000	−0.0005	0.1116	0.9964	0.0765
K. senegalensis	−3.4053	0.3660	0.0000	−1.4998	0.3928	0.0002	0.2249
Q. nigra	−5.5567	0.1794	0.0000	0.7648	0.2221	0.0007	0.2222
P. x hispanica	−4.9857	0.2699	0.0000	−0.5632	0.3251	0.0856	0.3752
A. hippocastanum	−5.3079	0.0997	0.0000	−0.4056	0.1306	0.0022	0.2028
T. cordata	−5.7364	0.1721	0.0000	−0.1285	0.1954	0.5119	0.1868

Parameter estimates (α_0, α_1) with standard errors (SE), significances (p), and the variance of the random effect (τ^2).

The effect of the degree of spatial confinement by soil sealing on the parameters diameter, height, and crown radius for the species *K. senegalensis* in Hanoi and *A. hippocastanum* in Munich is depicted in Figure 2. Thereby three classes of the variable *SCON*: no limitation (*SCON* = 0), 50% limitation (*SCON* = 0.5), and high limitation (*SCON* = 0.9) were considered.

For *A. hippocastanum* only a slight influence of the limitation factor can be seen. Still, the trees with the lowest degree of limitation show the highest growth levels. In contrast, for *K. senegalensis* a strong relationship between the non-paved area and the tree growth can be found. Thereby, all *K. senegalensis* trees were mid- to high-restricted. Due to this, only these two classes could be shown in Figure 2. To give an example of the influence of the restricted non-paved area, the tree height of a 100 year old high-limited tree is 60% of the tree height of a similar-aged medium-limited tree. For the stem diameter growth a value of 50% can be accounted. With increasing age, the growth of the trees is increasingly inhibited. These findings show that restricted *npa* influences tree growth, but that the effect is species-specific and can be weak, as in the case of *A. hippocastanum* in Munich, or strong, as in the case of *K. senegalensis* in Hanoi. These findings can be reported for five of the six investigated species (Table 6).

3.3. Biomass and Carbon Storage Estimation for Urban Trees

Finally, the aboveground woody biomass of the species was estimated by species-specific equations from literature. The amounts of produced above ground biomass clearly vary from species to species (Figure 3). A 100 year old *T. cordata* tree in Berlin has a biomass of 0.7 t compared to a similar-aged *Q. nigra* tree in Houston with 3.3 t. The differences in the total aboveground biomass of the tree species increase with the age of the trees. Lowest biomass productivity by the age of 160 years is shown by *T. cordata* in Berlin and *P. x hispanica* in Paris with 1.3 t and 2.5 t per tree, respectively. For the same age, *A. hippocastanum* and *K. senegalensis* have a biomass of 3.5 and 4.2 t, respectively. With values of 5.2 and 5.9 t per tree the highest biomass values are achieved by *A. cunninghamii* and *Q. nigra*.

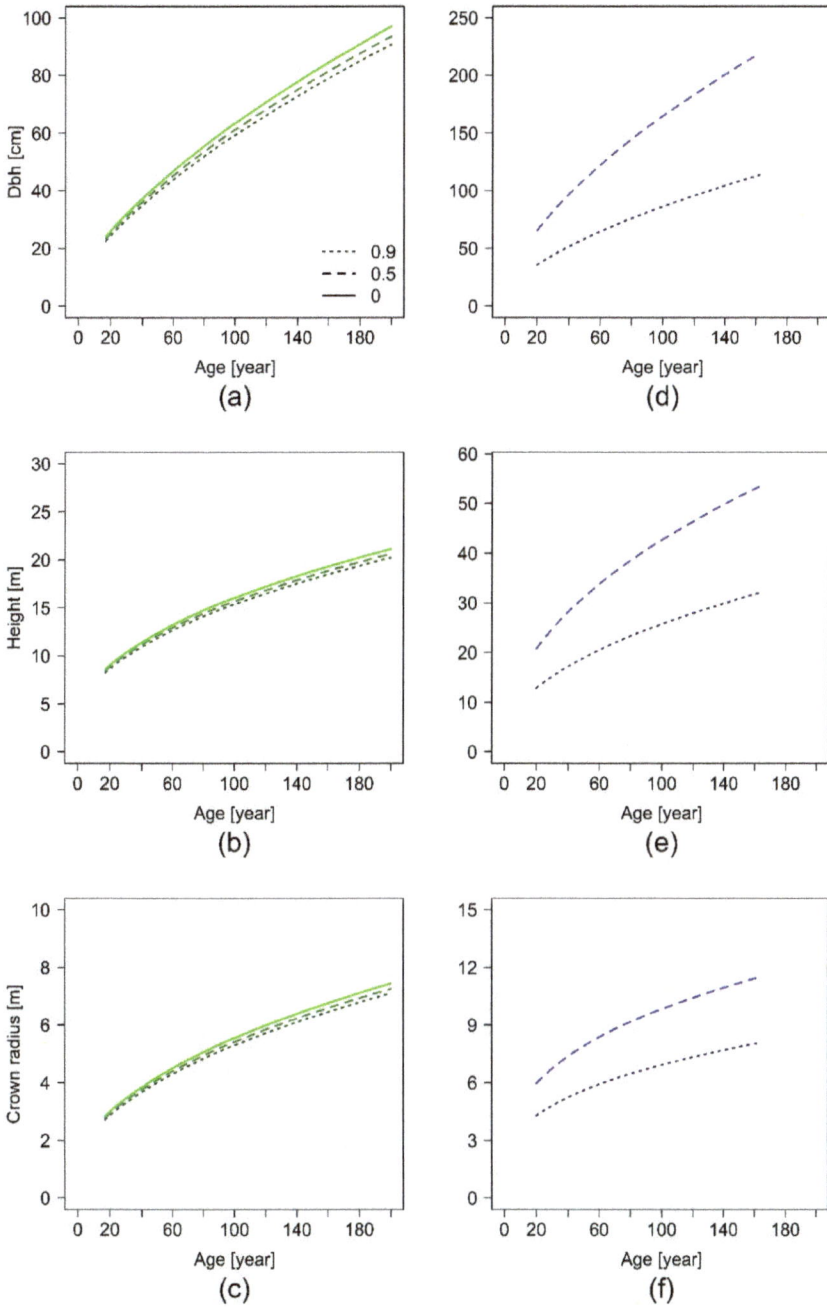

Figure 2. Effect of *SCON* on allometric growth relationships for diameter at breast height , tree height, and crown radius depending on age for the species *Aesculus hippocastanum* (panels **a–c**) and for *Khaya senegalensis* (panels **d–f**); solid line: no limitation, dashed line: 50% limited, and dotted line: highly limited (90%).

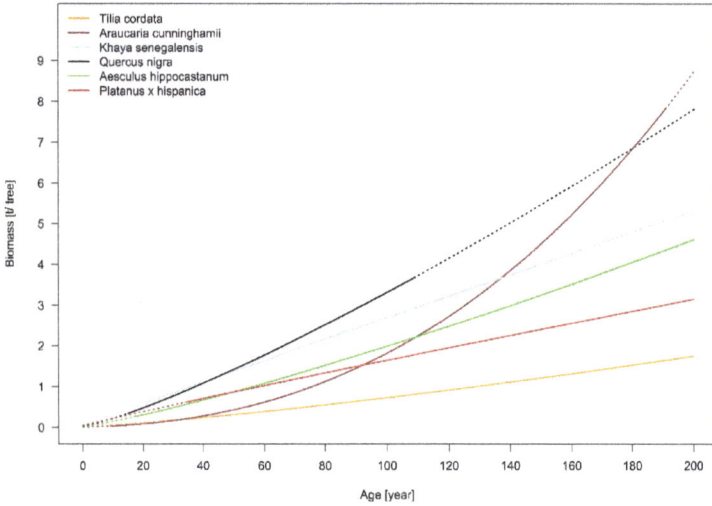

Figure 3. Estimated aboveground woody biomass for six tree species in urban environments, based on allometric equations. The data include the correction factor of 0.8 for transferring a stand biomass equation to urban trees. Dotted lines show projections which are not covered by measured data.

A comparison of the calculated carbon storage based on the measured tree parameters are shown in Figure 4, taking all available trees and the resulting average age into account (Tables 1 and 2). *A. hippocastanum* has the highest range from 122 to 4575 kg per tree and on average has 1106 kg per tree stored carbon; trees with lesser capacity include *P. x hispanica* (188–3038 kg per tree), *Q. nigra* (133–2145 kg per tree), and *K. senegalensis* (290–2161 kg per tree), the lowest values, as well as the smallest range, are attributed to *T. cordata* (71–1087 kg per tree) and *A. cunninghamii* (12–3531 kg per tree). The difference between the carbon storage (Figure 4) and the estimated biomass (Figure 3) is due to a variation in diameter at breast height (*dbh*)-range between the species, which can be seen in Table 2.

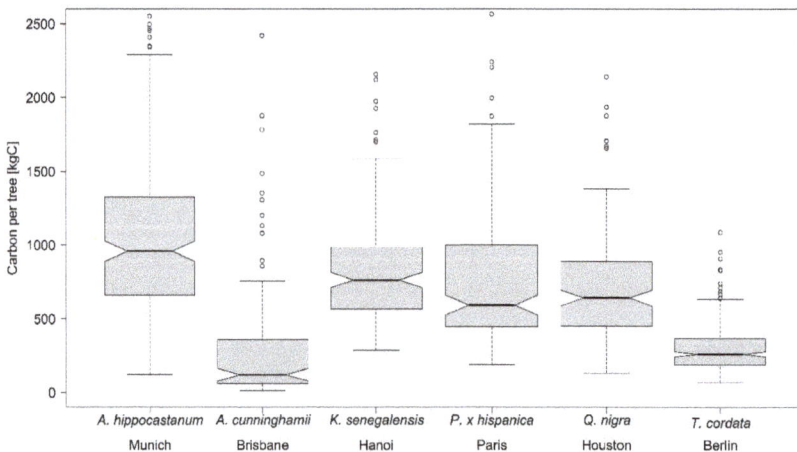

Figure 4. Carbon storage capacity of the investigated tree species in urban areas (cities), based on the measured data for diameter and tree height.

4. Discussion

4.1. Generalizabilty of the Identified Relationships

As mentioned in the introduction, this study cannot be interpreted as a usual species comparison, which would require a research design where the same set of species is observed under the same range of conditions. Our point is to compare urban tree growth and space requirements across a broad geographic climatic gradient based on a set of species that can be considered to be typical and highly adapted to the respective conditions. Some of the data we use are a subset of a data set that was used by Pretzsch *et al.* [10], where many urban tree species, including those dealt within the study at hand, are assigned to a small set of allometric types. This suggests that our findings can at least be roughly generalized for other species from the same allometric type and the same climate zone, and that they are representative for typical urban trees under the respective conditions.

In this context, the advantage of using allometric relations in contrast to other equation types is their strong foundation in biological theory. Thus, they are more than a mere description of a data set within the bounds of the sampled tree dimensions, as the parameter estimates transport direct biological information. An important consequence of this fact, allometric relationships can be extrapolated much more reliably than usual best-data-fit models. Species that show similar allometry can actually be considered to be comparable in terms of the variables involved. Furthermore, the allometry between tree size and size growth allows us to include temporal dynamics in our study, which are meaningful, albeit less precise, even when extrapolated due to their foundation in biological theory.

4.2. Growth and Space Requirements of Urban Trees Based on Allometric Relationships

Tree size and structure as described by tree allometry determine most functions and services of urban trees, such as shading, carbon fixation, or aesthetic embellishment. This underlines the relevance of a correct and species-specific quantification of tree allometry, which is in the focus of the study. Our continent-overarching study revealed the considerable extent to which allometry, and thus functions and services, can vary between species and change with proceeding tree ontogenesis. The study arrives at quantifying the allometry of six tree species common in urban environments. The quantification is relevant for practical planting guidelines for urban trees, for improving urban tree growth models, and parameterizing integrated tools for urban landscape planning.

The fundamental relationship of our analysis between basal area and basal area increment, which gives us the possibility of tracing back stem diameter growth and delivering us a sound scaling relationship, is based on increment cores. The results of our projected allometric relationships show that there are interspecific differences in urban tree allometry. The relationship of stem diameter to tree height shows the highest scaling parameter for *K. senegalensis* ($\beta_1 = 0.76$) in Hanoi and lowest for *Q. nigra* ($\beta_1 = 0.23$) in Houston. For β_0 only *K. senegalensis* ($\beta_0 = -0.13$) shows an unexpected negative value, all other species have positive values with the highest for *Q. nigra* ($\beta_0 = 1.81$). In comparison to this, allometric exponents for the relationship of stem diameter to crown radius are in a smaller range. The scaling factors are positive for all species and in a very narrow range ($\beta_1 = 0.43$ for *T. cordata* in Berlin, $\beta_0 = 0.71$ for *A. cunninghamii* in Brisbane). The lowest allometric exponent is $\beta_0 = -1.39$ for *A. cunninghamii* and is the highest for *T. cordata* with $\beta_0 = -0.03$. The results indicate that the growth of each species is individually based. Height growth, for example, increases from *A. hippocastanum* over *P. x hispanica* and *A. cunninghamii* to *K. senegalensis*, while crown radius expands from *A. cunninghamii* over *A. hippocastanum* and *P. x hispanica* to *Q. nigra*. These differences in growth dynamics for the specific species are highly relevant for urban forestry managers and planers. They need to know the space requirements of different species and how they will expand in the future in order to avoid potential conflicts with urban infrastructure.

Pretzsch *et al.* [22] state that interspecies differences in tree allometry exist in forest stands. They applied a similar method as in the present study and highlighted that competition can change plant morphology, and that allometric relationships represent structural configurations of forest stands on a species-specific or a general scaling level. Their analysis was based on stem slices of different tree species from forest stands in southern Germany.

The results of the present study show that differences in tree allometry are not only present in forest stands due to competition, but are similarly found in urban areas. Pretzsch *et al.* [10] also investigated crown allometry of different urban tree species, but without taking the temporal dynamic of the species into account. Their findings resulted in five allometric crown extension types, which also covered a distinct range of expansion and showed species-specific differences. These allometry types can also serve as a means of generalizing the results shown in this study.

4.3. Impact of Paved Surface Area on Urban Tree Growth

Our results show evidence that the restricted non-paved area limits tree growth in urban areas. Thereby, we could demonstrate that the level of growth limitation due to spatial confinement by sealed surfaces is species-specific.

This finding might be explained by water supply and the tree species' water storage capacity. Quigley [18] assumes that urban tree growth is slower than in forest stands due to the fact that a distinct proportion of the roots is below the pavement, and in this way is cut off from precipitation water. The reduction of water supply might be the main disadvantage of the restricted non-paved area affecting tree growth. The water storage capacity, which is very limited in the case of paved surfaces, is investigated for different tree species by Xiao and McPherson [13]. They state that this capacity is very species-specific and age-dependent.

Another study showed that the pavement type and profile design have an influence on tree growth, as well as the compaction, but that the pavements themselves do not cause a reduced growth rate [23]. We used the area of non-paved surface as a parameter for the influence of restricted area and found a trend for species-specific growth reduction with increasing restricted area.

The conflicts of roots with urban infrastructure are reviewed by Randrup *et al.* [35] in the way of analyzing the most relevant factors causing problems with tree roots in urban areas. As main factors they list among other things: species with a large maturity size, fast growing species, trees planted in restricted soil volumes, shallow irrigation, restricted distance between tree and sidewalk, and, in general, trees older than 15 to 20 years. Of these factors the restricted soil volume and shallow irrigation are likely to be given if the non-paved surface area is limited. Due to this limitation in non-paved area our analysis showed a significantly reduced growth, especially for *K. senegalensis*.

4.4. Comparison of Biomass and Carbon Storage Capacities of Different Urban Tree Species

McPherson *et al.* [7] investigated the differences in biomass between forest and urban trees and concluded that forest trees on average contain 20% more biomass than urban trees at a given height and stem diameter. Following their suggestion, we multiplied the outcomes obtained with the traditional biomass equations with a factor of 0.8 in order to achieve more realistic estimates for our urban trees. This factor is also considered in the i-Tree Eco/Urban Forest Effects (UFORE) model developed in the USA [36]. In contrast, Russo *et al.* [37] did not use such a factor for their study in Bolzano, Italy, due to their assumption that not all urban trees are open-grown. McHale *et al.* [6] also compared forest and urban biomass values and equations and stated that the value of 20% difference in biomass should be re-evaluated.

In Table 7 the results of two other urban tree studies on C storage [8,9] are compared with the values of the present study. For *T. cordata* the C storage values found by Moser *et al.* [8] are in a similar range to our findings. Both studies investigated *T. cordata* trees in German cities, which means that both studies took place within the same climate region. The minimum and mean C storage values of the present study are higher, due to the fact that less small trees were sampled and the mean diameter is higher. Again, Moser *et al.* [8] did not apply the correction factor of 0.8 and Yoon *et al.* [9] used it for avoiding an overestimation of the urban tree biomass for their estimations. Despite the fact that the two studies were done in different climate regions, and assuming that the two studied *Platanus* species (*P. x hispanica* and *P. orientalis*) show the same growth patterns, the C storage of the tree individuals fit together when based on their diameter.

Table 7. Comparison of carbon storage capacity based on the estimation of above ground woody biomass for trees in the present study with other studies in urban areas and the related diameter range.

Species	C Storage (kg/tree)			d (cm)
	Min	Mean	Max	Min–Max
Platanus x hispanica	259	702	2956	40–144
Platanus orientalis [1]	83	-	406	23–48
Tilia cordata	71	303	1087	25–81
Tilia cordata [2]	8	196	1341	6–107

[1] Data from Yoon *et al.* [9]; [2] Data from Moser *et al.* [8].

These results show that although a comparison of different studies is difficult due to different diameter ranges and different equations that might be used for the estimation, still the findings indicate that a tree species performs in a similar range on different sampling locations.

4.5. Carbon Storage Capacity of T. Cordata on City Scale

The present study estimates the temporal dynamic of urban tree growth and expansion. On this basis an estimation of the annual net carbon fixation of urban trees' woody biomass for a whole city is possible. Different studies [37–39], which quantified above-ground carbon storage on a city-wide scale, show that more research in this field is needed. For deriving a plausible estimation on the city scale the diameter distribution of the tree collective has to be known. For *T. cordata* a sample of 61,000 trees in Berlin [40] delivered us this distribution. In total, *T. cordata* trees amounted to 155,000 in 2014, by having a share of 35% of all street trees in Berlin [40]. By using the above mentioned biomass equation for *T. cordata* [33], which takes the diameter and the tree height into account, the carbon storage of this species sums up to 31.17 million tons C. On the base of the mean tree age of the respective diameter class the annual net average C-fixation of one single lime tree was calculated with 2.8 kg C year^{-1}, which results in an annual C-fixation for all lime trees in Berlin of 430 t C year^{-1}. Assuming a per capita C consumption of 1.5 t C year^{-1} for the year 2012 for Berlin [41], the annual carbon-fixation of lime trees corresponds to a carbon consumption of 284 inhabitants of Berlin.

Comparing the average value of 2.8· kg C year^{-1} of the present study to a similar study from Italy by Russo *et al.* [37] shows that with 2.8 kg C year^{-1} our estimation is out of his range, which lies between 9.79 to 43.06 kg C year^{-1}. It has to be mentioned though that *T. cordata* represents the lowest level of biomass production and carbon sequestration from our investigated species. This means that the other investigated species are within the range of Russo's values, which represent diverse tree species. Further, the differences between the studies might be due to the fact that the study by Russo *et al.* [37] included trees from urban areas as well as forest stands.

Schreyer *et al.* [42] derived the carbon storage per tree of different structure types in the City of Berlin. They list a value of 1,028,427 t carbon for the entire city of Berlin. In comparison, we quantified the value of 31,170 t C for 155,000 *Tilia* trees. If we take into account that *T. cordata* has a low C fixation rate compared to other species, and that just one third of Berlin's vegetation is represented with our selection, the total values of the two studies might fit together.

5. Conclusions

In summary, the results show the species-specific allometries of a set of typical urban trees worldwide, under the conditions they are typically found and best adapted to, by taking the temporal dynamic into account. The enlargement of diameter, tree height, and crown radius size are very species-specific and vary between the six investigated urban tree species. It can also be shown that the impact of paved surface inhibits tree growth. A significant impact of the non-paved area was found for the species *A. hippocastanum* in Munich and *K. senegalensis* in Hanoi. The estimations for above ground biomass are also species-specific and show the highest values for the deciduous species *Q. nigra* in Houston and the lowest for *T. cordata* in Berlin.

Our study gives statistically solid quantitative information about how typical urban tree species develop worldwide with specific regard given towards space requirements, which we deem useful for urban managers. Linking this information with the allometric species types as identified in Pretzsch *et al.* [10], the applicability of our study is beyond the species investigated here. With this knowledge better adaption of treatments for urban trees, which are in potential conflict with urban infrastructure, is possible; this information might be useful in supporting the benefits of urban trees. As a result, the additional costs for tree maintenance and care, which are well reported [2], can be reduced and a more effective way of managing might be possible.

Future research in this field is needed for an extended understanding of the growth behavior of urban trees and its temporal dynamic, specifically in relation to anthropogenic decisions, like paving surfaces. Last but not least, further investigations on urban tree biomass and carbon storage worldwide are important against the background of rising interest in urban air quality.

We based our study on both countries with long experience in urban tree research and countries that presently have begun caring for and investigating urban trees. As such, we tried to contribute to a continuous improvement of the appreciation, research, and knowledge base of urban trees worldwide.

Acknowledgments: Thanks to the AUDI Environmental Foundation for funding the project "Response of urban trees on climate change" and the several city ministries and forest services for the allowance of coring and measuring trees and for supporting the search for the trees. This work was supported by the German Research Foundation (DFG) and the Technical University of Munich (TUM) in the framework of the Open Access Publishing Program. The authors also want to thank two anonymous reviewers for their constructive and helpful comments on previous versions of the manuscript.

Author Contributions: Hans Pretzsch, Thomas Rötzer, and Enno Uhl conceived and designed the experiment; Jens Dahlhausen performed the experiments and analyzed the data; Peter Biber contributed statistical analysis tools; Jens Dahlhausen and Peter Biber wrote the paper. Thomas Rötzer and Enno Uhl revised the manuscript.

Conflicts of Interest: The authors declare no conflict of interest. The founding sponsors had no role in the design of the study; in the collection, analyses, or interpretation of data; in the writing of the manuscript, and in the decision to publish the results.

Abbreviations

The following abbreviations are used in this manuscript:

awb	aboveground woody biomass
ba	basal area
cb	crown base
cr	mean crown radius
cpa	crown projection area
hcb	height to crown base
iba	basal area increment
npa	non-paved area
OLS	Ordinary least squares
SCON	Spatial confinement of a tree

Appendix A

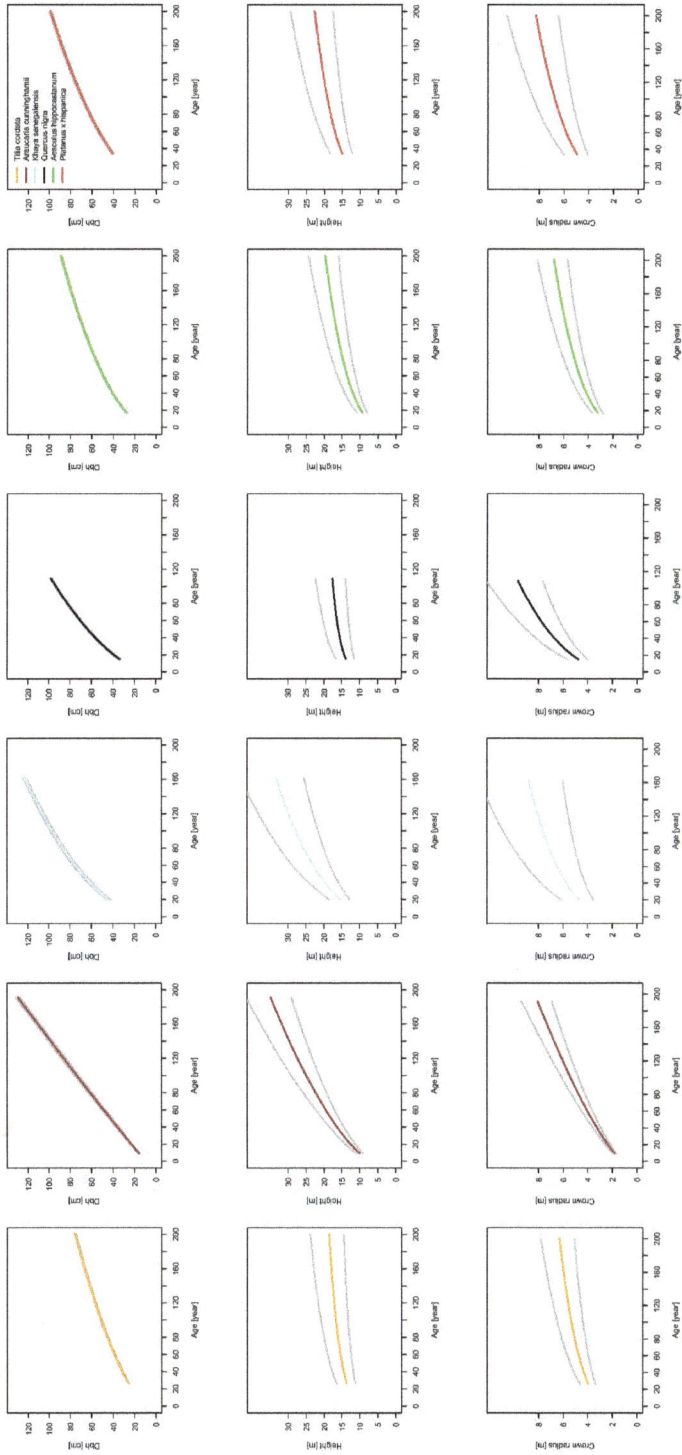

Figure A1. Allometric relationships for diameter at breast breast height (upper line), tree height (middle line) and crown radius (lower line) based on allometric model parametrizations for all six tree species (based on values of Tables 3–5). The coloured lines show the best-fit lines, and the grey lines indicate the visualized standard error of β_1.

References

1. Escobedo, F.J.; Kroeger, T.; Wagner, J.E. Uban forests and pollution mitigation: Analyzing ecosystem services and disservices. *Environ. Pollut.* **2011**, *159*, 2078–2087. [CrossRef] [PubMed]
2. Mullaney, J.; Lucke, T.; Trueman, S.J. A review of benefits and challenges in growing street trees in paved urban environments. *Landsc. Urban Plan.* **2015**, *134*, 157–166. [CrossRef]
3. McPherson, E.G.; Muchnick, J. Effect of street tree shade on asphalt concrete pavement performance. *J. Arboric.* **2005**, *31*, 303.
4. Livesley, S.J.; McPherson, G.M.; Calfapietra, C. The Urban Forest and Ecosystem Services: Impacts on Urban Water, Heat, and Pollution Cycles at the Tree, Street, and City Scale. *J. Environ. Qual.* **2016**, *45*, 119–124. [CrossRef] [PubMed]
5. Quigley, M.F. Fanklin Park: 150 years of changing design, disturbance, and impact on tree growth. *Urban Ecosyst.* **2002**, *6*, 223–235. [CrossRef]
6. McHale, M.R.; Burke, I.C.; Lefsky, M.A.; Peper, P.J.; McPherson, E.G. Uban forest biomass estimates: Is it important to use allometric relationships developed specifically for urban trees? *Urban Ecosyst.* **2009**, *12*, 95–113. [CrossRef]
7. McPherson, G.E.; Nowak, D.J.; Rowntree, R.A. *Cicago's Urban Forest Ecosystem: Results of the Chicago Urban Forest Climate Project*; United States Department of Agriculture: Radnor, PA, USA, 1994.
8. Moser, A.; Rötzer, T.; Pauleit, S.; Pretzsch, H. Structure and ecosystem services of small-leaved lime (*Tilia cordata* Mill.) and black locust (*Robinia pseudoacacia* L.) in urban environments. *Urban For. Urban Green.* **2015**, *14*, 1110–1121. [CrossRef]
9. Yoon, T.K.; Park, C.-W.; Lee, S.J.; Ko, S.; Kim, K.N.; Son, Y.; Lee, K.H.; Oh, S.; Lee, W.-K.; Son, Y. Allometric equations for estimating the aboveground volume of five common urban street tree species in Daegu, Korea. *Urban For. Urban Green.* **2013**, *12*, 344–349. [CrossRef]
10. Pretzsch, H.; Biber, P.; Uhl, E.; Dahlhausen, J.; Rötzer, T.; Caldentey, J.; Koike, T.; van Con, T.; Chavanne, A.; Seifert, T.; *et al.* Crown size and growing space requirement of common tree species in urban centres, parks, and forests. *Urban For. Urban Green.* **2015**, *14*, 466–479. [CrossRef]
11. Gill, S.E.; Handley, J.F.; Ennos, A.R.; Pauleit, S. Adapting cities for climate change: The role of the green infrastructure. *Built Environ.* **2007**, *3*, 115–133. [CrossRef]
12. McPherson, E.G.; Nowak, D.; Heisler, G.; Grimmond, S.; Souch, C.; Grant, R.; Rowntree, R. Quantifying urban forest structure, function, and value: The Chicago Urban Forest Climate Project. *Urban Ecosyst.* **1997**, *1*, 49–61. [CrossRef]
13. Xiao, Q.; McPherson, E.G. Srface water storage capacity of twenty tree species in Davis, California. *J. Environ. Qual.* **2016**, *45*, 188–198. [CrossRef] [PubMed]
14. Scharenbroch, B.C.; Morgenroth, J.; Maule, B. Tree species suitability to bioswales and impact on the urban water budget. *J. Environ. Qual.* **2016**, *45*, 199–206. [CrossRef] [PubMed]
15. Escobedo, F.J.; Clerici, N.; Staudhammer, C.L.; Corzo, G.T. Scio-ecological dynamics and inequality in Bogotá, Colombia's public urban forests and their ecosystem services. *Urban For. Urban Green.* **2015**, *14*, 1040–1053. [CrossRef]
16. Anderson, S.T.; West, S.E. Oen space, residential property values, and spatial context. *Reg. Sci. Urban Econ.* **2006**, *36*, 773–789. [CrossRef]
17. Peper, P.J.; Alzate, C.P.; McNeil, J.W.; Hashemi, J. Allometric equations for urban ash trees (*Fraxinus* spp.) in Oakville, Southern Ontario, Canada. *Urban For. Urban Green.* **2014**, *13*, 175–183. [CrossRef]
18. Quigley, M.F. Sreet trees and rural conspecifics: Will long-lived trees reach full size in urban conditions? *Urban Ecosyst.* **2004**, *7*, 29–39. [CrossRef]
19. Sanders, J.R.; Grabosky, J.C. 20 years later: Does reduced soil area change overall tree growth? *Urban For. Urban Green.* **2014**, *13*, 295–303. [CrossRef]
20. Blunt, S.M. Trees and pavements—Are they compatible? *Arboric. J.* **2008**, *31*, 73–80. [CrossRef]
21. Kirkpatrick, J.B.; Davison, A.; Daniels, G.D. Rsident attitudes towards trees influence the planting and removal of different types of trees in eastern Australian cities. *Landsc. Urban Plan.* **2012**, *107*, 147–158. [CrossRef]
22. Pretzsch, H.; Dauber, E.; Biber, P. Species-Specific and Ontogeny-Related Stem Allometry of European Forest Trees: Evidence from Extensive Stem Analyses. *For. Sci.* **2013**, *59*, 290–302. [CrossRef]

23. Morgenroth, J. Root growth response of Platanus orientalis to porous pavements. *Arboric. Urban For.* **2011**, *37*, 45.

24. Assmann, E. *Waldertragskunde: Organische Produktion, Struktur, Zuwachs und Ertrag von Waldbeständen;* BLV Verlagsgesellschaft: München, Germany, 1961.

25. Preuhsler, T. Ertragskundliche Merkmale oberbayerischer Bergmischwald-Verjüngungsbestände auf kalkalpinen Standorten im Forstamt Kreuth. *Forstwiss. Cent.* **1981**, *100*, 313–345. [CrossRef]

26. Niklas, K.J. Pant allometry: Is there a grand unifying theory? *Biol. Rev.* **2004**, *79*, 871–889. [CrossRef] [PubMed]

27. Enquist, B.J.; Niklas, K.J. Ivariant scaling relations across tree-dominated communities. *Nature* **2001**, *410*, 655–660. [CrossRef] [PubMed]

28. West, G.B.; Enquist, B.J.; Brown, J.H. A general quantitative theory of forest structure and dynamics. *Proc. Natl. Acad. Sci. USA* **2009**, *106*, 7040–7045. [CrossRef] [PubMed]

29. Jenkins, J.C.; Chojnacky, D.C.; Heath, L.S.; Birdsey, R.A. Ntional-scale biomass estimators for United States tree species. *For. Sci.* **2003**, *49*, 12–35.

30. Eamus, D.; McGuinness, K.; Burrows, W.W. *Review of Allometric Relationships for Estimating Woody Biomass for Queensland, the Northern Territory and Western Australia;* Technical Report No. 5a; Australian Greenhouse Office: Canberra, Australia, 2000.

31. Clément, J. Estimation des volumes et de la productivité des formations mixtes foresiéres et graminées tropicales. *Bois For. Trop.* **1982**, *198*, 35–58.

32. Clark, A.I.; Phillips, D.R.; Frederick, D.J. *Wight, Volume, and Physical Properties of Major Hardwood Species in the Gulf and Atlantic Coastal Plains;* United States Department of Agriculture: Radnor, PA, USA, 1985.

33. Čihák, T.; Hlásny, T.; Stolariková, R.; Vejpustková, M.; Marušák, R. Functions for the aboveground woody biomass in Small-leaved lime (Tilia cordata Mill.)/Funkce pro hodnocení biomasy nadzemních částí lípy maloListé (Tilia cordata Mill.). *For. J.* **2014**, *60*, 150–158. [CrossRef]

34. Penman, J.; Gytarsky, M.; Hiraishi, T.; Krug, T.; Kruger, D.; Pipatti, R.; Buendia, L.; Miwa, K.; Ngara, T.; Tanabe, K.; *et al. Good Practice Guidance for Land Use, Land-Use Change and Forestry;* Penman, J., Gytarsky, M., Hiraishi, T., *et al.*, Eds.; IPCC National Greenhouse Gas Inventories Programme: Kanagawa, Japan, 2003.

35. Randrup, T.B.; McPherson, E.G.; Costello, L.R. A review of tree root conflicts with sidewalks, curbs, and roads. *Urban Ecosyst.* **2001**, *5*, 209–225. [CrossRef]

36. Nowak, D.J.; Crane, D.E. The Urban Forest Effects (UFORE) Model: Quantifying Urban Forest Structure and Functions. Available online: http://www.nrs.fs.fed.us/pubs/gtr/gtr_nc212/gtr_nc212_714.pdf (accessed on 22 March 2016).

37. Russo, A.; Escobedo, F.J.; Timilsina, N.; Schmitt, A.O.; Varela, S.; Zerbe, S. Assessing urban tree carbon storage and sequestration in Bolzano, Italy. *Int. J. Biodivers. Sci. Ecosyst. Serv. Manag.* **2014**, *10*, 54–70. [CrossRef]

38. Davies, Z.G.; Edmondson, J.L.; Heinemeyer, A.; Leake, J.R.; Gaston, K.J. Mpping an urban ecosystem service: Quantifying above-ground carbon storage at a city-wide scale: Urban above-ground carbon storage. *J. Appl. Ecol.* **2011**, *48*, 1125–1134. [CrossRef]

39. Lee, J.-H.; Ko, Y.; McPherson, E.G. Te feasibility of remotely sensed data to estimate urban tree dimensions and biomass. *Urban For. Urban Green.* **2016**, *16*, 208–220. [CrossRef]

40. GRIS. Straßenbäume in Berlin—Bestand nach Hauptgattungen in den Berliner Bezirken. Available online: http://www.stadtentwicklung.berlin.de/umwelt/stadtgruen/stadtbaeume/de/daten_fakten/downloads/ausw_139.pdf (accessed on 31 December 2014).

41. Dank, H. Statistischer Bericht E IV 4–j/07 Energie-und CO_2-Bilanz im Land Brandenburg 2007. Available online: https://opus4.kobv.de/opus4-slbp/frontdoor/index/index/year/2010/docId/2218 (accessed on 16 March 2016).

42. Schreyer, J.; Tigges, J.; Lakes, T.; Churkina, G. Using Airborne LiDAR and QuickBird Data for Modelling Urban Tree Carbon Storage and Its Distribution—A Case Study of Berlin. *Remote Sens.* **2014**, *6*, 10636–10655. [CrossRef]

forests

MDPI

Article

How Do Urban Forests Compare? Tree Diversity in Urban and Periurban Forests of the Southeastern US

Amy Blood [1], Gregory Starr [1], Francisco Escobedo [2], Art Chappelka [3] and
Christina Staudhammer [1,*]

[1] Department of Biological Sciences, The University of Alabama, Tuscaloosa, AL 35401, USA;
    ablood@crimson.ua.edu (A.B.); gstarr@ua.edu (G.S.)
[2] Faculty of Natural Sciences and Mathematics, Biology Program, Universidad del Rosario, Bogotá D.C.,
    Colombia; franciscoj.escobedo@urosario.edu.co
[3] School of Forestry and Wildlife Sciences, Auburn University, Auburn, AL 36849, USA; chappah@auburn.edu
*   Correspondence: cstaudhammer@ua.edu; Tel.: +1-205-348-1538

Academic Editor: Timothy A. Martin
Received: 29 April 2016; Accepted: 3 June 2016; Published: 9 June 2016

Abstract: There is a need to understand how anthropogenic influences affect urban and periurban forest diversity at the regional scale. This study aims to compare urban and periurban tree composition along a geographic gradient, and test hypotheses about species composition and ecological homogeneity. We paired urban forest (UF) data from eight cities across the southeastern US with periurban forest (PF) data from the USDA Forest Service Forest Inventory and Analysis program. We found that tree diversity, as well as both observed and estimated species richness values were greater in UF *versus* PF. Community size structure analysis also indicated a greater proportion of large trees and greater numbers of non-native, invasive, and unclassified tree species in the UF *versus* the PF, regardless of location. Both forest type and ecological province had a significant effect on community species composition, with forests closer together in space being more similar to each other than those more distant. While land use change and management has been associated with ecological homogenization in human dominated landscapes, we found that species composition was more dissimilar along latitudinal lines than compared to between forest types, refuting this hypothesis, at least in terms of tree diversity.

Keywords: urban forest composition; regional diversity; forest inventory and analysis; ecological homogenization

1. Introduction

Rapid urbanization and land use change in proximity to urban areas has led to the alteration of structure and composition of forests [1]. Novel ecosystem assemblages have developed in both urban and periurban forests in response to land use change, as well as species introductions, ecological disturbance, and sociopolitical and economic shifts [2–5]. As natural landscapes are altered by urbanization, there is a gap in our understanding of the implications these changes might have on regional urban and periurban tree diversity. For example, have anthropogenic influences resulted in a homogenization of species composition within urban forests across regional scales? Are human-dominated landscapes providing adequate areas for native tree species? Is the species composition of urban areas more or less resistant to climate change, as compared to adjacent periurban forests?

There is mounting evidence suggesting that more diverse ecosystems have increased resistance to pests and disease [5–8]. Urban forests with low tree diversity may be at substantial risk in terms of potential alteration from ecological disturbances. For example, Dutch elm disease caused the mortality of millions of elm trees [5], greatly changing the urban forest composition and structure in cities such as Boulder City, NV and Chicago, IL in the United States (US) [9,10]. Since then, urban forester managers and planners have placed a higher value on urban tree diversity to limit future disturbances [11].

Because tree diversity is associated with enhanced ecosystem resistance to disturbances, tree diversity can also potentially increase resistance and adaptation to the impacts of a changing climate [8]. More diverse populations have greater adaptive capacity and thus are able to better withstand changes in ambient biophysical conditions [12,13]. Climate change has already caused shifts in tree species ranges and phenology [14,15]; if changes continue as predicted, this may lead to future unsuitability of species in both urban and periurban forests.

Invasive woody plants have had well-documented negative effects on both urban and periurban landscapes [16–18]. The presence of invasive plant species has been associated with reduced biodiversity and increased biotic homogenization [4,16], changes in forest structure, altered natural disturbance regimes [19,20], and subsequently modified ecosystem processes [21,22]. Moreover, innocuous non-native species may become invasive; worldwide, an estimated 62% of invasive species were introduced for horticulture and 13% for forestry [4]. For example, *Pyrus calleryana* Decne. has escaped cultivation in many areas of the US as a result of hybridization, forming dense homogenous plant communities and outcompeting local species [23,24]. *Pinus pinaster* Aiton and *P. halepensis* Mill. have negatively impacted water resources and biodiversity, as well as increased fire severity in Cape Town, South Africa [17]. Additionally, with changing climate, species that have not yet previously posed an invasive threat may become so with range shifts into areas more suitable for the non-native species [25].

Because non-native tree species (e.g., *P. calleryana* and *Acer platanoides* L. in the US) often have a broader niche that allows them to thrive in urban locales, these species have been historically planted intentionally throughout different regions of the world. These human planting preferences have led to the hypothesis that human dominated landscapes are undergoing ecological homogenization [3]. However, few studies have quantified similarities in tree species composition across urban and rural or periurban forest gradients.

Given this lack of information, our study aims to characterize community structure and composition of urban forests in the southeastern United States (SE US), and compare these metrics against adjacent periurban forests. For this study, we defined urban forests (UF) as a collection of trees within urban boundaries and its mixed land uses as well as in privately and publicly owned properties [26]. UF therefore included high- and low-density residential areas, as well as commercial, institutional, and industrial land uses. Periurban forests (PF) were defined as non-urban, forested areas located within 25 km of an urban city center. The SE US is an ideal location for this type of study due to its recognition as a tree diversity hotspot within North America and classification as a priority ecoregion for conservation [27,28]. We hypothesize that greater tree diversity exists in UF, with a greater number of both non-native and native species [29]. Additionally, a greater prevalence of human disturbance (as is the case in urban areas *versus* periurban areas) will result in a greater number of unintentional introductions, resulting in a prevalence of non-native and invasive species in UF *versus* PF [4]. We also expect that trees in UF will be smaller than those in adjacent PF. Moreover, we hypothesize that the species community composition among the UF of the different localities sampled will be more closely linked to geography than forest type (UF *vs.* PF).

2. Materials and Methods

2.1. Data Acquisition

Urban forest data were collected by independent research groups in eight localities (Figure 1; Table 1) between 2008 and 2014. The data were compiled into the Southeastern Urban Tree Database as part of a collaborative project sponsored by several universities and the Southern Research Station of the US Department of Agriculture (USDA) Forest Service. All UF data were collected using, or based on, the i-Tree Eco protocol [30], a standard sampling and inventory protocol that has been used extensively for sampling urban forests throughout the world [31,32]. In Gainesville, FL, plots were established randomly across varied urban land uses, such as commercial, vacant, and residential [33]. (See [22,30–36] for specific information regarding land use definition and the percentage of these in the different localities.) Four Virginia locations (Charlottesville, Abingdon, Winchester, and Roanoke City; [36]), as well as east Orlando, Florida were sampled using stratified random sampling, stratifying by land use [22]. One location in Virginia (Falls Church) was sampled using a randomized grid. Plots were selected in the City of Atlanta (Fulton and DeKalb counties) as part of a sub-study of the Proctor Creek Watershed (a Federal Urban Waters Project, [37]). Plots were selected via random generation of GIS locations from three land use class strata (residential, transportation, and all other) in proportion to US Geographical Survey land use land cover classifications. All urban localities in the study area were sampled using randomly selected 0.04 ha plots established within city limit boundaries (with the exception of east Orlando, where plots were placed within a 200 km^2 area around an eddy flux tower).

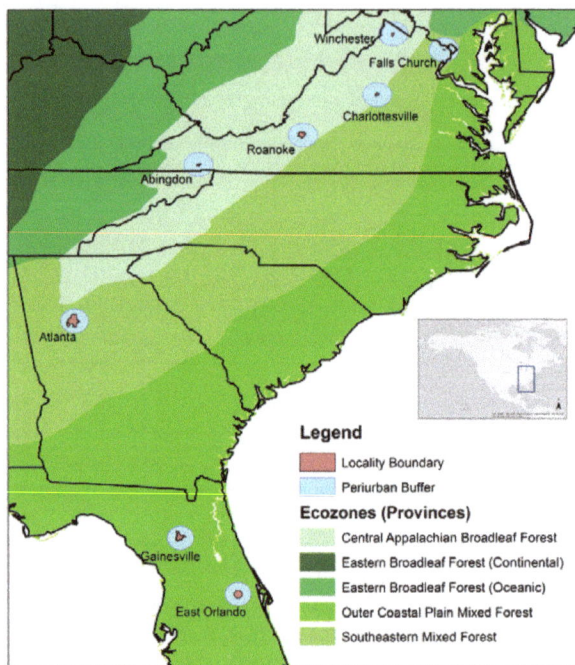

Figure 1. Eight urban forest locations sampled in the southeastern United States (SE US). Inset map shows where this region is relative to North America.

Each tree or palm with diameter at breast height (DBH; measured at 1.37 m from the ground) greater than 2.5 cm was measured and its species name recorded. Other measurements included height, land use, crown width, and crown light exposure. If a tree had multiple stems below DBH, it was counted as a singular tree, and the largest diameters (up to six) were recorded. For specific measurement methods, refer to [22,30,33–36]. For multi-stemmed trees, we calculated the quadratic mean diameter to represent the DBH underlying a tree with the same total tree cross-sectional area.

Table 1. Locality and plot characteristics for sampling locations and urban* and periurban** forest types in the southeastern US study area. Numbers of trees were not expanded to reflect lower sampling intensities in periurban microplots. Ecological province was defined by USDA Forest Service ecozones [38]. Population density is from 2010 U.S. Census Data [39]. CABF = Central Appalachian Broadleaf Forest-Coniferous Forest-Meadow, SMF = Southeastern Mixed Forest, OCP = Outer Coastal Plain.

Locality	Location Center	Ecological Province	Average Annual Rainfall (cm)	Urban Population Density (Person/km²)	Forest Type	Elevation Range (m)	Area Sampled (ha)	Number of Treed Plots	Number of Trees
Abingdon, VA	36°42'35" N, 81°58'32" W	CABF	116	392	Urban	590–709	1.56	39	279
					Periurban	414–1419	3.83	57	1970
Atlanta, GA	33°45'18" N, 84°23'24" W	SMF	126	1218	Urban	159–331	11.92	298	1548
					Periurban	145–513	2.02	30	659
Charlottesville, VA	38°1'47.64" N, 78°28'44.4" W	CABF	108	1640	Urban	92–198	2.36	59	409
					Periurban	68–964	3.16	47	1205
East Orlando, FL	28°35'33.54" N, 81°12'0.34" W	OCP	129	1023	Urban	1–29	3.16	79	732
					Periurban	–1–39	1.75	26	603
Falls Church, VA	38°52'56" N, 77°10'16" W	SMF	113	2382	Urban	82–134	1.24	31	178
					Periurban	–25–174	1.28	19	357
Gainesville, FL	29°39'7.19" N, 82°19'29.97" W	OCP	120	783	Urban	15–61	2.04	51	659
					Periurban	11–61	3.43	51	1429
Roanoke, VA	37°16'0" N, 79°56'0" W	CABF	105	880	Urban	271–531	4.56	114	1627
					Periurban	234–1197	2.76	41	1044
Winchester, VA	39°11'0" N, 78°10'0" W	CABF	97	1096	Urban	194–287	2.24	56	336
					Periurban	92–864	1.95	29	633

* Urban forest types sampled using i-Tree plots; ** Periurban forest types sampled using FIA plots.

All periurban data were from forested areas sampled with the USDA Forest Inventory and Analysis (FIA) protocol which were located within 25 km of an urban city center. Data were obtained from the USDA Forest Service datamart [40]. FIA data encompass areas that range from suburban to rural [2], but include only "forest land", which is defined as having an area of at least 0.4 ha with at least 10% canopy cover of live tree species of any size, either at the time of sampling or in the past, where the land is not subject to non-forest use which would prevent normal tree regeneration and succession (e.g., regular mowing, or grazing) [28]. All FIA plots are measured on a cyclic basis throughout the United States by the USDA Forest Service, with the intention of collecting data on an estimated 1 in 2428 hectares of land in the country to monitor the forest resources of the US. We extracted FIA data within a radius of 25 km around each urban forest location using ArcMap 10.1. The FIA plot locations are "fuzzed" and do not report exact spatial coordinates to comply with privacy issues; thus locations are between 0.8 and 1.6 km of the actual plot. Accordingly, we analyzed results on a community scale rather than by plot [41,42], so we estimated location error to be small. We obtained data from the years 2010–2013 for Virginia, and 2009–2013 for Georgia and Florida to obtain the maximum number of tree measurements while excluding re-measured trees. In three instances, data were also extracted from surrounding states because the 25 km buffer extended past state lines (Figure 1). Years extracted in those states were between 2009 and 2013, excluding 2009 if a re-measurement occurred in 2013.

The FIA plots consist of groups of four subplots that cover an area of approximately 0.067 hectares, with a microplot approximately 0.001 hectares in area located within each subplot. Woody plants with DBH values less than 12.7 cm but greater than or equal to 2.54 cm are measured only in the microplots. We used an expansion factor of 12.46 for the tree counts within microplots to adjust for their smaller sample plot. Tree data collected includes condition, species, DBH, height, and location within plot (for more information on FIA data collection, see [43]). In contrast to the i-Tree protocol, under the FIA protocol a tree that splits below DBH is recorded as more than one tree. Since trees that split between 0.3 m and 1.37 m are given the same cardinal distance and direction, we were able to combine multiple stems using the quadratic mean diameter to match the two protocols in these cases. We could not, however, differentiate between trees that split less than 0.3 meters from the ground and those which had not been split. Thus, we performed analyses with measures of stem density, as well as with basal area to account for this potential error.

2.2. Statistical Analyses

Unless otherwise stated, statistical analyses were carried out in R version 2.2-1 [44] with the *vegan* package [45]. We characterized species richness for each location using a bootstrap method to estimate the expected number of species in an area, using rare species to quantify how many species were likely unobserved. Species accumulation curves were created by graphing the area sampled rather than by the number of plots to account for different plot sizes. Curves were visually analyzed to verify species richness to avoid confounding results from locations with differing sampling intensities and plot sizes. Diversity by location was quantified using both the Shannon-Weaver (H') and the Inverse Simpson's (λ^{-1}) Indices:

$$H' = -\sum p_i \ln p_i \tag{1}$$

$$\lambda^{-1} = \left(\sum p_i^2\right)_{\lambda^{-1}} \tag{2}$$

where: p_i is the proportion of individuals in the *i*th species for a particular location. We did not quantify plot-level diversity because of differences in data collection methods between urban and periurban plots, such as different plot sizes and shapes. Additionally, PF plots contained, on average, many more trees than UF plots because all plots were classified as "forest land." Diversity metrics (species richness, Shannon-Weaver, and Inverse Simpson's) were analyzed with all species, as well as with only native and naturalized species to answer questions about species which could be expected to persist without further human influence. We defined "naturalized" species as those which are invasive or have been recorded as invasive anywhere within the US [46].

To assess community structure, we used graphical as well as statistical means. We focused on DBH-based measures of structure rather than other measures (e.g., tree height or biomass) due to the strong correlation between DBH and both biomass and tree height, and the lack of local and UF-specific biomass equations for most UF species. We first analyzed DBH distributions graphically according to forest type and location, and investigated significant differences between PF and UF as well as among locations using the Log-Rank and Wilcoxon tests via the SAS (9.3) procedure PROC LIFETEST [47].

Species composition was investigated by categorizing tree species into native, non-native, invasive, and unclassified. Native status was defined by classification in i-Tree [48], supplemented by the USDA PLANTS database [49]. Invasive species classification was defined using invasive plant species lists by state [50–52]. Species with any level of invasive status (e.g., threat level categorizations) within the state in which they were sampled were considered to be invasive. Trees were unclassified if they were either measured only to the genus-level or were cultivars with unspecified origin. To further understand community similarities or dissimilarities, we used a permutational ANOVA (PERMANOVA) [45] to determine whether differences in community structure were due to urbanization, geographic area, or both [53,54]. We included ecological province and land use (urban *vs.* periurban) in the model and used 999 permutations, following Anderson and Walsh [50]. Community differences were quantified using the Raup-Crick metric, which measures the similarities or dissimilarities between communities [55,56]. The Raup-Crick metric allows for comparisons between communities with varied numbers of species and sampling sizes, whereas other common similarity metrics such as Jaccard's could be skewed due to dissimilarities in species richness [55–57]. Using a matrix of comparisons between all pairs of associations, the Raup-Crick index compares observed number of species with the distribution of co-occurrences generated from 200 Monte Carlo random replicates [55]. The computed index ranges from a value of 1.0 indicating no similarity, to 0 indicating identical similarity. To further visualize the results, we created a dendrogram from a Raup-Crick dissimilarity metric to compare each site [55], utilizing presence/absence data to prevent bias from the differences in plot sizes between UF and PF samples.

3. Results

3.1. Tree Diversity Comparisons

In all localities except for Abingdon, VA, species richness values (both observed and estimated) were greater in UFs (Figure 2). The range of estimated number of species per location was 63–124 in UFs *versus* 35–64 in the PFs. Abingdon had the greatest number of species in its PF and the least number of species in its UF; however, species richness was not significantly different between its PF and UF (63.6 ± 2.5 and 63.2 ± 3.5, respectively; Figure 2). Roanoke's UF had the greatest number of species overall (124 and 106; estimated and observed), while East Orlando's PF had the fewest number of species (35 and 30; estimated and observed). In six of the eight locations, species richness, as illustrated by species accumulation curves (Figure 3 and Figure S1), was clearly greater in the UF. Differences were more difficult to distinguish between the PF and UF of Abingdon and Atlanta due to discrepancies in sample size.

Tree diversity as described with both the Shannon and Inverse Simpson's indices was greater in UF (Figure 4). These indices indicate that the most diverse PFs were less diverse than the least diverse UFs regardless if analyses were based on biomass or stem counts. Gainesville (both PF and UF) had the lowest diversity with both indices (2.3 and 3.2, for Shannon; 5.8 and 17.0 for Inverse Simpson's, for PF and UF, respectively), when compared to other UF and PF. As measured by the Shannon Index, Charlottesville's UF had the greatest diversity; as measured by the Inverse Simpson's Index, and Falls Church's UF has the highest diversity. Overall, the PF had greater relative abundance of the most common species than the UF (PF ranged from 15% to 36%, whereas UF ranged from 7% to 15%). This measure serves as an indicator of diversity as it is significantly related to the Shannon Diversity Index, with lower relative abundance of the most common species attributed to greater diversity [11]. Gainesville's PF had the greatest relative abundance of the most common species (36%, *Pinus elliottii* Engelm.).

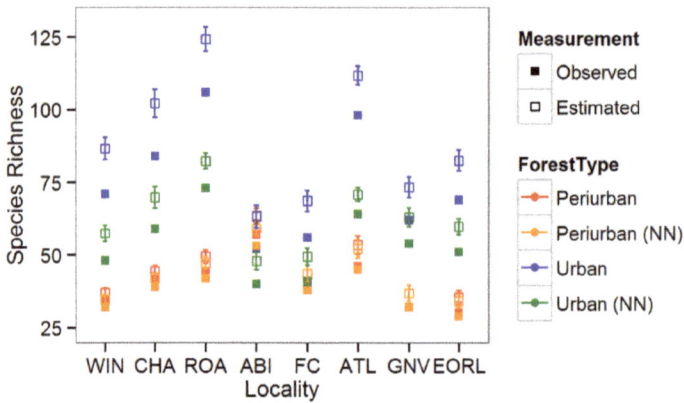

Figure 2. Number of tree species estimated by bootstrap methods and observed via collection. Locations are listed North-South, with the exception of Falls Church which is listed more southerly to reflect its ecological province. Secondary colors (orange and green) represent the species richness of only the native and naturalized species, whereas primary colors (red and blue) include all species. (ATL = Atlanta, GA; CHA = Charlottesville, VA; EORL = East Orlando, FL; FC = Falls Church, VA; GNV = Gainesville, FL; ROA = Roanoke, VA; ABI = Abingdon, VA; WIN = Winchester, VA, NN = Native and Naturalized).

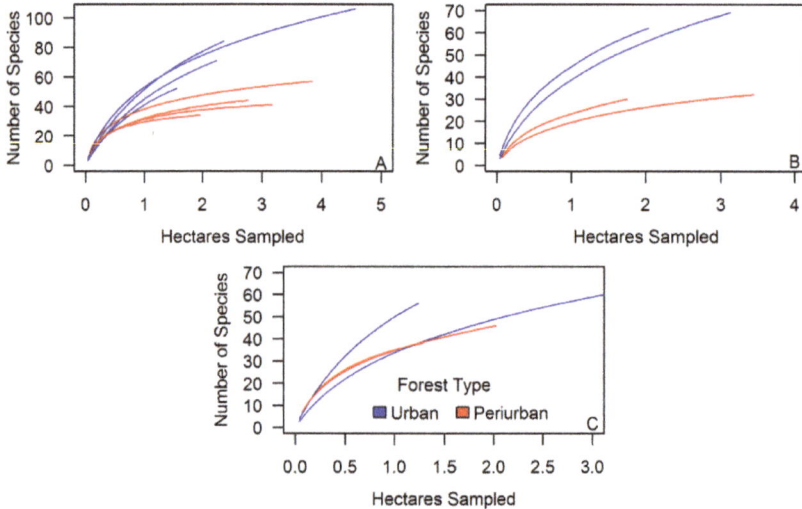

Figure 3. Species accumulation curves by ecoregion in the SE US. Regions are (**A**) Central Appalachian Broadleaf Forest; (**B**) Outer Coastal Plain; and (**C**) Southeastern Mixed Forest (Atlanta urban forest (UF) is truncated in image C for better visualization).

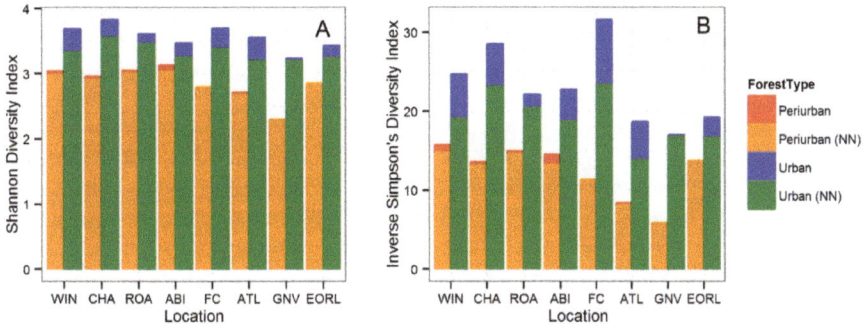

Figure 4. Diversity by location and forest type based on stem counts as defined by (**A**) Shannon Index and (**B**) Inverse Simpson's Index. Diversity indices for all species were overlaid with diversity indices including only the native and naturalized species; secondary colors (orange and green) represent native and naturalized species diversity, primary colors (red and blue) plus secondary colors represent diversity for all species. Locations are listed North-South, with the exception of Falls Church which is listed more southerly to reflect its ecological province. (ATL = Atlanta, GA; CHA = Charlottesville, VA; EORL = East Orlando, FL; FC = Falls Church, VA; GNV = Gainesville, FL; ROA = Roanoke, VA; ABI = Abingdon, VA; WIN = Winchester, VA, NN = Native and Naturalized).

3.2. Community Structure and Composition

The UFs and PFs presented mixed results in regards to community structure and composition. Community size structure was similar regardless of location or forest type. All locations had reverse J distributions (Figure 5), with a larger presence of small trees and fewer large trees. A significantly greater proportion of trees from PF had smaller diameters than trees from UF (Wilcoxon $p < 0.0001$). UF had a greater proportion of large (<100 cm) trees, though this difference was not significant, indeed, only three of eight PFs had trees in that size category (Log-Rank $p = 0.1057$).

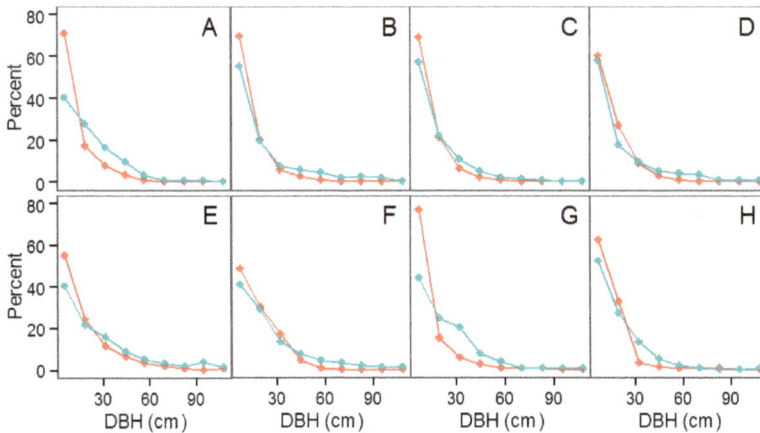

Figure 5. Distribution of trees by diameter at breast height (DBH) and location. Urban forest is in blue, periurban forest is red. (**A**) Winchester, VA; (**B**) Charlottesville, VA; (**C**) Roanoke, VA; (**D**) Abingdon, VA; (**E**) Falls Church, VA; (**F**) Atlanta, GA; (**G**) Gainesville, FL; (**H**) East Orlando, FL.

Altogether, there were 181 species that were present in UF but absent in PF, and 23 that were present in PF but absent in UF. The most widespread species was *Acer rubrum* L., which was present in every site (regardless of PF or UF location) and was within the three most common species for five of the eight PF locations and one of the eight UF locations (Table S1). Other widespread species include *Carya glabra* (Mill.) Sweet (present in all PF and six of the eight UF) and *Prunus serotina* Ehrh. (present in seven of the eight PF and UF). There were many species which were widespread in UF but absent in PF, such as *Lagerstroemia indica* L. and *P. calleryana*. Both were recorded in seven of the eight UF but not in any PF. The UF and PF had similar proportions of palms, with 1.0% in the UF and 0.8% in the PF.

There were greater numbers of non-native, invasive, and unclassified species in the UF *versus* the PF, regardless of location (Figure 6). Winchester and Falls Church UF had a greater number of non-native species than native species (36 *versus* 25 for Winchester, and 20 *versus* 31 for Falls Church). There was no common pattern describing the proportion of native species by forest type. By counts of individuals, PF had a substantially greater proportion of native trees in every location (Figure S2). Fewer than 50% of the trees in the UF of Winchester were native, but all other locations had communities where at least 50% of the trees were native. There were six invasive species throughout all PFs, and every invasive species found in PF was also found in UF. There were 19 additional invasive species found across our UF sample, with 10 found in Atlanta alone. However, based on numbers of trees, the UF in Atlanta had a smaller proportion of invasive trees than that of PF (Figure S2).

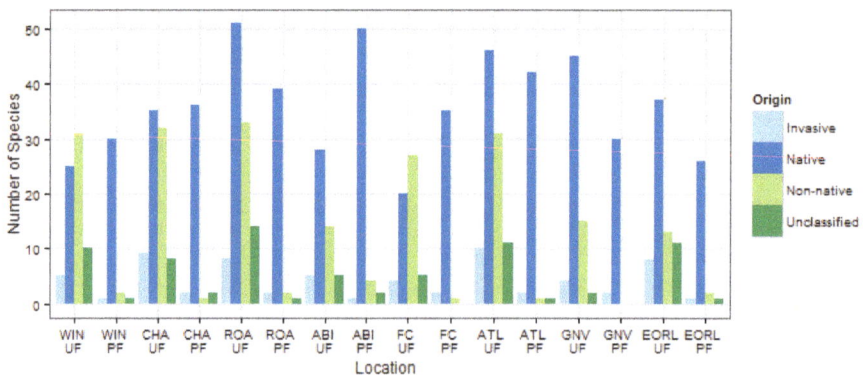

Figure 6. Number of observed species by origin and location. (ATL = Atlanta, GA; CHA = Charlottesville, VA; EORL = East Orlando, FL; FC = Falls Church, VA; GNV = Gainesville, FL; ROA = Roanoke, VA; ABI = Abingdon, VA; WIN = Winchester, VA; UF = Urban forests, PF = Periurban forests.) Locations are listed North-South, with the exception of Falls Church which is listed more southerly to reflect its ecological province.

There was a significant effect (PERMANOVA, $p < 0.05$) of both forest type (Pseudo-$F_{(1,15)}$ = 5.003, $p = 0.001$) and ecological province (Pseudo-$F_{(2,15)}$ = 4.845, $p = 0.001$) on community species composition (Table S2). The Raup-Crick metric indicated that community species composition was more dissimilar between localities in Florida and Virginia than they were between UF and PF (Figure 7). Gainesville and East Orlando, regardless of forest type, had very similar communities. In Virginia, PF communities were more similar to each other than UF communities. The more northern Virginia UF communities with lower elevation gradients (Winchester and Falls Church) were more similar to each other than they were to the more southern, more elevated Virginia communities. Atlanta's UF was more similar to its and Virginia's PF than to any other UF.

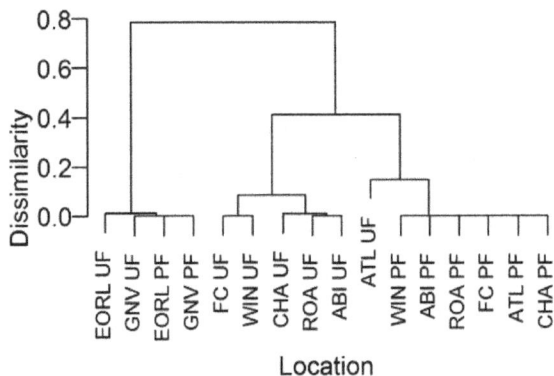

Figure 7. Cluster dendrogram grouping locations based on their similarities in species composition via the Raup-Crick Dissimilarity index. PF = Periurban Forest, UF = Urban Forest, ATL = Atlanta, GA; CHA = Charlottesville, VA; EORL = East Orlando, FL; FC = Falls Church, VA; GNV = Gainesville, FL; ROA = Roanoke, VA; ABI = Abingdon, VA; WIN = Winchester, VA.

4. Discussion

Tree species richness and diversity was greater in UF, which was expected (Figures 2–4) [29]. Moreover, in two cases (Atlanta and Abingdon), the species accumulation curves show that sampling efforts were vastly different. On the other hand, the PF curve in Atlanta, when viewed at its full extent, shows a lessened slope indicative of fewer remaining (uncounted) species (Figure S1). While it could be argued that a large urban center like that of Atlanta may require a buffer larger than our chosen distance of 25 km, this distance (although arbitrary) represented a balance to limit geoclimatic variability and link to direct urban impacts. The use of estimated numbers of species rather than actual sampled numbers of species mitigates the effects of varying sampling efforts, but does carry with it uncertainty that should be addressed in future studies.

In forested ecoregions of North America, an estimated 1/3 of urban trees are planted; the rest originate from remnant or regenerated forest [58]. Therefore, UF often contain similar forest communities to PF while also containing highly altered planted communities, making it unsurprising that species found in PF are almost all in UF. The additional species richness in UF (181 species in UF not found in PF) is no doubt due to human planting or introductions of non-native species [4] and the maintenance of otherwise unfit species through fertilization and irrigation [3]. The UF had similar numbers of native species as the PF, but proportionally many more non-native species (Figure 6). Across all locations, all but one UF (Gainesville) had proportionally more non-native species than that of the 15% reported for riparian forests in urban areas of Manaus, Brazil [59] for example, and half of our UF study sites had more than twice this proportion, supporting a hypothesis of a management effect.

In North America, 77% of invasive species introductions are a result of horticulture plantings and 14% are via forestry [4]. This unintentional introduction of invasive species was noticeable in the UF, where numbers of invasive species were greater than in PF (Figure 6). Interestingly, *P. calleryana*, which was found in all but one of the urban locations and has been widely planted throughout the region, is now considered to be an invasive species [11,50–52]. Although less than 1% of woody plant introductions become invasive, this underscores the need to monitor non-native species [4]. Likely, the greater proportion of invasive trees in Atlanta's PF is due to the large amount of human activity within the area. Indeed, the percent of developed land was positively correlated with the number of invasive species for both the PF and UF (data not shown). Atlanta's PF was more urbanized than the other cities in this study, with a greater proportion of its land cover area classified as developed [60].

Unsurprisingly, the two invasive species found in the PF of Atlanta (*Ailanthus altissima* (Mill.) Swingle and *Albizia julibrissin* Durazz.) commonly invade disturbed areas [61].

There is a growing body of research on the structural and ecological dynamics of UFs using permanent field plot and forest inventory methods similar to the ones presented in this study [2,3,5,22,33,35,58] (and citations therein). The field measurement protocols in these disparate studies are, however, often different because of (for example) land use definitions, the stem size criteria for differentiating tree and shrub growth forms, and the presence of multi-stemmed tall woody shrubs in certain biomes (e.g., mangroves, shrublands) [2,30,33]. However, this study's use of the same UF and PF inventory protocols and tree size criteria eliminated many of these issues. Similarly, we found that PFs were found to have a greater proportion of smaller DBH trees when compared to UF (Figure 5). This was not surprising, considering that PFs are likely to have an understory component containing saplings, due to planting or natural regeneration. On the other hand, only UF plots falling in vacant areas, parks or remnant forest patches would be likely to have comparable understories [58]. However, we did not measure the age nor assess the origin of sampled trees in our study.

Overall, we found greater species-level homogeneity in PFs than in UFs as evidenced by a greater relative abundance of the most common species in the region's PFs. The PF of Gainesville was especially homogenous and other PF areas with low diversity were made up of very few species; greater than 50% of the trees in the PF of Gainesville were pines (primarily *Pinus taeda* L. and *P. elliottii*). This is likely a result of human intervention and legacy effects of past land use practices favoring pine plantations [1,27,62], and also a result of the generally lower tree species richness in the naturally pine-dominated landscapes sampled in this study. While hardwood-dominated depressional wetlands of this region contain higher tree diversity, our sample included few PF forests of this type. The more northern locations (those in Virginia), with their prevalence of non-native species within their UF (Figure 6), had more distinct PF and UF communities (Figure 7). Conversely, Florida locales had fewer non-native tree species, which could explain the similarities between the PF and UF in those locations. This is reflective of the lower natural tree species diversity in pine-dominated outer coastal plain landscapes sampled in this study, *versus* broadleaf and mixed forests, which are dominated by hardwoods and other species [27,28].

More importantly, species composition analyses indicated that urban forests were more similar to their periurban counterparts within eco-zone than were urban forests to each other, indicating that UFs are strongly influenced by the natural diversity patterns in the local vegetation. This result refutes the hypothesis of ecological homogeneity across urban areas, at least in terms of tree diversity. If homogeneity were to hold true in this region, we would have expected the urban locations to be more similar to each other, regardless of ecoregion. Instead, there was a clear distinction due to ecoregion. The results of the PERMANOVA showed that species composition of urban forest was significantly different than periurban, and significant differences also were found among urban ecological provinces.

The increased diversity in UF was a likely result of the greater diversity of the Central Hardwood forests, as well as greater landscape heterogeneity, which has been shown to result in more diverse communities [63]. However, the PF in Abingdon had an elevation gradient of greater than 1000 meters (Table 1), which could support a larger range of plant communities [64]. Abingdon's PF had species richness which was comparable to not only its UF, but also that of Falls Church. The smaller species richness values in Abingdon's UF may be a result of Abingdon's less urbanized character (with a population density of <400 people/km^2) and correspondingly fewer opportunities for planted species and less heterogeneity.

5. Conclusions

Our results indicate that urban tree species distributions are not geographically homogenous throughout the SE US. Urban forests in this region have maintained a fair amount of native species, and thus are a reflection of their respective periurban tree communities. Our conclusions are, however, somewhat limited by different plot size and measurement techniques used in the urban (i-Tree) and periurban (FIA) protocols. On the other hand, with the relatively recent (2014) implementation of urban FIA sampling, there is a need to develop techniques and metrics to compare the data that are available in order to further investigate hypotheses about the impacts of urbanization on trees further into the future.

Utilizing existing SE US urban forest data collected under standardized inventory and monitoring protocols, as well as available data from the USDA Forest Service FIA program, we used a novel approach to investigate urban forest community composition relative to adjacent periurban forests along a geographic and urban to rural gradient. Comparing community composition and structure between urban forests and their periurban counterparts could be used further to better assess the ecological stability and sustainability of an area in the face of climate change and urbanization. Overall, our findings indicate that urban forests, with their great diversity, should fare well in the face of future global changes. However, care should be taken to monitor non-native and invasive species and to ensure that the species and size composition support sustainability of both urban and periurban forests. In the future, more comprehensive studies should be undertaken that include more global cities to encompass a larger variety of land uses and management histories, as well as a wider range of vegetation types, to gain a more complete understanding of regional anthropogenic influences on diversity in PFs and UFs.

Supplementary Materials: The following are available online at www.mdpi.com/1999-4907/7/6/120/s1, Table S1: Common species by location, Table S2: PERMANOVA results, Figure S1: Species accumulation curves by location, Figure S2: Origins of trees by locations.

Acknowledgments: The authors would like to thank Dudley Hartel at USDA Forest Service-Urban Forestry South for funding and direction. Funding for this project and for publishing of this manuscript was provided by a grant from the US Forest Service (Forest Service Agreement Number 13-CS-11330144-061, titled "Regional Urban Forest i-Tree Eco Inventory Study"). Data for VA studies were generously provided by Dr. P. Eric Wiseman (Virginia Tech College of Natural Resources and Environment), whose funding was provided in part by USFS UCF formula funding via VA Dept. of Forestry, and in part by USDA National Institute of Food and Agriculture McIntire-Stennis formula funding. We would also like to thank multiple municipal and county employees and students who collected data, and the Staudhammer and Starr labs for valuable feedback on earlier versions of the manuscript.

Author Contributions: C.S. and A.B conceived and designed the experiments; A.B. analyzed the data; F.E. and C.S. contributed reagents/materials/analysis tools; C.S., A.B., F.E, A.C, and G.S. wrote the paper.

Conflicts of Interest: The authors declare no conflict of interest. USDA Forest Service—Urban Forestry South was responsible for the collection of data in Atlanta, GA; however, this sponsor had no role in the design of the study; in the analyses, or interpretation of data; in the writing of the manuscript, and in the decision to publish the results.

Abbreviations

The following abbreviations are used in this manuscript:

DBH	Diameter at breast height
FIA	Forest Inventory and Analysis
PERMANOVA	Permutational Analysis of Variance
PF	Periurban forests
SE US	Southeastern United States
UF	Urban forests
US	United States
USDA	United States Department of Agriculture

References

1. Wear, D.N. Forecasts of Land Uses. In *The Southern Forest Futures Project: Technical Report*; Greis, J.G., Ed.; USDA Forest Service: Washington, DC, USA, 2013; pp. 45–72.
2. Conway, T.M.; Bourne, K.S. A comparison of neighborhood characteristics related to canopy cover, stem density and species richness in an urban forest. *Landsc. Urban Plan.* **2013**, *113*, 10–18. [CrossRef]

3. Groffman, P.M.; Cavender-Bares, J.; Bettez, N.D.; Grove, J.M.; Hall, S.J.; Heffernan, J.B.; Hobbie, S.E.; Larson, K.L.; Morse, J.L.; Neill, C.; *et al.* Ecological homogenization of urban USA. *Front. Ecol. Environ.* **2014**, *12*, 74–81. [CrossRef]
4. Richardson, D.M.; Rejmanek, M. Trees and shrubs as invasive alien species—A global review. *Divers. Distrib.* **2011**, *17*, 788–809. [CrossRef]
5. Raupp, M.; Cumming, A.; Raupp, E. Street Tree Diversity in Eastern North America and Its Potential for Tree Loss to Exotic Borers. *Arboric. Urban For.* **2006**, *32*, 297–304.
6. Subburayalu, S.; Sydnor, T.D. Assessing street tree diversity in four Ohio communities using the weighted Simpson index. *Landsc. Urban Plan.* **2012**, *106*, 44–50. [CrossRef]
7. Alvey, A.A. Promoting and preserving biodiversity in the urban forest. *Urban For. Urban Green.* **2006**, *5*, 195–201. [CrossRef]
8. Thompson, I.; Mackey, B.; McNulty, S.; Mosseler, A. Forest Resilience, Biodiversity, and Climate Change. In *A Synthesis of the Biodiversity/Resilience/Stability Relationship in Forest Ecosystems*; Technical Series No. 43; Secretariat of the Convention on Biological Diversity: Montreal, Canada, 2009; p. 67.
9. McPherson, E.; Simpson, J.R. Carbon dioxide reduction through urban forestry: Guidelines for professional and volunteer tree planters. *Gen. Tech. Rep.* **1999**, *171*, 237.
10. Dreistadt, S.H.; Dahlsten, D.L.; Frankie, G.W. Urban Forests and Insect Ecology. *Bioscience* **2015**, *40*, 192–198. [CrossRef]
11. Kendal, D.; Dobbs, C.; Lohr, V.I. Global patterns of diversity in the urban forest: Is there evidence to support the 10/20/30 rule? *Urban For. Urban Green.* **2014**, *13*, 411–417. [CrossRef]
12. Chapin, F.S., III; Zavaleta, E.S.; Eviner, V.T.; Naylor, R.L.; Vitousek, P.M.; Reynolds, H.L.; Hooper, D.U.; Lavorel, S.; Sala, O.E.; Hobbie, S.E.; *et al.* Consequences of changing biodiversity. *Nature* **2000**, *405*, 234–242. [CrossRef] [PubMed]
13. Cumming, G.S.; Olsson, P.; Chapin, F.S.; Holling, C.S. Resilience, experimentation, and scale mismatches in social-ecological landscapes. *Landsc. Ecol.* **2013**, *28*, 1139–1150. [CrossRef]
14. Walther, G.-R.; Post, E.; Convey, P.; Menzel, A.; Parmesan, C.; Beebee, T.J.C.; Fromentin, J.-M.; Hoegh-Guldberg, O.; Bairlein, F. Ecological responses to recent climate change. *Nature* **2002**, *416*, 389–395. [CrossRef] [PubMed]
15. Cleland, E.E.; Chuine, I.; Menzel, A.; Mooney, H.A.; Schwartz, M.D. Shifting plant phenology in response to global change. *Trends Ecol. Evol.* **2007**, *22*, 357–365. [CrossRef] [PubMed]
16. Motard, E.; Muratet, A.; Clair-Maczulajtys, D.; MacHon, N. Does the invasive species *Ailanthus altissima* threaten floristic diversity of temperate peri-urban forests? *Comptes Rendus-Biol.* **2011**, *334*, 872–879. [CrossRef] [PubMed]
17. Gaertner, M.; Larson, B.M.H.; Irlich, U.M.; Holmes, P.M.; Stafford, L.; van Wilgen, B.W.; Richardson, D.M. Managing invasive species in cities: A framework from Cape Town, South Africa. *Landsc. Urban Plan.* **2016**, *151*, 1–9. [CrossRef]
18. Staudhammer, C.L.; Escobedo, F.J.; Holt, N.; Young, L.J.; Brandeis, T.J.; Zipperer, W. Predictors, spatial distribution, and occurrence of woody invasive plants in subtropical urban ecosystems. *J. Environ. Manag.* **2015**, *155*, 97–105. [CrossRef] [PubMed]
19. Vitousek, P.; D'Antonio, C.; Loope, L.L.; Westbrooks, R. Biological invasions as global environmental change. *Am. Sci.* **1996**, *84*, 468–478.
20. Crossman, N.D.; Bryan, B.A.; Cooke, D.A. An invasive plant and climate change threat index for weed risk management: Integrating habitat distribution pattern and dispersal process. *Ecol. Indic.* **2011**, *11*, 183–198. [CrossRef]
21. Escobedo, F.J.; Kroeger, T.; Wagner, J.E. Urban forests and pollution mitigation: Analyzing ecosystem services and disservices. *Environ. Pollut.* **2011**, *159*, 2078–2087. [CrossRef] [PubMed]
22. Horn, J.; Escobedo, F.J.; Hinkle, R.; Hostetler, M.; Timilsina, N. The Role of Composition, Invasives, and Maintenance Emissions on Urban Forest Carbon Stocks. *Environ. Manag.* **2015**, *55*, 431–442. [CrossRef] [PubMed]
23. Schierenbeck, K.A.; Ellstrand, N.C. Hybridization and the evolution of invasiveness in plants and other organisms. *Biol. Invasions* **2009**, *11*, 1093–1105. [CrossRef]
24. Vincent, M.A. On the Spread and Current Distribution of *Pyrus calleryana* in the United States. *Castanea* **2005**, *70*, 20–31. [CrossRef]

25. Thuiller, W. Biodiversity: Climate change and the ecologist. *Nature* **2007**, *448*, 550–552. [CrossRef] [PubMed]

26. Konijnendijk, C.C.; Ricard, R.M.; Kenney, A.; Randrup, T.B. Defining urban forestry—A comparative perspective of North America and Europe. *Urban For. Urban Green.* **2006**, *4*, 93–103. [CrossRef]

27. Ricketts, T.; Imhoff, M. Biodiversity, urban areas, and agriculture: Locating priority ecoregions for conservation. *Ecol. Soc.* **2003**, *8*, 1.

28. Currie, D.J.; Paquin, V. Large-scale biogeographical patterns of species richness of trees. *Nature* **1987**, *329*, 326–327. [CrossRef]

29. McKinney, M.L. Effects of urbanization on species richness: A review of plants and animals. *Urban Ecosyst.* **2008**, *11*, 161–176. [CrossRef]

30. Nowak, D.; Crane, D.; Stevens, J.C.; Hoehn, R.E.; Walton, J.T.; Bond, J. A Ground-Based Method of Assessing Urban Forest Structure and Ecosystem Services. *Arboric. Urban For.* **2008**, *34*, 347–358.

31. Baró, F.; Chaparro, L.; Gómez-Baggethun, E.; Langemeyer, J.; Nowak, D.J.; Terradas, J. Contribution of ecosystem services to air quality and climate change mitigation policies: The case of urban forests in Barcelona, Spain. *Ambio* **2014**, *43*, 466–479. [CrossRef] [PubMed]

32. Martin, N.A.; Chappelka, A.H.; Keever, G.J.; Loewenstein, E.F. A 100 % Tree Inventory Using i-Tree Eco Protocol: A Case Study at Auburn University, Alabama, U.S. *Arboric. Urban For.* **2011**, *37*, 207–212.

33. Lawrence, A.B.; Escobedo, F.J.; Staudhammer, C.L.; Zipperer, W. Analyzing growth and mortality in a subtropical urban forest ecosystem. *Landsc. Urban Planing* **2012**, *104*, 85–94. [CrossRef]

34. Escobedo, F.J.; Adams, D.C.; Timilsina, N. Urban forest structure effects on property value. *Ecosyst. Serv.* **2015**, *12*, 209–217. [CrossRef]

35. Timilsina, N.; Staudhammer, C.L.; Escobedo, F.J.; Lawrence, A. Tree biomass, wood waste yield, and carbon storage changes in an urban forest. *Landsc. Urban Planing* **2014**, *127*, 18–27. [CrossRef]

36. Wiseman, P.E.; King, J. *i-Tree Ecosystem Analysis: Town of Abingdon, Urban Forest Effects and Values*; Virginia Tech College of Natual Resources and Environment: Blacksburg, VA, USA, 2012.

37. Urban Waters Partnership: Proctor Creek Watershed/Atlanta (Georgia). Available online: https://www.epa.gov/urbanwaterspartners/proctor-creek-watershedatlanta-georgia (accessed on 5 May 2016).

38. Bailey, R.G. Identifying ecoregion boundaries. *Environ. Manag.* **2005**, *34*, 14–26. [CrossRef] [PubMed]

39. US Census Bureau. TIGER/Line Shapefiles (machine-readable data files), 2012. Available online: https://www2.census.gov/geo/pdfs/maps-data/data/tiger/tgrshp2010/TGRSHP10SF1.pdf (accessed on 6 April 2016).

40. Forest Inventory and Analysis Database. Available online: http://apps.fs.fed.us/fiadb-downloads/datamart.html (accessed on 6 April 2016).

41. Delphin, S.; Escobedo, F.J.; Abd-Elrahman, A.; Cropper, W.P. Urbanization as a land use change driver of forest ecosystem services. *Land Use Policy* **2016**, *54*, 188–199. [CrossRef]

42. McRoberts, R.; Holden, G.; Nelson, M.D.; Liknes, G.C.; Moser, W.K.; Lister, A.J.; King, S.L.; Lapoint, E.; Coulston, J.W.; Smith, W.B.; *et al.* Estimating and circumventing the effects of perturbing and swapping inventory plot locations. *J. For.* **2005**, *103*, 275–279.

43. Woudenberg, S.W.; Conkling, B.L.; O'Connell, B.M.; LaPoint, E.B.; Turner, J.A.; Waddell, K.L.; Boyer, D.; Christensen, G.; Ridley, T. FIA database description and users manual for Phase 2. Version 5.1. US Department of Agriculture: Washington, DC, USA, 2011.

44. R Core Team. *R: A Language and Environment for Statistical Computing 2013*; R Foundation for Statistical Computing: Vienna, Austria, 2014.

45. Vegan: Community Ecology Package, Version 2.3-5. Available online: https://cran.r-project.org/web/packages/vegan/vegan.pdf (accessed on 3 March 2016).

46. Invasive Plant Atlas of the United States. Available online: http://www.invasiveplantatlas.org/ (accessed on 7 June 2016).

47. SAS/STAT®. SAS Institute Inc: Cary, NC, USA, 2011. Available online: http://www.sas.com/en_sg/software/analytics/stat.html (accessed on 7 June 2016).

48. Nowak, D.; Crane, D. The Urban Forest Effects (UFORE) Model: Quantifying urban forest structure and functions. In *Integrated Tools for Natural Resources Inventories in the 21st Century*; Hansen, M., Burk, T., Eds.; US Department of Agriculture, Forest Service, North Central Forest Experiment Station: St. Paul, MN, USA, 2000; pp. 714–720.

49. USDA, N. The PLANTS Database. Available online: http://plants.usda.gov (accessed on 20 March 2016).

50. FLEPPC List of Invasive Plant Species 2015. Available online: www.fleppc.org (accessed on 7 June 2016).
51. GEPPC List of Non-native Invasive Plants in Georgia. Available online: http://www.gainvasives.org/species/weeds/ (accessed on 1 February 2016).
52. Hefferman, K.; Engle, E.; Richardson, C. *Virginia Invasive Plant Species List*; Virginia Department of Conservation and Recreation: Richmond, VA, USA, 2014.
53. Anderson, M.J.; Walsh, D.C.I. PERMANOVA, ANOSIM, and the Mantel test in the face of heterogeneous dispersions: What null hypothesis are you testing? *Ecol. Monogr.* **2013**, *83*, 557–574. [CrossRef]
54. Anderson, M.J. A new method for non-parametric multivariate analysis of variance. *Austral Ecol.* **2001**, *26*, 32–46.
55. Chase, J.M.; Kraft, N.J.B.; Smith, K.G.; Vellend, M.; Inouye, B.D. Using null models to disentangle variation in community dissimilarity from variation in α-diversity. *Ecosphere* **2011**, *2*. [CrossRef]
56. Raup, D.M.; Crick, R.E. Measurement of faunal similarity in paleontology. *J. Paleontol.* **1979**, *53*, 1213–1227.
57. Vellend, M.; Verheyen, K.; Flinn, K.M.; Jacquemyn, H.; Kolb, A.; Van Calster, H.; Peterken, G.; Graae, B.J.; Bellemare, J.; Honnay, O.; *et al.* Homogenization of forest plant communities and weakening of species—Environment relationships via agricultural land use. *J. Ecol.* **2007**, *95*, 565–573. [CrossRef]
58. Nowak, D.J. Contrasting natural regeneration and tree planting in fourteen North American cities. *Urban For. Urban Green.* **2012**, *11*, 374–382. [CrossRef]
59. dos Anjos Santos, O.; Couceiro, S.R.M.; Rezende, A.C.C.; de Sousa Silva, M.D. Composition and richness of woody species in riparian forests in urban areas of Manaus, Amazonas, Brazil. *Landsc. Urban Plan.* **2016**, *150*, 70–78. [CrossRef]
60. Homer, C.; Dewitz, J.; Yang, L.; Jin, S.; Danielson, P.; Xian, G.; Coulston, J.; Herold, N.; Wickham, J.; Megown, K. Completion of the 2011 National Land Cover Database for the Conterminous United States—Representing a Decade of Land Cover Change Information. *Photogramm. Eng. Remote Sens.* **2015**, *81*, 345–354.
61. Landenberger, R.E.; Warner, T.A.; McGraw, J.B. Spatial patterns of female ailanthus altissima across an urban-to-rural land use gradient. *Urban Ecosyst.* **2009**, *12*, 437–448. [CrossRef]
62. Huggett, R.; Wear, D.N.; Li, R.; Coulston, J.W.; Liu, S. Forecasts of Forest Conditions. In *The Southern Forest Futures Project; Technical Report*; Wear, D.N., Greis, J.G., Eds.; USDA Forest Service: Washington, DC, USA, 2013; pp. 73–102.
63. Rosenzweig, M.L. *Species Diversity in Space and Time*; Cambridge University Press: Cambridge, UK, 1995.
64. Stevens, G.C. The Elevational Gradient in Altitudinal Range: An Extension of Rapoport's Latitudinal Rule to Altitude. *Am. Nat.* **2012**, *140*, 893–911. [CrossRef] [PubMed]

forests

MDPI

Article

Spatio-Temporal Changes in Structure for a Mediterranean Urban Forest: Santiago, Chile 2002 to 2014

Francisco J. Escobedo [1,*], Sebastian Palmas-Perez [2], Cynnamon Dobbs [3], Salvador Gezan [2] and Jaime Hernandez [4]

[1] Biology Program—Functional and Ecosystem Ecology Unit, Faculty of Natural Sciences and Mathematics, Universidad del Rosario, Kr 26 No 63B-48, Bogotá D.C. 111221492, Colombia
[2] School of Forest Resources and Conservation, IFAS-University of Florida, Gainesville, FL 32611, USA; spalmas@ufl.edu
[3] Departamento de Ecosistemas y Medio Ambiente, Facultad de Agronomia e Ingenieria Forestal, Pontificia Universidad Catolica de Chile, Av. Vicuna Mackenna 4860, Macul, Santiago 7820436, Chile; cdobbsbr@gmail.com
[4] Laboratorio de Geomática y Ecología del Paisaje, Facultad de Ciencias Forestales y de la Conservación de la Naturaleza, Universidad de Chile, Santiago 7820436, Chile; jhernand@uchile.cl
* Correspondence: franciscoj.escobedo@urosario.edu; Tel.: +57-1-297-0200 (ext. 3359)

Academic Editor: Timothy A. Martin
Received: 15 March 2016; Accepted: 2 June 2016; Published: 11 June 2016

Abstract: There is little information on how urban forest ecosystems in South America and Mediterranean climates change across both space and time. This study statistically and spatially analyzed the spatio-temporal dynamics of Santiago, Chile's urban forest using tree and plot-level data from permanent plots from 2002 to 2014. We found mortality, ingrowth, and tree cover remained stable over the analysis period and similar patterns were observed for basal area (BA) and biomass. However, tree cover increased, and was greater in the highest socioeconomic stratum neighborhoods while it dropped in the medium and low strata. Growth rates for the five most common tree species averaged from 0.12 to 0.36 cm·year^{-1}. Spatially, tree biomass and BA were greater in the affluent, northeastern sections of the city and in southwest peri-urban areas. Conversely, less affluent central, northwest, and southern areas showed temporal losses in BA and biomass. Overall, we found that Santiago's urban forest follows similar patterns as in other parts of the world; affluent areas tend to have more and better managed urban forests than poorer areas, and changes are primarily influenced by social and ecological drivers. Nonetheless, care is warranted when comparing urban forest structural metrics measured with similar sampling-monitoring approaches across ecologically disparate regions and biomes.

Keywords: basal area; urban forest biomass; spatial analysis; urban tree growth; urban forest mortality

---

1. Introduction

Urban forests are characterized by unique soils and urban morphologies, heterogeneous vegetation structure and composition, and novel assemblages of native and exotic tree species [1–3]. The spatial and temporal characteristics of urban forest structure are driven by biophysical factors such as topography, climate, biogeochemical cycles, and disturbances such as drought [2,4,5]. They are also affected by socioeconomic factors such as management and planning regimes, people's preferences, and socio-political budgets and directives [6–8]. As all these drivers alter the structure and composition of an urban forest, so too are ecological processes affected and subsequently the provision of ecosystem services [6]. A few studies have examined urban forests, their functions, and mortality with some examples from Mediterranean climate urban forests in California (United States, US), Spain, and Italy [9–12]. However, other than these few studies, there is little information on how and why urban forest structure in South America, the global south, and Mediterranean climates changes across space and time.

Urban forests are a key component in cities and provider of ecosystem services as they influence the well-being of urban inhabitants [11,13]. They are able to store and sequester carbon, regulate hydrologic cycles, ameliorate climate, remove air pollutants, provide habitat for fauna and space for recreation and spiritual enjoyment, among other services [13,14]. However, multi-scale economic and socio-ecological drivers can alter urban forest structure, thus understanding the spatio-temporal dynamics of urban forest structure and composition can provide insights on their management and planning [15].

Tree planting preferences and management decisions by both private individuals and communities can influence urban forest cover and density, growth, mortality, and distribution [6]. Similarly, species composition, age diversity, condition, site characteristics, and socio-ecological disturbances have an effect on growth of the urban forest [2,4,7,8], affecting several ecosystem processes, disservices and services. The type of land use and building density (*i.e.*, urban morphology) and socioeconomic status in particular, affect the structure and distribution of the urban forest, leading to changes in biomass with consequences for carbon storage and sequestration [5,6,10,12]. Growth, mortality, and regeneration in both urban and natural forests depend on environmental factors such as rainfall, temperature, and soil conditions; however, urban tree growth and biomass have been shown to be often greater in urban than natural landscapes [16]. Other anthropogenic factors affecting growth and mortality include exposure to pollution, artificial irrigation, and vegetation management and maintenance practices [7,8,16,17].

Irrigation becomes extremely influential in Mediterranean semi-arid and arid cities where frequent droughts and water scarcity during growing periods can cause reduced growth, basal area, and biomass, and can increase mortality rates [15]. These structural effects can be measured in growth rates, biomass, basal area, and mortality rates, and are key for analyzing temporal patterns of carbon sequestration and storage [17], tree wood waste biomass [5], and overall effects of climate on urban forest structure [6]. Research on urban tree growth, biomass, and basal area is scarce and, to our knowledge, limited to a number of species mainly from North America [10,16,18] and a few studies from other regions [19–21]. Further information on drivers of urban tree growth and mortality can improve decision-making on tree selection, urban forest carbon accounting, and benefit estimates [10,17,22,23].

Temporal studies of urban forest ecosystems are not common and are mainly based on coarse resolution information derived from satellite imagery [7,24–27]. Most research for monitoring the urban forest has been based on land use/land cover change analyses, with several examples from the US [5,7,16], Europe [24], and Asia [25], with few examples from Latin America [26] and Africa [27]. Research based on inventories and field data are particularly scarce, aside from monitoring of planted street trees [10,20] and citation therein], and a few examples from the subtropics [5,8,16,28] and humid temperate areas of North America [7] exist. However, most of these mentioned studies are located in temperate and industrialized regions, with a scarce few studies from elsewhere [2,15] such as those in South America and from cities in Mediterranean climates. The use of permanent urban forest plots for monitoring the structure and composition of the urban forest is increasingly being used for not only recognizing necessary changes in planning and management goals, but also for distinguishing the most effective practices that maximize the provision of ecosystem services [4,7]. Nevertheless, such research and information is rare for urban forests in South America and in Mediterranean climates, such as south and western Australia [15], southern Europe [20,23], and southern Africa [27].

The aim of this study is to analyze the spatio-temporal dynamics of a South American urban forest with a Mediterranean climate using tree and plot-level data from permanent monitoring plots.

Additionally, to our knowledge, such an approach would make this study one of the first of its kind in South America. Our specific study objectives are to analyze temporal changes from 2002 to 2014 and spatial differences in urban forest structural characteristics from Santiago, Chile including: (1) overall tree population growth, mortality, and ingrowth; (2) basal area, biomass, and tree cover change dynamics; (3) correlates of structural change; and (4) spatial patterns across the study area. We propose to address these research questions using statistical and spatial analyses of field data collected during 2002 and 2014. This type of study can be used to better understand the effects of urbanization and land use change on the structure of not only Latin American but Mediterranean climate urban forests as well. Findings can also be compared to similar plot based studies from other urban forests. Similarly, structural information, such as growth rates and biomass change, is key towards assessing ecosystem services and disservices that are frequently being used in urban planning, sustainability and climate change initiatives, and land management decisions.

2. Methods

2.1. Study Area and Field Sampling

Santiago, Chile is located in the middle of the Chilean Mediterranean climate zone (33°27′ S–70°41′ W). According to the Köppen climate classification systems, Santiago has a cool, semi-arid climate with warm, dry, hot, summers (November to March; [29]). Temperatures vary from an average of 20 °C in January to 8 °C in June–July with an annual average of 14.4 °C, while mean rainfall is 312.5 mm per year ([29]). The city lies in the center of a valley surrounded by a coastal mountain range to the west and the foothills of the Andes to the east. The elevation varies from 400 to 900 m with an average of 540 m. In 2002 its population was 5.3 million inhabitants with a population density of 10,000 people/km^2 and 55,700 ha of built up area, which increased to 61,679 ha by 2009 [30]. The study area of 967 km^2 is within the Santiago Metropolitan area and its 2014 population of approximately 7 million inhabitants, and encompasses multiple land use, land covers, and tenures. The eastern higher elevation portion of the study area is located in Andean piedmont shrublands while the western portion, once an *Acacia* spp. and grass dominated alluvial plain, has now mostly been altered to agricultural and urban land covers [12,30].

A total of 200 stratified random 400 m^2 plots, originally measured during January and February 2002 using criteria outlined in [12], were allocated across all of the Santiago Metropolitan area's 36 *comunas* and an additional four unincorporated ones (*i.e.*, Colina, Lampa, Puente Alto, and San Bernardo). A *comuna* in Chile is a geographically and administratively delineated municipality. Of the 200 originally measured plots in the study area, 192 were relocated and re-measured during October 2014 to January 2015 (Figure 1).

Re-measured plots were assigned land use classes based on existing land use and land cover types originally defined in 2002 ([12]; Table 1). We also classified the 2002 pervious surface cover types on each plot into percent: maintained grass, herbaceous vegetation, and bare soil. Percent impervious surface covers were also measured and include: cement, buildings, and paved surfaces. Plots were also post-stratified into low, medium, or high socioeconomic strata based on the *Asociación Chilena de Empresas de Investigación de Mercado* and an approach described in [12]; socioeconomic strata being defined as a combination of average annual income, education, vehicle ownership, and house services (e.g., fixed telephone). Due to the low number of individual trees per species found on the plots, we grouped tree species into five tree growth form types based on Lawrence *et al.*'s [16] classes: conifers, broadleaf-evergreen, broadleaf-deciduous, shrubs, and palms.

Figure 1. Plot locations within the Santiago de Chile study area in 2002 (**left**) and 2014 (**right**). Plots located on high, medium, and low socioeconomic strata are represented in red, gray, and blue circles, respectively. Vegetated land covers are green in color, bare soil areas are in pink, while orange and brown depict sparsely vegetated areas. High to low urban density areas are dark to light magenta, respectively, and dark magenta, linear features are major transportation rights of way. Linear light magenta areas in the southeast and far east in the Andean foothills are a river flood plain and barren rock, respectively. Note: images are from Landsat-7 RGB 742 taken on February 2002 and January 2014.

Table 1. 2002 land use/cover classes analyzed in Santiago, Chile.

Land Use Classes	Land Use/Land Covers	Number of Plots
Residential	Low to high density residential, multi-family, mixed residential	86
Commercial/Industrial	Commercial shopping areas, industrial areas, public buildings, airports, athletic stadiums	26
Green Areas	Vacant areas, shrub lands, plazas, parks, cemeteries, golf courses, athletic fields	36
Agriculture	Agricultural areas	38
Transportation	Highways, major transportation rights-of-way	6

2.2. Plot Matching and Data Analysis

During 2014, plot locations were relocated using 2002 measured field reference data (e.g., Geographical Information System plot centroid coordinates, aerial photographs, plot center photos, and plot sketches). Once true plot center was identified using azimuth and distance to permanent reference object measurements (e.g., utility poles, manhole covers, building corners), we identified trees, palms, and shrubs originally measured in 2002. A tree, palm, or shrub is defined as a woody plant or palm with a diameter at breast height (DBH) of 2.5 cm or greater at 1.5 m above the ground surface. Plot data in 2002 recorded each tree's distance and direction to the plot center. This same information was available in 2014 to locate and identify individuals that were measured in 2002. If species matched and DBH was similar to the 2002 measurement, or wherever diameter was indicated to have been originally measured in the case of forked or deformed stems; trees were re-measured. Missing trees that were originally measured in 2002, but were not located in 2014, were further investigated to determine if stumps were present to confirm mortality or removal. Trees in 2014 that were not present in 2002 were also recorded. See [16,28] for details on specific plot and tree matching approaches.

Urban forest structure changes were analyzed using tree mortality and ingrowth variables as previously done in studies from North America [16,28]. Specifically, we defined mortality as 2002 trees that could not be matched to 2014 trees. As such, mortality does not distinguish between tree removal due to maintenance or land clearing activities. Ingrowth was defined as the presence of a new tree in 2014, not measured in 2002 that could be the result of planting (*i.e.*, existing trees that grew into the 2.5 DBH criteria), or natural regeneration. Average annual tree diameter growth (ΔDBH; cm· year^{-1}) was estimated for matched trees by calculating the change in DBH divided by the total number of days between measurements and then annualized using 365 days per year [16]. Plot-level basal area change (ΔBA; m^2· year^{-1}) was also estimated using average annualized tree diameter growth for all trees on a plot [28].

We estimated individual tree biomass (kg) using aboveground allometric biomass equations from [19] and from the GlobAllomeTree database [31]. If species-specific equations were not available, we used equations from the same genus, family, or tree type (*i.e.*, conifer and hardwoods) following [32] and [33]'s approaches. Plot-level biomass was obtained as the sum of individual tree biomass on a specific plot; temporal changes in biomass (ΔBiomass kg· year^{-1}) were estimated by subtracting 2014 biomass from 2002 and dividing by the time since measurements. We did not include new biomass from 2002 due to ingrowth or biomass from 2014 lost to mortality. Plot-level basal area and biomass estimates were then converted to a per hectare basis for subsequent analyses.

2.3. Statistical Analysis

We tested for differences in DBH, BA, and biomass change at the species-level for both the five most frequent tree species as well as tree classification groups based on their form. We also analyzed differences in 2014 basal area and biomass and 2002 to 2014 ΔDBH, ΔBA, and ΔBiomass by *comuna*, socioeconomic strata, soil surface cover types, and land use classification. The 2014 biomass data was transformed using a logarithmic function for subsequent statistical analysis. We also analyzed for statistical differences in tree cover change as well as mortality and ingrowth according to socioeconomic strata, soil surface cover, and land use.

To test for statistical differences in mortality and ingrowth according to tree form, we used a five-sample test for equality of proportions without continuity correction using the R procedure *prop.test* [34] and alternative two-sided hypotheses. To better identify the plot-level surface cover factors, or correlates, that can possibly be driving the changes in structure, we determined the correlation between surface cover types and DBH, BA, and biomass (B). Specifically, we calculated Spearman correlation coefficients between 2002 to 2014 change in basal area and biomass with 2014 surface cover data using the *cor* procedure. Residuals were evaluated using Q-Q plots to assess their distribution and statistical differences were determined using Analyses of Variance with the *aov* function. This function was also used to test for differences in plot-level tree cover among strata while a paired t-test was used to determine city-wide tree cover differences during the analysis period.

Spatial correlations were determined using the Spatial Dependence: Weighting Schemes, Statistics and Model (*spdep*), and Data Analysis in Ecology (*pgirmess*) packages with the R *correlog* function. Finally, to better visually assess spatial patterns in BA and biomass we mapped urban tree biomass using spatial interpolation with an Inverse Distance Weighting (IDW) method using QGIS 2.10 and a distance coefficient of 0.05. Spatial autocorrelation was determined using Moran's *I* with the *correlog* procedure in the *pgirmess* package. All analyses were done using R version 3.1.3 [34].

3. Results

Of the original 200 plots from 2002, only eight could not be re-measured due to lack of access and permission. Given Santiago's semi-arid environment, high building density, and low tree cover; our overall sample size and number of matched trees (Table 2) was low compared to other urban forest ecosystem studies from the subtropics [16,28,33] and more humid temperate areas [7]. Table 2 below provides an overview of the basal area, ingrowth, and mortality in Santiago, Chile's urban forest from 2002 to 2014.

Table 2. Basal area and average annual urban forest percent mortality and ingrowth for different tree forms in Santiago, Chile during 2002 to 2014.

Tree Forms	Re-Measured Trees	Total Basal Area $(m^2 \cdot ha^{-1})$	% Mortality (SE)	% Ingrowth (SE)
Broadleaf-Deciduous	476	2.29	2.99 (0.18)	2.94 (0.18)
Broadleaf-Evergreen	210	0.76	2.98 (0.27)	3.13 (0.27)
Conifer	43	0.65	3.29 (0.64)	2.71 (0.59)
Palm	20	0.59	2.92 (0.88)	3.33 (0.91)
Shrub	86	0.14	3.1 (0.43)	2.81 (0.42)

Note: SE is Standard error.

3.1. Spatial Differences in Tree Biomass, Cover, and Basal Area

We found that 2014 tree biomass $(kg \cdot ha^{-1})$ was greater in the northeastern sections of the study area (Figure 2) and significant differences were found among socioeconomic strata ($p < 0.01$; Table 3). In general, wealthier *comunas* (*i.e.*, the northeastern section of the study area) had greater tree biomass than lower income ones. We observed that city-wide plot-level tree cover was not significantly different ($p = 0.45$) during the analysis period, with a value of 18.2% ($+/- 1.6\%$) in 2002 that dropped to 16.6% ($+/- 1.6\%$) in 2014. Tree cover in the high socioeconomic stratum was significantly greater than the other two strata and increased during the period of analysis from 17.3% to 23.1%. Conversely, tree cover dropped from 16.8% to 13.1% in the medium strata, and from 20.1% to 12.8% in the low strata (Table 3). We also observed a high variability in ΔBA and ΔBiomass, but no significant differences were found across socioeconomic strata ($p = 0.4$, Table 3) or land uses ($p = 0.71$, Table 4). Overall, we found no statistically significant differences in both ΔBA and ΔBiomass, and we observed almost a static biomass accumulation and BA increases during the analyzed period (Table 4). Overall, biomass ($p = 0.42$) and BA ($p = 0.37$) were not significantly different across land uses (Table 4). However, commercial/industrial areas did present lower average biomass and BA values than other land use types.

Table 3. Mean aboveground tree biomass in 2014, change in aboveground tree biomass (ΔBiomass), and mean percent tree cover change (ΔTree Cover%) according to socioeconomic strata during 2002 to 2014 in Santiago, Chile.

Socioeconomic Strata	Plots	2014 Biomass $(Mg \cdot ha^{-1})$ (SE)	ΔBiomass 2002 to 2014 $(Mg \cdot ha^{-1} \cdot year^{-1})$ (SE)	ΔTree Cover% (2002 to 2014)
High	69	21.66 (5.20)	0.29 (0.31)	5.80 (3.53) [a]
Medium	61	8.32 (3.24)	−0.05 (0.18)	−3.70 (3.67) [a,b]
Low	62	5.98 (2.05)	−0.15 (0.17)	−7.30 (3.84) [b]

Note: SE is Standard error and different letter superscripts between strata in ΔTree cover represent significant statistical difference ($p < 0.05$).

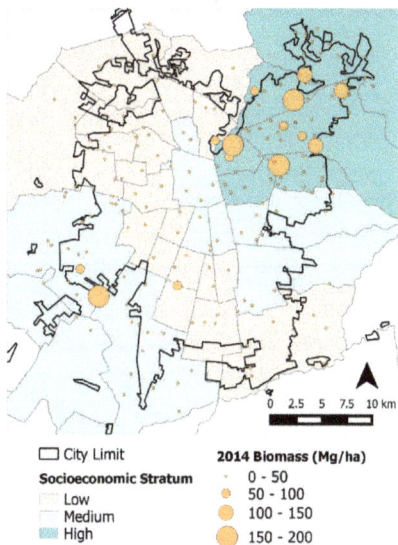

Figure 2. Aboveground tree biomass distribution in Santiago, Chile in 2014. Larger sized circles represent plots with higher biomass estimates.

Table 4. Mean tree biomass, change in tree biomass (ΔBiomass) during 2002 to 2014, mean basal area, and change in basal area (ΔBA) during 2002 to 2014 in Santiago, Chile.

Land Use Class	Plots	2014 Biomass $(Mg \cdot ha^{-1})$ (SE)	ΔBiomass $(Mg \ ha^{-1} \cdot year^{-1})$ (SE)	2014 Basal Area $(m^2 \cdot ha^{-1})$ (SE)	ΔBA $(m^2 \cdot ha^{-1} \cdot year^{-1})$ (SE)
Agriculture	38	7.91 (5.01)	−0.09 (0.18)	2.10 (0.83)	0.00 (0.06)
Commercial/Industrial	26	4.85 (1.56)	−0.38 (0.32)	1.83 (0.45)	−0.04 (0.05)
Green Areas	36	11.90 (4.77)	0.22 (0.32)	2.72 (0.81)	0.03 (0.05)
Residential	86	16.84 (4.10)	0.16 (0.25)	3.96 (0.89)	0.07 (0.06)
Transportation	6	11.90 (7.68)	−0.13 (0.70)	5.50 (3.10)	−0.06 (0.17)
Santiago	192	12.35 (2.29)	0.04 (0.14)	3.12 (0.47)	0.03 (0.03)

Note: SE is Standard error.

3.2. Changes in DBH, BA, and Biomass According to Species and Tree Form

The five most frequent tree species represented approximately 25% of all our matched trees, and their overall growth rate by species ranged from 0.12 to 0.36 cm· $year^{-1}$ (Table 5). These species also had an increase in biomass between 0.46 to 6.47 kg· $year^{-1}$ during the analyzed period; with *R. pseudoacacia* having the highest growth rates and *P. ceracifera* the greatest biomass change (Table 5). Actual ΔDBH ($p < 0.0001$) and ΔBiomass ($p = 0.03$) during the analysis period were significantly different among the five most frequent tree species (Table 5).

Table 5. Growth diameter rate (ΔDBH), change in biomass (ΔBiomass) and basal area for the five most frequent tree species in Santiago, Chile during 2002 to 2014.

Species	Measured Trees	ΔDBH (cm· year^{-1}) (SE)	ΔBiomass (kg· year^{-1}) (SE)	Total Basal Area (m^2· ha^{-1}) (SE)
Acer negundo	15	0.12 (0.15)	0.46 (0.40)	0.20 (0.02)
Citrus limon	12	0.26 (0.12)	2.49 (1.47)	0.02 (0.00)
Prunus amygdalus	14	0.13 (0.05)	1.09 (0.46)	0.04 (0.00)
Prunus ceracifera	20	0.21 (0.11)	6.47 (4.33)	0.12 (0.01)
Robinia pseudoacacia	27	0.36 (0.05)	2.61 (1.29)	0.30 (0.01)

Growth diameter rates (ΔDBH) were greater for conifers, while conifers and broadleaf deciduous trees exhibited the greatest change in biomass (Table 6). The large standard errors found for conifers and palms are likely due to a very small sample size, stem shrinkage or swelling, and the palm's stem sheath and fronds that makes DBH re-measurements unreliable [4] (Table 6).

Table 6. Annual average change in growth diameter rate (ΔDBH) and aboveground biomass for different tree forms in Santiago Chile during 2002 to 2014.

Tree Form	Trees	ΔDBH (SE) (cm· year^{-1})	ΔBiomass (SE) (kg· year^{-1})
Broadleaf-Deciduous	149	0.47 (0.06)	9.92 (2.34)
Broadleaf-Evergreen	56	0.49 (0.41)	7.35 (4.03)
Conifer	12	1.94 (0.65)	*
Palm	5	*	1.86 (1.01)
Shrub	25	0.36 (0.14)	6.91 (2.80)

* Standard errors were much too large to report value.

3.3. Surface Cover Correlates

We assessed the use of plot-level surface cover as an indicator for urban forest structural changes and found that both BA and biomass were poorly correlated with all measured surface covers (Table 7). Increases in building cover indicate lower BA and biomass, however, low building cover was not indicative of BA and Biomass. We only found that cement coverage was correlated to 2002 Basal Area ($r = 0.43$) and 2002 and 2014 Biomass ($r = 0.41$ and $r = 0.36$, respectively; Table 7).

Table 7. Spearman correlation coefficients, r, for basal area (BA), basal area change (ΔBA), biomass (kg), and biomass change (ΔBiomass) according to 2002 and 2014 surface covers in Santiago, Chile.

Surface Cover	BA m^2· ha^{-1} 2002	BA m^2· ha^{-1} 2014	ΔBA m^2· ha^{-1}	Biomass (kg) 2002	Biomass (kg) 2014	ΔBiomass (kg· year^{-1})
Impervious	0.198	0.179	−0.021	0.186	0.147	−0.018
Asphalt	0.111	0.163	0.088	0.099	0.179	0.082
Building	0.161	0.147	−0.036	0.165	0.108	−0.068
Cement	0.433	0.39	−0.066	0.405	0.362	−0.061
Pervious	−0.077	−0.012	0.021	−0.082	0.017	0.008
Bare soil	0.020	0.027	−0.017	0.039	0.032	−0.013
Grass	0.287	0.318	0.026	0.231	0.291	−0.024
Herbaceous	−0.134	−0.160	−0.091	−0.106	−0.104	−0.068

3.4. Spatio-Temporal Changes and Patterns in Biomass and Basal Area

Similar to the previous statistical analyses, our spatial analyses also showed greater tree biomass towards the northeastern section of the city and a cluster of plots with greater biomass in the peri-urban, southwest *comunas* of the study area (Figures 2 and 3). These same plots had greater basal area. Spatial interpolation by IDW of ΔBiomass also exhibited a similar pattern to the plot-level biomass distribution

(Figures 2 and 4). Plots and areas in the central, northwest, and southern sections of Santiago showed losses in both ΔBA and ΔBiomass (Figure 3). Figure 4 maps biomass change but should be interpreted carefully since the IDW spatial interpolation based on the plot influence distance for urban areas has high uncertainty, given the non-continuous heterogeneous character of cities. Similarly, we found some significantly high spatial autocorrelation according to Moran's I statistic in plots that were less than 4 km in proximity relative to each other.

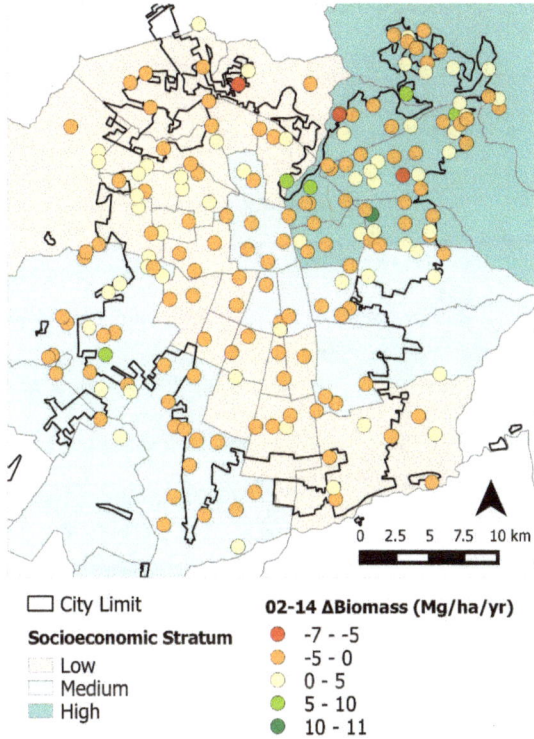

Figure 3. Plot-level urban tree biomass change in Santiago, Chile from 2002 to 2014 based on 192 re-measured plots.

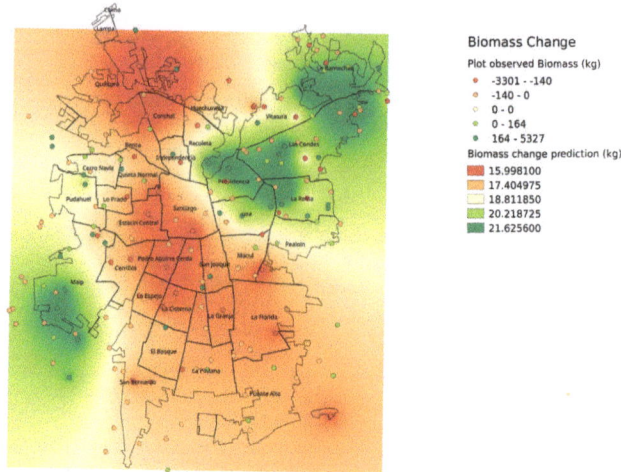

Figure 4. Urban tree biomass change prediction map for Santiago, Chile using Inverse Distance Weighting interpolation using the 192 re-measured plots.

4. Discussion

Although no clear evidence can be observed in the 2002 and 2014 Landsat imagery (Figure 1), the Santiago Metropolitan Area has undergone noticeable urbanization in the form of infill and land use changes since 2002 [30]. Rapid economic development and recent large-scale infrastructure projects such as new transportation rights of way and hubs, tunnels, and building-housing projects [35] has negatively affected the structure of Santiago's urban forest. These changes have occurred for most of the city with the exception of the most affluent areas, the northeast and few areas in the oldest sections of the city center [35]. Santiago's semi-arid climate and relative drought related water scarcity explains its low tree cover; however, even small changes can have noticeable effects in terms of the urban forest structure and subsequent functions and services. The tree scarcity in semi-arid Santiago resulted in different patterns than previous studies that used similar field sampling approaches from wet and humid urban forests in temperate and subtropical North America (e.g., Syracuse, USA [7]; Gainesville, USA [16]; Orlando, USA [33] and San Juan, Puerto Rico [28]); making this study unique.

These socio-ecological and urbanization effects were observed in the results from both our statistical and spatial analyses. This is particularly evident from the differences in tree cover, losses in biomass throughout the study area, and BA changes among socioeconomic strata (Table 3). More affluent *comunas* are mostly located in the northeastern section of the city and, because of greater resources, have more tree cover, larger trees (*i.e.*, greater BA), more abundant trees, and therefore, greater ΔBiomass (0.29 Mg$\cdot$ha$^{-1}\cdot$year^{-1}) [6,15,36]. Conversely, plots on the medium and lower socioeconomic strata had losses in overall tree cover and negative ΔBiomass. These plots generally correspond to areas with higher building densities and *comunas* with lesser economic resources ([12]; Figure 4). Spatially, plots in the northeastern section of the city had greater tree biomass; however, in the far southwest section of study area we also found high biomass areas corresponding to agricultural land from the peri-urban section (Figure 2). Rainfall gradients across the study areas could also be playing a role in structural differences as upper elevation plots in the Andean piedmont receive more rain, but management and maintenance regimes (*i.e.*, irrigation and fertilization) could be masking the effects of precipitation on tree growth and mortality.

Poor correlations between surface cover and BA and biomass might be a result of having small sample sizes and a number of plots with both high cement cover and high biomass and others with high grass cover and low biomass. High cement cover and high biomass are typical of large street trees near building areas. Using the sum of the different vegetation cover (horizontal) and height (vertical) estimates might increase the correlations, as observed by [33]. We noted that cover types are regularly used as indicators of urbanization, planting space, and urban forest structure and function [7,16], but given our semi-arid and high building densities, this might have resulted in different patterns.

We acknowledge that our sample size, in terms of plot density and tree numbers, was low relative to similar urban forest studies using fixed area plots [7,16,28,33]. Factors—not analyzed in this study—such as climate change and socio-ecological dynamics—could have also driven these changes. Additionally, our use of forest grown allometric equations on urban trees can often lead to errors of up to 40% in biomass estimates [5]. Similarly, the sample size for conifers and palms was very low, seasonal stem shrinkage and swelling is likely [4,16], and this was also confounded by the difficulties in measuring palms using measurement techniques developed for single stem temperate trees [4,16,28]. However, we note that all these cited studies had the same limitations. Indeed, given the context of our study area's: size, socio-political dynamics, and limited infrastructure that characterizes cities from developing countries, we feel that our results do provide a better understanding of urban forests outside the frequently studied areas of the United States and Canada, giving more insights to understudied areas such as Latin America [1,13,32], Mediterranean climates [12,28,32], and Australia [3,15].

When comparing urban forest structure results such as annual mortality to other studies using inventory data, we can distinguish the effect of climate and management. Mortality of urban trees was apparently more related to management than to climate; Santiago showed much lower mortality rates (3.0%) than subtropical San Juan, Puerto Rico [28], with values between 30%–40%. Conversely a subtropical city in the United States, Gainesville [16] had an average mortality rate between 10%–19%. Looking at growth rates, climate appears to be a stronger driver than management; values for subtropical cities of San Juan and Gainesville are closer to each other (0.7–1.0 cm· year^{-1}), while Santiago had lower average values between 0.1 to 0.4 cm· year^{-1}. In terms of the temporal changes to urban forest structure, there are no similarities to relevant studies [7,28].

Studies that use permanent random plots often report number of trees per unit area—as opposed to basal area—as a measure of tree density [7]; however, multi-stemmed trees with a shrub form or secondary tropical forests and mangroves can confound comparisons, thus our preference for basal area. Values for tree density can vary from 222–328 trees per ha in San Juan in 2001 and 2010, respectively [28]. While there is an average of 34 trees per ha in a temperate city such as Syracuse, US [7], in Santiago, a Mediterranean shrubland biome, this value reached 64 trees per ha [12]. Meanwhile, there was an overall annual net loss of approximately four trees per ha in subtropical Gainesville [16]. Tucker-Lima *et al.* report a basal area of 4.6 m^2· ha^{-1} in San Juan, which is very similar to our 2014 estimate of 4.8 m^2· ha^{-1}. Changes in basal area in Santiago of 0.1 m^2· ha^{-1} were, however, much lower than in San Juan (1.0 m^2· ha^{-1}) [28].

The growing body of literature using similar sized, long-term monitoring plots as utilized in this study can be used to compare trends for urban forests across different regions of the globe [5,6,16,28,33,36]. The characteristics of urban forest growth, mortality, and the effects of site characteristic such as irrigation, and ecological disturbance on these, could be analyzed against field measured plot and site correlates [5,16,20,21]. However, the presence of palms, multi-stemmed tall shrubs, and size criteria for tree-shrub differentiation across different biomes can confound some of these comparisons. Care is also warranted when comparing shrubland dominated biomes such as Santiago, to dense subtropical secondary and mangrove forests or temperate forests. Similarly, because of the difficulty in sampling large, heterogeneous, urbanized areas, sampling intensities can be low and result in larger uncertainties. This is particularly true in Santiago's densely built, semi-arid, urban context, which also resulted in a reduced sample of measured and matched trees.

Our study provides one of the few comparative insights into how a South American and Mediterranean urban forest changes across space and time. Future research could analyze spatio-temporal changes in urban forest composition and its subsequent effects such as the spatial dynamics of ecosystem service provision and disservice hotspots [11,13,21,22,24–26]. Plot level structure-function information could also be used to test land management and planning scenario effects on the demand and supply of services [4,7,14,33]. Quantitative analyses could also determine the socio-ecological causal factors behind, and the drivers related to, changes in urban forests, their processes, and services [1,3,6,23,36]. More basic research could use these monitoring sites for better understanding floral diversity and the occurrence of invasives and understory dynamics in urbanized forests, as well as for developing allometric equations [36–39]. Permanent plots could also facilitate dendrochronological analysis of different tree species and their growth and mortality as affected by climate change, pollution, maintenance practices, and community preferences.

5. Conclusions

Overall, we found that urban forest mortality, ingrowth, and tree cover in the greater Santiago area remained stable during 2002 to 2014. We also noticed slight losses in basal area and biomass change across the study area. However, there were some noticeable trends during the analysis period in that tree cover increased and was greater in the highest socioeconomic stratum; however, it decreased in the medium and low strata. Similarly, the less affluent central, northwest, and southern plots, in particular, exhibited losses in BA and biomass from 2002 to 2014.

As previously mentioned, other studies have used plot-level data to analyze changes in an urban forest but, to our knowledge, most of these studies are from North America. Here, we describe for the first time the spatio-temporal dynamics of a South American urban forest in a Mediterranean climate. Overall, we found that these urban forests follow similar trends as in other parts of the world. For example, affluent areas tend to have more and better condition urban forests than poorer areas. Thus, barring substantial ecological disturbance (e.g., storms, drought, pests, or urbanization), changes in urban forest structure will primarily depend on human management (e.g., maintenance and irrigation) and biophysical factors (*i.e.*, growing space and climate). Findings could also be used to identify which tree species perform better in terms of growth and basal area. Accordingly, this information could be used to identify tree functional traits that are most associated with ecosystem service provision (e.g., carbon offsets) and disservice minimization (e.g., allergenic tree locations).

However, Latin America and the rest of the developing world are generally characterized by unplanned land uses and marked socioeconomic inequities. Thus, studies such as ours can be used for targeting specific sites for improved management and setting monitoring and evaluation standards for municipalities. Conversely, areas that are maintaining their urban forest and standards—despite limited budgets—could also be identified. Most importantly, issues of environmental justice and resilience to the effects of climate and socio-political changes could be better addressed with this long-term data. Indeed, the growing body of literature calls for the development of an available global urban forest monitoring database that could be used for research by the global scientific community.

Acknowledgments: We would thank the Chilean *Fondo Nacional de Desarrollo Científico y Tecnológico* (FONDECYT 1140319) project "Vegetation Knowledge-based Indicators for Urban Sustainable Planning "for funding this study. Research was also partly funded by FONDECYT 3150352 "Provision of urban ecosystem services, exploring the effects of planning, urbanization, climate and environmental conditions on the urban forest of Santiago and La Serena". We are also very grateful to the three reviewers whose suggestions greatly improved this manuscript.

Author Contributions: All authors contributed equally to this work.

Conflicts of Interest: The authors declare no conflict of interest.

References

1. Aronson, M.F.J.; La Sorte, F.A.; Nilon, C.H.; Katti, M.; Goddard, M.A.; Lepczyk, C.A.; Warren, P.S.; Williams, N.S.G.; Cilliers, S.; Clarkson, B.; *et al.* A global analysis of the impacts of urbanization on bird and plant diversity reveals key anthropogenic drivers. *Proc. R. Soc. Lond. B Biol. Sci.* **2014**, *281*. [CrossRef] [PubMed]

2. Avolio, M.; Pataki, D.E.; Gillespie, T.; Jenerette, G.D.; McCarthy, H.R.; Pincetl, S.; Weller-Clarke, L. Tree diversity in southern california's urban forest: The interacting roles of social and environmental variables. *Front. Ecol. Evol.* **2015**, *3*. [CrossRef]
3. Kendal, D.; Williams, N.S.G.; Williams, K.J.H. Drivers of diversity and tree cover in gardens, parks and streetscapes in an Australian city. *Urban For. Urban Green.* **2012**, *11*, 257–265. [CrossRef]
4. Brandeis, T.J.; Escobedo, F.J.; Staudhammer, C.L.; Nowak, D.J.; Zipperer, W. *San Juan Bay Estuary Watershed Urban Forest Inventory*; USDA-Forest Service, Southern Research Station: Asheville, NC, USA, 2014; p. 44.
5. Timilsina, N.; Staudhammer, C.L.; Escobedo, F.J.; Lawrence, A. Tree biomass, wood waste yield, and carbon storage changes in an urban forest. *Landsc. Urban Plan.* **2014**, *127*, 18–27. [CrossRef]
6. Zhao, M.; Escobedo, F.J.; Staudhammer, C. Spatial patterns of a subtropical, coastal urban forest: Implications for land tenure, hurricanes, and invasives. *Urban For. Urban Green.* **2010**, *9*, 205–214. [CrossRef]
7. Nowak, D.J.; Hoehn, R.E.; Bodine, A.R.; Greenfield, E.J.; O'Neil-Dunne, J. Urban forest structure, ecosystem services and change in Syracuse, NY. *Urban Ecosyst.* **2013**, 1–23. [CrossRef]
8. Staudhammer, C.L.; Escobedo, F.J.; Lawrence, A.B.; Duryea, M.; Smith, P.; Merritt, M. Rapid assessment of change and hurricane impacts to houston's urban forest structure. *Arboricult. Urban For.* **2011**, *37*, 60.
9. Manes, F.; Marando, F.; Capotorti, G.; Blasi, C.; Salvatori, E.; Fusaro, L.; Ciancarella, L.; Mircea, M.; Marchetti, M.; Chirici, G.; *et al.* Regulating ecosystem services of forests in ten Italian metropolitan cities: Air quality improvement by PM_{10} and O_3 removal. *Ecol. Indic.* **2016**, *67*, 425–440. [CrossRef]
10. Roman, L.A.; Battles, J.; McBride, J.R. *Urban Tree Mortality: A Primer on Demographic Approaches*; US Department of Agriculture, Forest Service, Northern Research Station: Newtown Square, PA, USA, 2016; p. 24.
11. Cariñanos, P.; Casares-Porcel, M.; Quesada-Rubio, J.M. Estimating the allergenic potential of urban green spaces: A case-study in Granada, Spain. *Landsc. Urban Plan.* **2014**, *123*, 134–144. [CrossRef]
12. Escobedo, F.J.; Nowak, D.J.; Wagner, J.E.; De la Maza, C.L.; Rodríguez, M.; Crane, D.E.; Hernández, J. The socioeconomics and management of Santiago de Chile's public urban forests. *Urban For. Urban Green.* **2006**, *4*, 105–114. [CrossRef]
13. Dobbs, C.; Kendal, D.; Nitschke, C.R. Multiple ecosystem services and disservices of the urban forest establishing their connections with landscape structure and sociodemographics. *Ecol. Indic.* **2014**, *43*, 44–55. [CrossRef]
14. Escobedo, F.J.; Kroeger, T.; Wagner, J.E. Urban forests and pollution mitigation: Analyzing ecosystem services and disservices. *Environ. Pollut.* **2011**, *159*, 2078–2087. [CrossRef] [PubMed]
15. Dobbs, C.; Kendal, D.; Nitschke, C. The effects of land tenure and land use on the urban forest structure and composition of Melbourne. *Urban For. Urban Green.* **2013**, *12*, 417–425. [CrossRef]
16. Lawrence, A.B.; Escobedo, F.J.; Staudhammer, C.L.; Zipperer, W. Analyzing growth and mortality in a subtropical urban forest ecosystem. *Landsc. Urban Plan.* **2012**, *104*, 85–94. [CrossRef]
17. Escobedo, F.; Varela, S.; Zhao, M.; Wagner, J.E.; Zipperer, W. Analyzing the efficacy of subtropical urban forests in offsetting carbon emissions from cities. *Environ. Sci. Policy* **2010**, *13*, 362–372. [CrossRef]
18. Peper, P.J.; McPherson, E.G.; Mori, S.M. Equations for predicting diameter, height, crown width, and leaf area of San Joaquin Valley street trees. *J. Arboricult.* **2001**, *27*, 306–317.
19. Dobbs, C.; Hernández, J.; Escobedo, F. Above ground biomass and leaf area models based on a non destructive method for urban trees of two communes in central Chile. *Bosque (Valdivia)* **2011**, *32*, 287–296. [CrossRef]
20. Semenzato, P.; Cattaneo, D.; Dainese, M. Growth prediction for five tree species in an italian urban forest. *Urban For. Urban Green.* **2011**, *10*, 169–176. [CrossRef]
21. Rahman, M.A.; Armson, D.; Ennos, A.R. A comparison of the growth and cooling effectiveness of five commonly planted urban tree species. *Urban Ecosyst.* **2014**, *18*, 371–389. [CrossRef]
22. Xie, Q.; Zhou, Z.; Chen, F. Quantifying the beneficial effect of different plant species on air quality improvement. *Environ. Eng. Manag. J.* **2011**, *10*, 959–963.
23. Haase, D.; Larondelle, N.; Andersson, E.; Artmann, M.; Borgström, S.; Breuste, J.; Gomez-Baggethun, E.; Gren, Å.; Hamstead, Z.; Hansen, R.; *et al.* A quantitative review of urban ecosystem service assessments: Concepts, models, and implementation. *AMBIO* **2014**, *43*, 413–433. [CrossRef] [PubMed]
24. Strohbach, M.W.; Haase, D. Above-ground carbon storage by urban trees in leipzig, germany: Analysis of patterns in a European city. *Landsc. Urban Plan.* **2012**, *104*, 95–104. [CrossRef]

25. Gong, C.; Yu, S.; Joesting, H.; Chen, J. Determining socioeconomic drivers of urban forest fragmentation with historical remote sensing images. *Landsc. Urban Plan.* **2013**, *117*, 57–65. [CrossRef]
26. Sperandelli, D.; Dupas, F.; Dias Pons, N. Dynamics of urban sprawl, vacant land, and green spaces on the metropolitan fringe of São Paulo, Brazil. *J. Urban Plan. Dev.* **2013**, *139*, 274–279. [CrossRef]
27. Cilliers, S.; Cilliers, J.; Lubbe, R.; Siebert, S. Ecosystem services of urban green spaces in African countries—Perspectives and challenges. *Urban Ecosyst.* **2012**, *16*, 681–702. [CrossRef]
28. Lima, J.M.T.; Staudhammer, C.L.; Brandeis, T.J.; Escobedo, F.J.; Zipperer, W. Temporal dynamics of a subtropical urban forest in San Juan, Puerto Rico, 2001–2010. *Landsc. Urban Plan.* **2013**, *120*, 96–106. [CrossRef]
29. Peel, M.C.; Finlayson, B.L.; McMahon, T.A. Updated world map of the Köppen-Geiger climate classification. *Hydrol. Earth Syst. Sci.* **2007**, *11*, 1633–1644. [CrossRef]
30. Romero, H.; Vásquez, A.; Fuentes, C.; Salgado, M.; Schmidt, A.; Banzhaf, E. Assessing urban environmental segregation (UES). The case of Santiago de Chile. *Ecol. Indic.* **2012**, *23*, 76–87. [CrossRef]
31. Henry, M.; Bombelli, A.; Trotta, C.; Alessandrini, A.; Birigazzi, L.; Sola, G.; Vieilledent, G.; Santenoise, P.; Longuetaud, F.; Valentini, R.; Picard, N. GlobAllomeTree: International platform for tree allometric equations to support volume, biomass and carbon assessment. *iFor. Biogeosci. For.* **2013**, *6*, 326. [CrossRef]
32. Escobedo, F.J.; Clerici, N.; Staudhammer, C.L.; Corzo, G.T. Socio-ecological dynamics and inequality in Bogotá, Colombia's public urban forests and their ecosystem services. *Urban For. Urban Green.* **2015**, *14*, 1040–1053. [CrossRef]
33. Horn, J.; Escobedo, F.J.; Hinkle, R.; Hostetler, M.; Timilsina, N. The role of composition, invasives, and maintenance emissions on urban forest carbon stocks. *Environ. Manag.* **2014**, *55*, 431–442. [CrossRef] [PubMed]
34. R Core Team. *R: A Language and Environment for Statistical Computing*; R Foundation for Statistical Computing: Vienna, Austria, 2008.
35. Figueroa, O.; Rodriguez, C. Urban Transport, Urban Expansion and Institutions and Governance in Santiago, Chile. Available online: http://unhabitat.org/wp-content/uploads/2013/06/GRHS.2013.Case_.Study_.Santiago.Chile_.pdf (accessed on 20 May 2016).
36. Staudhammer, C.L.; Escobedo, F.J.; Holt, N.; Young, L.J.; Brandeis, T.J.; Zipperer, W. Predictors, spatial distribution, and occurrence of woody invasive plants in subtropical *Urban Ecosyst. J. Environ. Manag.* **2015**, *155*, 97–105. [CrossRef] [PubMed]
37. Loeb, R.E. Arboricultural introductions and long-term changes for invasive woody plants in remnant urban forests. *Forests* **2012**, *3*, 745. [CrossRef]
38. Pretzsch, H.; Biber, P.; Uhl, E.; Dahlhausen, J.; Rötzer, T.; Caldentey, J.; Koike, T.; van Con, T.; Chavanne, A.; Seifert, T.; *et al.* Crown size and growing space requirement of common tree species in urban centres, parks, and forests. *Urban For. Urban Green.* **2015**, *14*, 466–479. [CrossRef]
39. He, R.; Yang, J.; Song, X. Quantifying the impact of different ways to delimit study areas on the assessment of species diversity of an urban forest. *Forests* **2016**, *7*, 42. [CrossRef]

forests

MDPI

Article

Do Indigenous Street Trees Promote More Biodiversity than Alien Ones? Evidence Using Mistletoes and Birds in South Africa

Charlie Shackleton

Department of Environmental Science, Rhodes University, Grahamstown 6140, South Africa;
c.shackleton@ru.ac.za; Tel.: +27-46-603-7001

Academic Editors: Francisco Escobedo, Stephen John Livesley and Justin Morgenroth
Received: 19 April 2016; Accepted: 29 June 2016; Published: 13 July 2016

Abstract: Trees in urban landscapes provide a range of ecosystem services, including habitat, refugia, food, and corridors for other fauna and flora. However, there is some debate whether the richness and abundance of other biodiversity supported is influenced by the provenance of trees, i.e., native or non-native. This study assessed the presence of mistletoes and birds (and nests) in 1261 street trees. There were marked differences between native and non-native street trees, with the former having a significantly higher prevalence of birds (and nests) and supporting more species and in greater densities, whilst the latter supported a higher prevalence of mistletoes. Additionally, for birds, the proximity to green space, tree size and species were also important, whilst for mistletoes, the proximity to green space, slope aspect, and tree species were significant. Preference ratios indicated that some tree species had a higher than random occurrence of birds or mistletoes, whilst others had a low abundance. The indigenous tree species, *Acacia karroo* Hayne was the only reasonably abundant street tree species that was important for birds, nests, and mistletoes. At the street scale, there was a positive relationship between street tree species richness and bird species richness. These results emphasise the importance of selecting appropriate tree species if biodiversity conservation is a core outcome.

Keywords: biodiversity; connectivity; preference ratio; street trees; tree size; urban

1. Introduction

Trees in public urban spaces provide a variety of provisioning, regulating, cultural, and supporting ecosystem services to urban residents and necessary ecological processes [1,2]. Consequently, there is a growing emphasis on the greening of urban environments to maximise the provision of such ecosystem services [3]. Street trees are a particularly important component of urban greening because (i) they are located throughout the urban matrix; (ii) although streets are narrow, their combined area is often much larger than that of formal parks and green spaces; and (iii) a considerable proportion of the time that urban residents spend outdoors is spent on the streets, according to Todorova et al. [4] a considerable proportion of the time that urban residents spend out of doors in spent on the streets. Thus, ensuring that the streets are attractive, safe, and as functional as possible for urban residents is an import consideration for city planners and parks officials [5].

Besides all the benefits that urban residents receive from street and other urban trees, they also provide supporting services in the form of food, habitat, shelter, refugia, nesting materials, and breeding sites for many other species occurring in towns and cities. For example, the early work of Tzilkowsjki et al. [6] revealed that one-third of street trees had birds in them and that the prevalence differed markedly between different tree species, whilst Kubista & Bruckner [7] reported that urban trees provided 50% of the roost sites for several species of bats.

As a means of promoting the benefits of trees as habitat and food for other biodiversity, it is often stated that indigenous (native) trees should be encouraged above alien (exotic/non-native) species, because the services that they provide are already part of the local ecology, and other native species will have co-adapted with them [8–11]. These works present several examples of local biodiversity plans or urban ward authorities specifically regulating for native species over alien ones. Thus, from a biodiversity conservation perspective, indigenous trees are often seen as more beneficial for other native species than are alien ones. However, there has been limited empirical evaluation of this claim in urban settings [8,11]. This is particularly so for street trees.

Carthew et al. [12] recorded the presence of hollows in trees in six parks in Adelaide, Australia, and their occupation by brushtail possums (*Trichosurus vulpecula*). They reported that only one-third of possum dens were in indigenous trees, and the statistical analysis indicated that indigenous tree species were less likely to be used for den sites than the most common alien tree species. Blanchon et al. [13] found that an urban forest dominated by alien woody plants had 50% more species of ground beetles than a smaller, nearby urban forest dominated by indigenous plants. In contrast, bird and butterfly species richness was more diverse within and between green patches in Singapore dominated by native vegetation relative to those dominated by non-native species [14]. Gariola et al. [15] found that the alien tree species *Melia azedarach* L. was a common host for mistletoes in urban parks in Durban (South Africa).

Whilst such comparisons are valuable, there are confounding variables—most notably, the size of the forest or park studied and the structural diversity of the broader site. Two studies with birds reduced these confounding factors to some degree by examining visitation rates of native birds to individual trees. French et al. [16] and Gray and van Heezik [17] both compared visitation rates of birds to selected indigenous and alien tree species, the former in Australia and the latter in New Zealand. French et al. [16] found that, whilst birds visited all four tree/shrub species, the rate of visitation was significantly higher for the indigenous ones. The results of Gray and van Heezik [17], who monitored six species, were more equivocal and varied with season, but they concluded that exotic trees can sustain native birds. The same cannot be said for epiphytes, as there is a scarcity of literature on epiphytes in urban trees. The commentary of Johnston et al. [10] and the recent review by Chalker-Scott [11] also concluded that alien tree species can provide the same services to other biodiversity as native tree species do. However, if this is to be situated within a conservation debate, the question needs to be more nuanced, i.e., not whether exotic trees species can sustain native biodiversity, but rather can they sustain more, either in species richness or abundance. If the exotic species were removed, would it result in a loss of certain native biodiversity that was dependent upon them? If yes, would it be at levels about which managers and conservationists should be concerned? Dickie et al. [18] describe several examples of where removal of alien tree species was resisted or halted due to their perceived importance in providing food or habitat to charismatic or endangered native fauna, but can the same food or habitat be provided by indigenous species?

From the above, it is apparent that current debates on the relative merits of native or alien trees in supporting other indigenous biodiversity are fraught with conflicting positions and sometimes equivocal results confounded by differences in patch size, the presence of other species in the patch, variable structure between patches, and the nature of the surrounding matrix. Using street trees as the sample unit reduces or eliminates the confounding issues. Within the context of the above, this paper reports on a study, the aim of which was to ascertain whether there is any difference in the use of native or alien street trees by other biodiversity, using street trees as the sample unit and birds and mistletoes as components of supported biodiversity.

2. Materials and Methods

2.1. Study Site

Grahamstown (33°18′ S; 26°33′ E) is a medium-sized town in the Eastern Cape province of South Africa, with a population of approximately 70,000 people. It is the administrative centre of the Makana local municipality. Having been founded as a military base during the colonial frontier wars of the early 1800s, it is now a well-known educational centre, with a university and numerous private and state schools. Grahamstown is located at an altitude of 650 m.a.s.l and has a moderate climate with an average seasonal temperature ranging from 9.8 to 23.1 °C. The hottest months are December to March, and the coldest months are June and July [19]. It receives, on average, 669 mm of rainfall annually [20], with bimodal peaks in October–November and again in March–April, largely as frontal rain showers. The city is situated within a region of high biodiversity as it lies in the convergence zone of four major biomes, namely, fynbos, grassland, thicket, and karoo [21].

The more affluent, western suburbs are well greened, both in terms of formal green spaces and street trees, whilst the poorer eastern suburbs are not [22,23]. The inequitable distribution is a legacy of South Africa's racially discriminatory past [22], current developmental budgets favouring basic infrastructure over environmental or what are deemed luxury concerns [24], and high rates of vandalism and livestock damage to trees in some areas [25]. Consequently, the study was restricted to the western suburbs. Mean housing density varies from 4.3 ha^{-1} in the western suburbs to 32.3 ha^{-1} in the newly constructed low-cost state housing areas (reserved for the indigent) in the east [22]. Kuruneri-Chitepo & Shackleton [23] reported that approximately 60% of the street trees are not indigenous to South Africa and that the three most common species are *Grevillea robusta* A.Cunn, *Jacaranda mimosifolia* D.Don, and *Brachychiton acerifolium* (A.Cunn.) F. Muell., all alien species.

2.2. Field Methods

Eight residential suburbs were selected on the basis of their high street tree abundance [23] (namely, Currie Park, Hill 60, Kingswood, Oatlands, Oatlands North, Somerset Heights, Sunnyside, and West Hill), allowing for the full range of aspects (north-, east-, south-, and west-facing). In each suburb, all streets longer than 50 m and containing at least ten street trees taller than 2 m, running parallel to the prevailing slope, were sampled. Sampling continued until 100 trees with mistletoes were recorded (a total of 38 streets). Sampling was done in early to mid-morning and again in mid to late afternoon; rainy or windy days were avoided.

Within each street, all street trees taller than 2 m on both sides of the road were inventoried. Since not all properties have boundary fences, it is possible that some of the sampled trees were not public street trees but planted by the property owners, but that has no effect on the objectives and results of this study. For each tree, the following information was recorded: (i) the species; (ii) the basal diameter at approximately 35 cm above ground level; (iii) the number and species of any mistletoes; (iv) a visual estimate of the proportion of the tree canopy occupied by mistletoes; (v) the number of bird's nests; and (vi) any birds (number and species noted) in the tree or any that flew into or out of the tree as it was approached. The bird observation time was approximately 6 min per tree, within the usual 5 or 10 min typical for bird surveys (which usually survey a much larger area, such as within a 25 m radius). The author's presence was unlikely to have had much effect on the birds since, being on streets, there was already a measure of human activity with pedestrians, bicycles, and vehicles. The size of each mistletoe was visually estimated as small, medium, or large, roughly corresponding to <0.5 m, 0.5–1.0 m, and >1.0 m diameter, respectively. If a given tree was multi-stemmed, the diameter of only the largest stem was measured. If a tree was branching at the measurement height, the diameter was measured above the branching or swelling. Sampling was done at the end of winter (August and September 2015) to optimise the visibility of mistletoes and nests, as this was the time of lowest leaf abundance on trees and when all trees were in the same phenophase to eliminate the effects of differential timing of flowers, fruiting, or seeding on bird presence. It is therefore likely that most nests observed were old, being from the previous summer. Coniferous species (a negligible proportion of street trees in Grahamstown [23]) were omitted because of their evergreen nature and, for several species, very dense canopy, which made detection of mistletoes and nests almost impossible. That period of the year was prior to the arrival of summer migratory bird species, which will have reduced the frequency of bird encounters and species richness to some extent, but does not undermine the comparative basis of the study. Any trees that could not be identified due to the absence of leaves were revisited three months later in the spring, when leaves and flowers were available. The linear distance from the mid-point along the street to the nearest public green area (formal or informal) with at least 10% woody plant cover was measured using Google Earth images (2015).

2.3. Data Analyses

Preference ratios per common tree species were determined as the percentage that a given tree species contributed to all trees infected with mistletoes to the percentage contribution of that same species to all street trees sampled. The same was done for birds and bird's nests. A preference ratio of greater than 1 signifies a rate of occurrence of mistletoes, birds, or nests in a particular tree species greater than would be expected if their presence was random. A preference ratio of less than 1 signifies active avoidance, and a preference ratio of close to 1 indicates a more or less random presence. Differences in the proportion of indigenous and alien street trees with mistletoes, birds, and bird's nests were tested via chi-square tests. Binomial logistic regression was used to determine factors that predict the presence of mistletoes, birds, and nests. The factors included were tree provenance (alien or indigenous), tree circumference, species, street, aspect, and distance to the nearest green space. All data analyses were conducted in Statistica v12. (StatSoft, 213, Tulsa, OK USA).

3. Results

A total of 1261 street trees spanning close to 100 species were enumerated, of which almost two-thirds (64.6%) were alien species. All the three most common species were alien, namely, *Jacaranda mimosifolia*, *Schinus terebinthifolius* Raddi, and *Fraxinus* spp. (Table 1).

Table 1. The five most common alien and indigenous street tree species in Grahamstown (*n* = 1261).

Alien	% of All Trees	Indigenous	% of All Trees
Jacaranda mimosifolia	8.7	*Erythrina caffra*	6.0
Schinus terebinthifolius	8.6	*Celtis africana*	5.6
Fraxinus spp.	7.3	*Acacia karroo*	5.0
Brachychiton acerfolium	6.4	*Ekebergia capensis*	3.4
Grevillea robusta	5.9	*Harpephyllum caffrum*	3.2
Total number of tree species	61	Total number of tree species	40
Total proportion (%) of all street trees	64.6	Total proportion (%) of all street trees	35.4

Only one mistletoe species was found: *Viscum obscurum* Thunb. Across all sampled trees, 7.9% had mistletoes. Corresponding figures for bird's nests and for birds were 6.2% and 8.9%, respectively. There were significant differences in the presence rates between indigenous and alien street tree species for mistletoes, birds, and bird' nests (Figure 1). In the case of mistletoes, they were significantly more prevalent in alien tree species than indigenous tree species. For birds and bird's nests, the opposite pattern prevailed, with significantly more indigenous street tree species harbouring them than alien ones.

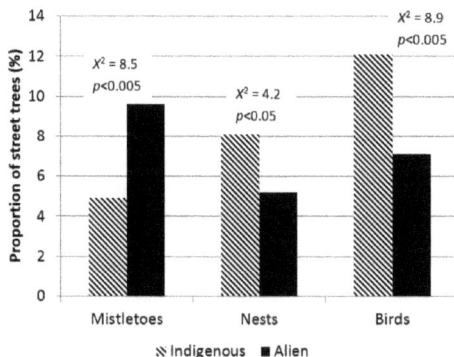

Figure 1. Prevalence (% of trees) of mistletoes, birds, and bird's nests in indigenous and alien street trees in Grahamstown (chi-square results indicate the significance of the difference in % of indigenous and alien street trees with each of mistletoes, birds, and nests).

A greater number of bird species were recorded in indigenous street trees than in alien ones—25 and 18, respectively (Figure 2)—even though there were almost 50% fewer indigenous street trees. The mean density of birds per tree also significantly favoured indigenous species over alien ones, being almost three times greater (Figure 2). Twenty-nine species of birds were recorded, of which 12 were recorded only in indigenous tree species, 11 were common to both indigenous and alien tree species, and 6 were recorded in alien trees only. The most commonly recorded bird species were the laughing dove (*Streptopelia senegalensis* L.), the redwing starling (*Onychognathus morio* L.), the cape weaver (*Ploceus capensis* L.), the cape white eye (*Zosterops virens* Sundevall), and the black-capped bulbul (*Pycnonotus barbatus* Desfontaines).

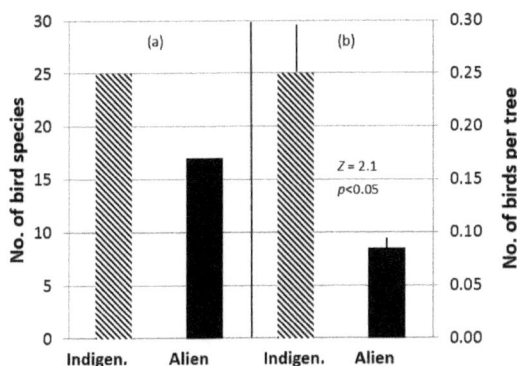

Figure 2. Bird species richness (**a**) and mean density per tree (**b**) for indigenous and alien street trees in Grahamstown.

Presence rates were not equal across tree species, with some being strongly favoured and some largely avoided. Of the 15 most abundant street tree species, *Acacia karroo* Hayne was favoured for birds and nests, with mistletoe infection being approximately proportional to *A. karroo* representation in the tree sample (Table 2). None of the common indigenous street tree species were avoided by birds. Most of the more common alien tree species had low PRs for mistletoes, birds, and nests, except *Fraxinus* spp., for which the PR was high for mistletoes and birds, and *Quercus robur* L., which had high PRs for nests and birds.

Table 2. Preference ratios of mistletoes, birds and bird's nests for the 15 most common street trees in Grahamstown (PR of close to 1 indicates more or less random occurrence in street tree species; >1 indicates positive association with the tree species; PR of <1 signifies under-representation or avoidance of that tree species).

Species	Origin	No. Sampled	Mistletoes	Nests	Birds
Acacia karroo	Indig	63	0.8	3.2	2.1
Celtis africana	Indig	70	0.2	1.9	1.0
Ekebergia capensis	Indig	43	0	0.4	2.1
Erythrina caffra	Indig	76	0	0.4	1.9
Harpephyllum caffrum	Indig	40	0	0	1.1
Podocarpus falcatus	Indig	34	0	0.5	1.3
Brachychiton acerfolium	Alien	81	0	0.2	0.4
Brachychiton populneum	Alien	24	0.5	0.7	0.5
Eucalyptus ficifolia	Alien	40	0	0	0.3
Fraxinus sp	Alien	92	6.4	1.9	0.4
Grevillea robusta	Alien	74	0	0.9	1.2
Jacaranda mimosifolia	Alien	110	0.1	0.3	0.7
Quercus robur	Alien	91	0.3	1.8	1.5
Schinus terebenthifolius	Alien	109	0	0.2	0.8
Tipua tipu	Alien	17	0	0	0.7

Considered against all factors measured, the origin of the tree as either indigenous or alien was the most significant predictor of the presence or absence of mistletoes, birds, and nests (Table 3). However, it was not the only one. Proximity to the nearest green space was also positively related with all three—most for mistletoes and least for nests. Overall, the presence of mistletoes was significantly related to distance, origin, species, and aspect. For the latter, north-facing aspects were relatively devoid of mistletoes. Birds were significantly associated with large indigenous trees and proximity to green spaces, whilst nests were associated with only indigenous trees and proximity.

Table 3. Significant predictors of the presence of mistletoes, birds, and nests in street trees in Grahamstown (*p* values; n.s = not siginficant).

Attribute	Mistletoes	Nests	Birds
Street	n.s	n.s	n.s
Distance to nearest green area	0.00002	0.0023	0.0116
Aspect	0.0257	n.s	n.s
Tree species	0.0048	n.s	n.s
Origin	0.0003	0.0361	0.0016
Tree circumference	n.s	n.s	0.00001

At a larger spatial scale, a positive relationship ($r^2 = 0.304$; $p < 0.0005$) was evident between bird species richness per street and trees species richness per street (Figure 3); thus, more trees species in a street resulted in more bird species. Bird species richness per street was also positively associated with tree density per street, albeit only weakly ($r^2 = 0.11$; $p < 0.05$).

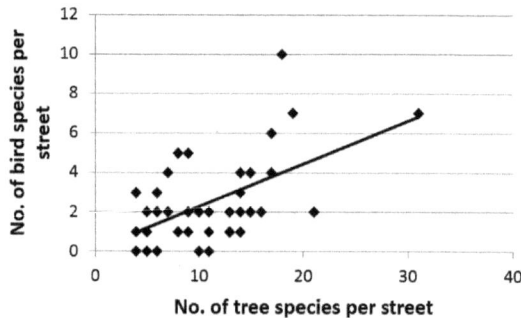

Figure 3. Bird species richness per street as a function of tree species richness per street ($r^2 = 0.304$; $p < 0.0005$).

4. Discussion

With respect to the research objective, the results are contradictory between the two taxonomic groups investigated, with bird richness, abundance, and breeding (indicated by nests) favoured by native street trees and mistletoes by alien street trees. The findings regarding birds tend to support previous studies, in that some bird species favour indigenous woody plants, whilst others are influenced more by density and structure of the vegetation than the provenance of the woody plant species. Nonetheless, the seminal study by Ikin et al. [26] illustrated markedly higher bird species richness and abundance in native tree species than exotic ones, irrespective of structure, in Canberra, Australia. For insectivorous birds, this may be a consequence of the higher abundance and richness of arthropods on indigenous trees than exotic ones as reported by Bhullar & Majer [27] in contrasting two indigenous and two exotic trees species. Chong et al. [14] showed higher beta diversity for birds across parks dominated by indigenous flora relative to those with exotic flora.

Most previous work in this debate has been on fauna, with limited consideration of flora [11] such as mistletoes and understorey flora; hence, there is limited opportunity for comparative analysis of the results. Mistletoes are important as they attract frugivorous birds which, in turn, disperse the mistletoe seeds as well as seeds of other fruiting plants [28]. Gairola et al. [15] reported mistletoes on 30 tree species in urban parks in Durban (South Africa), of which 16 were alien species. The alien tree species, *Melia azedarach* L., had the highest infestation rate. In Singapore, Izuddin & Webb [29] found 3937 epiphytes spanning 51 species in 1170 *Albizia saman* F.Muell. street trees. My results corroborate theirs in that tree size and proximity to green patches were positively related to epiphyte presence and richness.

The results show that the presence of birds was higher in large trees than in small ones, mirroring the findings from other settings [6,30,31]. This may be an active selection for large trees or may simply be a reflection of the size effect (larger trees have a larger volume and therefore a greater random likelihood of harbouring a bird). Nonetheless, it does show the importance of ensuring that there is a range of tree sizes available in the urban forest. The need for a diversity of trees sizes has recently been emphasised [32] because of the increasing tendency for city authorities to plant small statured species because they are cheaper and easier to manage and require less space when mature [33]. However, large trees do not only provide greater canopy volume, but also a greater diversity of habitats as they age [32]. For example, they typically contain more hollows [12], and both fungal density [34] and lichen prevalence [35] are higher in larger trees than smaller ones. Whilst my results did not show a relationship between tree size and the presence of mistletoes, Gairola et al.'s [15] did, with larger trees having more.

The positive relationship between the presence of mistletoes, birds, and nests in street trees and proximity to the nearest green space reveals the importance of connectivity. Whilst the abundance of street trees in the western section of Grahamstown is high [23], it is highly probable that not all resource needs are met from the street trees only, but that birds disperse laterally into private gardens and broader patches of formal and informal green space. Since mistletoes are bird-dispersed, they are likely to follow a similar pattern. Connectivity is important for a variety of organisms in human-transformed landscapes such as urban areas and agricultural lands [36,37]. These results suggest the importance of street trees in facilitating movement between urban green spaces. This is further emphasised through the positive relationship between street tree richness at the street scale and of bird species richness at the same scale.

This work corroborates previous conclusions [10,11] that alien tree species can provide ecosystem services in support of other biodiversity because mistletoes, birds, and nests were found in alien as well as indigenous street trees. However, from a conservation perspective, this conclusion is insufficient. As argued earlier, the question should not be whether alien tree species can provide the same services and support to other biodiversity as native species, but whether they support a greater suite of species or abundance of native flora and fauna for the same unit of abundance (such as canopy volume). If they do not, then there is no reason against favouring indigenous species from a conservation point of view. The preference ratios indicate strong associations of birds for a number of native street trees, but very few alien ones, which suggests that were there no alien street trees in the study area, there would probably be little negative effect on the richness of avifauna in the city. On the other hand, the abundance of mistletoes would likely be greatly reduced if there were fewer or no alien street trees. However, they would not be absent, as mistletoes were also recorded in indigenous street trees—most notably, *Acacia karroo*. The choice of whether to favour birds or mistletoes in selecting trees to plant is a value judgement. However, birds are inspiring organisms for many urban residents, and there are many different bird species making use of the street trees. In contrast, mistletoes do not instil such similar levels of respect and enjoyment, and, in Grahamstown, only one species was encountered. Moreover, heavy infections of mistletoes can result in branch loss, which adds to tree maintenance costs. However, epiphytes such as mistletoes can be important in providing habitat to invertebrate fauna [29], which requires investigation in South Africa.

5. Conclusions

In conclusion, this study has shown that native street trees show higher species richness and density of birds than do non-native street trees and that birds display a positive preference for several native tree species, but only one alien tree species. In contrast, mistletoe prevalence was markedly higher for alien tree species, notably *Fraxinus sp.*, than for indigenous tree species, but they were still found in native tree species. Overall, the results indicate that a policy of promoting the planting of only indigenous street trees is likely to favour birds without unduly effecting mistletoes.

Acknowledgments: This work was sponsored by the South African Research Chairs Initiative of the Department of Science and Technology and the National Research Foundation of South Africa. Any opinion, finding, conclusion, or recommendation expressed in this material is that of the author, and the NRF does not accept any liability in this regard.

Conflicts of Interest: The author declares no conflict of interest. The founding sponsors had no role in the design of the study; in the collection, analyses, or interpretation of data; in the writing of the manuscript; or in the decision to publish the results.

References

1. Niemelä, J.; Saarela, S.-R.; Söderman, T.; Kopperoinen, L.; Yli-Pelkonen, V.; Väre, S.; Kotze, D.J. Using the ecosystem services approach for better planning and conservation of urban green spaces: A Finland case study. *Biodiv. Conserv.* **2010**, *19*, 3225–3243. [CrossRef]
2. Roy, S.; Byrne, J.; Pickering, C. A systematic quantitative review of urban tree benefits, costs, and assessment methods across cities in different climatic zones. *Urban For. Urban Green.* **2012**, *11*, 351–363. [CrossRef]
3. Livesley, S.J.; McPherson, G.M.; Calfapietra, C. The urban forest and ecosystem services: Impacts on urban water, heat and pollution cycles at the tree, street and city scale. *J. Environ. Qual.* **2016**, *45*, 119–124. [CrossRef] [PubMed]
4. Todorova, A.; Asakawa, S.; Aikoh, T. Preferences for and attitudes towards street flowers and trees in Sapporo, Japan. *Landsc. Urban Plan.* **2004**, *69*, 403–416. [CrossRef]
5. Antupit, S.; Gray, B.; Woods, S. Steps ahead: Making streets that work in Seattle, Washington. *Landsc. Urban Plan.* **1996**, *35*, 107–122. [CrossRef]
6. Tzilkowski, W.M.; Wakeley, J.S.; Morris, L.J. Relative use of municipal street trees by birds during summer in state college, Pennsylvania. *Urban Ecol.* **1986**, *9*, 387–398. [CrossRef]
7. Kubista, C.E.; Bruckner, A. Importance of urban trees and building as daytime roost for bats. *Biologia* **2015**, *70*, 1545–1552. [CrossRef]
8. Kendle, A.D.; Rose, J.E. The aliens have landed! What are the justifications for 'native only' policies in landscape plantings? *Landsc. Urban Plan.* **2000**, *47*, 19–31. [CrossRef]
9. Smith, R.M.; Thompson, K.; Hodgson, J.G.; Warren, P.H.; Gaston, K.J. Urban domestic gardens (IX): Composition and richness of the vascular plant flora, and implications for native biodiversity. *Biol. Conserv.* **2006**, *129*, 312–322. [CrossRef]
10. Johnston, M.; Nail, S.; James, S. 'Natives versus aliens': The relevance of the debate to urban forest management in Britain. In *Trees, People and the Built Environment*, Proceedings of the Urban Trees Research Conference, Birmingham, UK, 13–14 April 2012; Johnston, M., Percival, G., Eds.; pp. 181–191.
11. Chalker-Scott, L. Nonnative, noninvasive woody species can enhance landscape biodiversity. *Arboric. Urban For.* **2015**, *41*, 173–186.
12. Carthew, S.M.; Yáñez, B-M.; Ruykys, L. Straddling the divide: Den use by brushtail possums (*Trichosurus vulpecula*) in urban parklands. *Urban Ecosyst.* **2015**, *18*, 525–538. [CrossRef]
13. Blanchon, D.; Pusateri, J.; Galbraith, M.; Thorpe, S. Sampling indigenous ground-living beetles in a stand of non-native tree privet (*Ligustrum lucidum*) in New Zealand raises new management questions. *Ecol. Manag. Restor.* **2011**, *12*, 234–236. [CrossRef]
14. Chong, K.Y.; Teo, S.; Kurukulasuriya, B.; Chung, Y.F.; Rajathurai, S.; Tan, H.T.W. Not all green is as good: Different effects of the natural and cultivated components of urban vegetation on bird and butterfly diversity. *Biol. Conserv.* **2014**, *171*, 299–309. [CrossRef]

15. Gairola, S.; Bhatt, A.; Govender, Y.; Baijnath, H.; Porcheş, S.; Ramdhani, S. Incidence and intensity of tree infestation by the mistletoe *Erianthemum dregei* (Eckl. & Zeyh.) V. Tieghem in Durban, South Africa. *Urban For. Urban Green.* **2013**, *12*, 315–322.

16. French, K.; Major, R.; Hely, K. Use of native and exotic garden plants by suburban nectarivorous birds. *Biol. Conserv.* **2005**, *121*, 545–559. [CrossRef]

17. Gray, E.R.; van Heezik, Y. Exotic trees can sustain native birds in urban woodlands. *Urban Ecosyst.* **2015**. [CrossRef]

18. Dickie, I.A.; Bennett, B.M.; Burrows, L.E.; Nunez, M.A.; Peltzer, D.A.; Porté, A.; Richardson, D.M.; Rejmánek, M.; Rundel, P.W.; van Wilgen, B.W. Conflicting values: Ecosystem services and invasive tree management. *Biol. Invasions* **2014**, *16*, 705–719. [CrossRef]

19. Climatedata.eu. Available online: http://www.climatedata.eu/climate.php?loc=sfzz0020&lang=en (accessed on 13 March 2013).

20. State of the Environment South Africa, 2007. Available online: http://soer.deat.gov.za/332.htm (accessed on 15 March 2015).

21. Mucina, L.; Rutherford, M.C. *The Vegetation of South Africa, Lesotho and Swaziland. Strelitzia 19*; South African National Biodiversity Institute (SANBI): Pretoria, South Africa, 2006; p. 816.

22. McConnachie, M.M.; Shackleton, C.M. Public green space inequality in small towns in South Africa. *Habit. Int.* **2010**, *34*, 244–248. [CrossRef]

23. Kuruneri-Chitepo, C.; Shackleton, C.M. The distribution, abundance and composition of street trees in selected towns of the Eastern Cape, South Africa. *Urban For. Urban Green.* **2011**, *10*, 247–254. [CrossRef]

24. Gwedla, N.; Shackleton, C.M. The development visions and attitudes towards urban forestry of officials responsible for greening in South African towns. *Land Use Policy* **2015**, *42*, 17–26. [CrossRef]

25. Richardson, E.; Shackleton, C.M. The extent, causes and local perceptions of street tree damage in small towns in the Eastern Cape, South Africa. *Urban For. Urban Green.* **2014**, *13*, 425–432. [CrossRef]

26. Ikin, K.; Knight, E.; Lindenmayer, D.B.; Fischer, J.; Manning, A.D. The influence of native versus exotic streetscape vegetation on the spatial distribution of birds in suburbs and reserves. *Divers. Distrib.* **2013**, *19*, 294–306. [CrossRef]

27. Bhullar, S.; Majer, J. Arthropods on street trees: A food resource for wildlife. *Pac. Conserv. Biol.* **2000**, *6*, 171–173. [CrossRef]

28. Watson, D.M. The relative contribution of specialists and generalists to mistletoe dispersal: Insights from a neotropical rain forest. *Biotropica* **2013**, *45*, 195–202. [CrossRef]

29. Izuddin, M.; Webb, E.L. The influence of tree architecture, forest remnants, and dispersal syndrome on roadside epiphyte diversity in a highly urbanized tropical environment. *Biodivers. Conserv.* **2015**, *24*, 2063–2077. [CrossRef]

30. Palomino, D.; Carrascal, L.M. Birds on novel island environments: A case study with the urban avifauna of Tenerife (Canary Islands). *Ecol. Res.* **2005**, *20*, 611–617. [CrossRef]

31. MacGregor-Fors, I. Relation between habitat attributes and bird species richness in western Mexico suburbs. *Landsc. Urban Plan.* **2008**, *84*, 92–08. [CrossRef]

32. Stagoll, K.; Lindenmayer, D.B.; Knight, E.; Fischer, J.; Manning, A.D. Large trees are keystone structures in urban parks. *Conserv. Lett.* **2012**, *5*, 115–122. [CrossRef]

33. Nagendra, H.; Gopal, D. Tree diversity, distribution, history and change in urban parks: Studies in Bangalore, India. *Urban Ecosyst.* **2011**, *14*, 211–223. [CrossRef]

34. Heilmann-Clausen, J.; Christensen, M. Does size matter? On the importance of various dead wood fractions for fungal diversity in Danish beech forests. *For. Ecol. Manag.* **2004**, *201*, 105–117. [CrossRef]

35. Johansson, P.; Erlén, J. Influence of habitat quantity, quality and isolation on the distribution and abundance of two epiphytic lichens. *J. Ecol.* **2003**, *91*, 213–221.

36. Vergnes, A.; Le Viol, I.; Clergeau, P. Green corridors in urban landscapes affect the arthropod communities of domestic gardens. *Biol. Conserv.* **2012**, *145*, 171–178. [CrossRef]

37. Braaker, S.; Ghazoul, J.; Obrist, M.K.; Moretti, M. Habitat connectivity shapes urban arthropod communities: The key role of green roofs. *Ecology* **2014**, *95*, 1010–1021. [CrossRef] [PubMed]

forests

MDPI

Article

City "Green" Contributions: The Role of Urban Greenspaces as Reservoirs for Biodiversity

Ian MacGregor-Fors [1,*,†], Federico Escobar [1,*,†], Rafael Rueda-Hernández [1],
Sergio Avendaño-Reyes [1], Martha Lucía Baena [2], Víctor M. Bandala [1], Santiago Chacón-Zapata [1],
Antonio Guillén-Servent [1], Fernando González-García [1], Francisco Lorea-Hernández [1],
Enrique Montes de Oca [3], Leticia Montoya [1], Eduardo Pineda [1], Lorena Ramírez-Restrepo [1],
Eduardo Rivera-García [1] and Elsa Utrera-Barrillas [1]

[1]   Instituto de Ecología, A.C. (INECOL), Carretera antigua a Coatepec 351, El Haya, Xalapa 91070, Veracruz,
      Mexico; rarh82@hotmail.com (R.R.-H.); sergio.avendano@inecol.mx (S.A.-R.);
      victor.bandala@inecol.mx (V.M.B.); santiago.chacon@inecol.mx (S.C.-Z.);
      antonio.guillen@inecol.mx (A.G.-S.); fernando.gonzalez@inecol.mx (F.G.-G.);
      francisco.lorea@inecol.mx (F.L.-H.); leticia.montoya@inecol.mx (L.M.); eduardo.pineda@inecol.mx (E.P.);
      bioramirez@gmail.com (L.R.-R.); eduardo.rivera@inecol.mx (E.R.-G.); elsa.maria@inecol.com (E.U.-B.)
[2]   Instituto de Investigaciones Biológicas, Universidad Veracruzana. Avenida Luis Castelazo Ayala s/n.,
      Xalapa 91190, Veracruz, Mexico; mbaena@uv.mx
[3]   Facultad de Biología, Universidad Veracruzana, Circuito Gonzalo Aguirre Beltrán s/n, Zona Universitaria,
      Xalapa 91070, Veracruz, Mexico; endamont@gmail.com
*   Correspondence: ian.macgregor@inecol.mx (I.M.-F.); federico.escobar@inecol.mx (F.E.);
      Tel.: +52-228-842-1800 (ext. 4322) (I.M.-F.); +52-228-842-1800 (ext. 4111) (F.E.)
†   These authors collaborated equally to this work.

Academic Editors: Francisco Escobedo, Stephen John Livesley and Justin Morgenroth
Received: 31 March 2016; Accepted: 6 July 2016; Published: 15 July 2016

Abstract: Urbanization poses important environmental, social, and ecological pressures, representing
a major threat to biodiversity. However, urban areas are highly heterogeneous, with some greenspaces
(e.g., urban forests, parks, private gardens) providing resources and a refuge for wildlife communities.
In this study we surveyed 10 taxonomic groups to assess their species richness and composition in
six greenspaces that differ in size, location, management, and human activities. Species richness
differed among taxonomic groups, but not all differed statistically among the studied greenspaces
(i.e., sac fungi, bats). Plants, basidiomycetous and sac fungi, and birds showed intermediate
assemblage composition similarity (<54%). The composition of assemblages of copro-necrophagous
beetles, grasshoppers, amphibians, and bats was related to the specific traits of greenspaces, mainly
size and location. The species richness contribution of each greenspace considering all studied
taxonomic groups was highest in the largest greenspace that is located at the southeastern border
of the city, while the lowest contribution was recorded in the smallest ones, all of them closer to the
city's center. Our results shed some light on the way in which different taxonomic groups respond to
an array of neotropical urban greenspaces, providing an important basis for future studies.

Keywords: urban ecology; urban forests; multi-taxonomic analysis; Neotropics; Mexico; urbanization;
species-area relationships; turnover rates; greenspace management; assemblage

1. Introduction

As the human population rises and global economic models drive people from rural areas to
urban centers, urbanization poses intensive demands with socio-economical and environmental
effects at multiple scales [1–3]. Such effects have been related to major components of global
change (i.e., land-use change, biogeochemical cycle shifts, climate change, biodiversity loss, biological

invasions [1]), with urbanization identified as one of the most important causes of native species endangerment [4,5]. Thus, the functioning, sprawling, and establishment of urban areas represent important ecological and health issues [1].

Novel urban systems are highly heterogeneous, including distinctive land uses that often reflect their origin and socio-economical role [6]. Among urban land uses, greenspaces—including a wide range of variants (e.g., private gardens, sport fields, parks, urban forests/preserves, golf courses, rights-of-way)—have been shown to shelter higher biodiversity than heavily developed land-uses (e.g., residential, commercial, industrial) [7–10]. However, not all greenspaces play similar roles for biodiversity, with active management practices and human activities (e.g., pruning, high visitation volume) reducing their ecological value [11,12], as well as other habitat and geographic variables (e.g., size, location) that mold their potential as biodiversity refuges [9,13]. The responses to such variables can be attributed to underlying factors such as the biology of species, their sensitivity to disturbance, species-area relationships, edge effects, habitat heterogeneity associated with greenspaces and the history of local and regional land uses [13–17].

A vast body of literature focused on the ecology of urban systems has drawn generalized biodiversity patterns [8]. However, there is a worrisome lack of knowledge regarding other biological groups, given that much evidence-based urban management and planning is heavily biased towards birds and vegetation [9]. Additionally, there is also a gap in our comprehension of urban ecology patterns and processes in developing countries, where biodiversity is often high and urbanization is expected to increase in the next decades [3,16–19]. Recently, urban ecologists have addressed multi-taxonomic responses of native, non-native, and invasive species in urban areas [20,21]. These studies are not only a baseline for biological inventories, but also show the type of information that several taxonomic groups can provide when assessing their ecological responses to urbanization.

Although urban greenspaces are intended to have important environmental functions, together with other economic, social, and health functions [22], little is known about the relative role that the arrangement of greenspaces in a city can have in sheltering biodiversity. In this study we assessed multi-taxonomic species richness and composition shifts, as well as the overall species contribution of six urban greenspaces with different sizes, management, and human activities in a neotropical city (Xalapa, Veracruz, southeast Mexico). We expected an overall increase of species richness in larger, peri-urban greenspaces with fewer management practices (e.g., pruning intensity, trail clearing) and human activities (e.g., visitors, noise). Regarding species composition, we predicted peri-urban greenspaces to have higher dissimilarity in relation to intra-urban sites due to their connectivity with extra-urban systems; however, we expected the biology of each group and its association with urbanization (e.g., positive for ants) also to drive this pattern. Finally, we expected larger urban greenspaces, farther from the city's geographic center and with fewer management practices, to shelter a greater average proportion of all the studied groups when compared to smaller, managed greenspaces located in the city's core.

2. Materials and Methods

2.1. Study Area

We conducted this study in the city of Xalapa-Enríquez (referred to hereafter as Xalapa), state capital of Veracruz de Ignacio de la Llave. Xalapa is located on the hillside of the easternmost Trans-Mexican Volcanic Axis, where Neartic and Neotropic biotas meet (19°32'2″ N, 96°55'8″ W, rainfall: 1100–1600 mm/year; elevation: 1100–1560 m above sea level). Xalapa is a small- to medium-size city, the urban continuum of which includes four municipalities (~64 km^2; [23]), and houses ~600,000 inhabitants [24]. Historically, predominant vegetation surrounding Xalapa included cloud forests in its central-southwestern section, tropical dry forests in its central-southeastern section, and temperate forests in the northern section [21]. Currently, the landscape in which Xalapa is embedded shows the urban land cover to be the most representative (49%), followed by agriculture

(37%), grassland (9%), and "forested" areas (5%) [24]. Within its limits, Xalapa has a significant amount of woody vegetation cover, basically comprised by parks, urban forests, private gardens, and vacant lots [25]. In fact, Xalapa has been considered a green city, making it a unique setting for urban ecological studies [21].

2.2. Urban Greenspaces

We surveyed six urban greenspaces in the central and southern parts of the city of Xalapa (Figure 1), two of them peri-urban, with low management and moderate visitor rates (i.e., Parque Natura, Santuario del Bosque de Niebla); the other three are intra-urban with moderately high management and human activities (i.e., Parque Ecológico Macuiltépetl, Parque Miguel Hidalgo, Parque de Los Tecajetes), and one is intra-urban without active management (i.e., Parque Ecológico Molinos de San Roque; Figure 1).

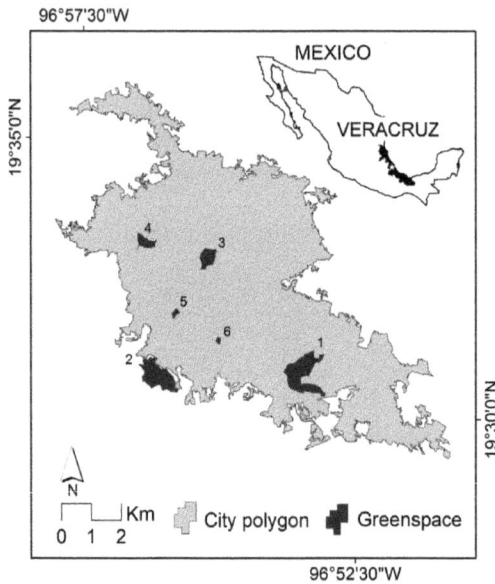

Figure 1. Greenspace location in Xalapa (light gray polygon). 1: Natura; 2: Santuario; 3: Macuiltépetl; 4: San Roque; 5: Tecajetes; 6: Berros.

Parque Natura (referred to hereafter as Natura; 95.5 ha; ~1300 m a.s.l.) is the largest urban greenspace of the city and part of its most extensive vegetation patch together with other greenspaces (e.g., El Tejar-Garnica). It is a State Natural Protected Area located in the southeastern end of the city. Predominant vegetation is represented by second-growth heavily-disturbed cloud-forest elements with an important calid influence, as well as the presence of post-disturbance colonizing deciduous species. Management and human activities are restricted to certain roads and recreational areas.

Santuario del Bosque de Niebla (referred to hereafter as Santuario; 76.8 ha; ~1340 m a.s.l.) is also a State Natural Protected Area (managed by INECOL) located in the southwestern section of the city, where dominant vegetation is second-growth cloud forest. This Natural Protected Area adjoins some urban settlements, forest remnants, and shade-grown coffee plantations. Management is limited to a gravel path and small trails for visitors.

Parque Ecológico Macuiltépetl (referred to hereafter as Macuiltépetl; 26.5 ha; ~1560 m a.s.l.) is also a State Natural Protected Area located near the geographic centroid of Xalapa. Vegetation in this inactive volcano includes second-growth cloud forests, although exotic ornamental species are well represented in the park. Management activities include gardening along main trails and specific areas, tree removal and pruning, as well as the maintenance of a paved road; human activities are considerable and include jogging, picnicking, group activities (e.g., dancing, martial arts), among others, that peak during weekends.

Parque Ecológico Molinos de San Roque (referred to as San Roque; 17.6 ha; ~1450 m a.s.l.) is also a State Natural Protected Area located in the northwest end of the city, mostly covered with second-growth cloud forest; firewood extraction is a common but illegal practice. Management is absent and human activities are low, with police patrolling along the trails due to increased insecurity.

Parque de Los Tecajetes (referred to hereafter as Tecajetes; 4.6 ha; ~1410 m a.s.l.) is a recreational greenspace near downtown Xalapa. It is characterized by a steep terrain and a perennial spring that supplies water to multiple fountains and fish tanks. Dominant tree species are *Platanus mexicana* and *Liquidambar styraciflua;* understory is scarce in half of the park but in the steeper areas understory resembles that of an old second-growth forest. Management includes intense gardening and landscaping. Human activities include sports (e.g., basketball, football, skating), picniking, and visiting on a regular basis.

Parque Miguel Hidalgo (referred to hereafter as Berros; 3.2 ha; ~1370 m a.s.l.) is a small recreational greenspace located southwest in downtown Xalapa with important ornamental value. Although most present tree species are planted, both native and exotic to Central Veracruz, the two dominant tree species are *Platanus mexicana* and *Liquidambar styraciflua*. Human activities are high throughout the week, including jogging, bicycling, skating, and even pony-back riding on weekends.

2.3. Sampling Methods

From June to November 2014, we sampled vascular plants, fungi (Ascomycetes and Basidiomycetes), ants, grasshoppers, copro-necrophagous beetles, butterflies, amphibians, birds, and bats in the six aforementioned greenspaces of Xalapa (Figure 1).

Plants: We sampled vegetation in the six greenspaces between July and October. We set five 10 m^2 plots (1 m × 10 m) per site which represented the variety of plant condition at each greenspace, which were dominated by shrubs (woody vegetation up to 3 m with various stems) and herb species (non-woody vegetation). For all species found in each quadrant, we obtained herbarium specimens following Lot and Chiang [26]. After species identification, we deposited specimens in the XAL herbarium.

Fungi: Ascomycetes (sac fungi): We delimited one quadrant of 10 m × 20 m (200 m^2) per greenspace that was visited three times between August and September to carry out random opportunistic paths, as recommended by Mueller et al. [27]. We collected sac fungi from fallen branches and trunks with the help of a hand lens [28,29]. We analyzed blade cuts made on the sporecarps and measured the asci and ascospores on a compound microscope. We dehydrated and labeled all specimens for further deposition in the fungi collection of the XAL herbarium. Basidiomycetes: We performed weekly opportunistic surveys for fructifications [27,30] of basidiomycetous macrofungi, focusing on Agaricales, Russulales, and Boletales. We performed collects with the same sample effort at each greenspace between May and October. We recorded and analyzed specimens macro- and microscopically [31–33]. For both fungi groups (i.e., sac, basidyomicetous), specialized literature was the basis for the taxonomic treatment of species [34–37].

Ants: We visited all sites between July and October. We outlined two perpendicular transects randomly at each park separated at least 200 m, with 10 sampling locations every 10 m (*n* = 20) and where bait bond papers (10 m × 10 cm), one with tuna fish and a second one with bee honey as attractants. We collected all individuals attracted by baited papers after one hour [38]. Additionally, we complemented our sampling by searching for foraging ants in the area (e.g., soil, vegetation) for 10 min per sampling site. Following capture, we identified all individuals with the aid of specialized keys [39,40] before final deposition in the entomological collection of the Instituto de Ecologia, Xalapa (IEXA).

Grasshoppers: We performed direct air net collection on four transects (36–40 m long × 1 m wide), in 12 sampling points per greenspace between July and September. Total travel time was ~10 min per sampling point [41,42].

Copro-necrophagous beetles: We sampled copro-necrophagous beetles at each greenspace in July and August. We used pitfall traps baited with decaying squid and human feces, using propylen-glycol as preserver; a modified version of the NecroTrap 80 (NTP-80) [43,44]. We set three traps with each bait type, separated at least 15 m apart from each other, across equivalent areas at each greenspace. We left the traps in site for three weeks and returned the collected material to the lab. We identified all collected individuals with specialized literature [44–46] and deposited them in the IEXA entomological collection.

Butterflies: We surveyed diurnal butterflies (Lepidoptera: Papilionoidea) in September and October, between 9 a.m. and 4 p.m. when environmental conditions were adequate (i.e., no rainfall, environmental temperature was ⩾20 °C [47]). We recorded all individuals seen or caught during a four-hour intensive search per greenspace. Following capture, we identified most individuals in the field with the aid of specialized local field guides [48,49]. We collected the unidentified specimens and deposited them in the IEXA entomological collection after their identification.

Amphibians: We sampled amphibians between September and November using a time-constraint technique [50]. Each greenspace was sampled once at nighttime between 8 p.m. to 1 a.m. for three hours by four well-trained field technicians (12 h/greenspace, 72 total h/person). We identified all individuals to species level and released them at the site of capture.

Birds: We surveyed landbirds during June at each greenspace using 25 m radius point counts (from 7 a.m. to 10 a.m., 5 min [51]). We performed 10 point-count repetitions at each greenspace and considered all birds seen or heard actively using the surveyed area in our analysis.

Bats: We surveyed aerial insectivorous bats in August by using ultrasound detection in 20 repetitions per greenspace [52]. We started our recordings 10 min after sunset in 5 min intervals. We digitized bat calls (250 kHz sampling rate, 8 bit) with an Ultramic 250kHz microphone (Dodotronic, Dodotronic di Ivano Pelicella, Castel Gandolfo, Italy). We inspected the recordings using sonographic software (BatSound Pro, Pettersson Elektronik AB, Uppsala, Sweden) and assigned calls to species by comparison with a reference collection of recordings obtained in the region. Finally, we quantified the "relative abundance" of bats as total number of bat detection instances or "passes" registered at each site [53].

2.4. Statistical Analysis

To allow multiple comparisons, we determined the statistical expectation of species richness (Sm) of every studied taxonomic group in all greenspaces using individual-based data calculation with the package iNEXT [54]. This software, allows the calculation of comparable, rarefied species richness values for multiple datasets by simultaneously extrapolating all samples up to two times their sampling effort, which avoids unstable results [55]. To establish statistical differences in species richness among greenspaces for each taxonomic group, we contrasted Sm 84% confidence intervals (CIs) and assumed statistical differences in non-overlapping intervals. We used 84% CIs as they precisely mimic 0.05 tests, as contrasting 95% ones can lead to uncertain interpretations (often resulting in Type II errors) when intervals overlap [56].

To assess taxonomic composition differences among greenspaces, we calculated β_{sim} [57]. This index quantifies the relative magnitude of shared species in relation to the sample with less unique species, making it useful for samples with different species richness [58,59]. We calculated β_{sim} using the vegan package in R [60] (formula in vegan: min(b,c)/(min(b,c)+a)) and generated matrices of taxonomic similarity among greenspaces for each studied group and represented it using average hierarchical clustering dendrograms using R [60]. In order to evaluate overall species contribution of each of the studied greenspaces to the total recorded diversity, we calculated the percentage of species pertaining to each studied taxonomic group recorded at each greenspace in relation to the overall recorded species for each group. We then computed a generalized linear model (GLM, family: Gaussian; transformation for percentages: arcsin$\sqrt[2]{x}/100$) to assess differences in the proportional contribution by taxonomic group at each of the studied greenspaces. Finally, we calculated post hoc general linear hypothesis tests for paired statistical comparisons in R [60].

3. Results

3.1. Species Richness

Due to the low number of recorded species of grasshoppers (in Tecajetes and Berros), as well as copro-necrophagous beetles and amphibians in all six greenspaces, we were not able to calculate CIs. Regardless of the latter, we found three main species richness patterns from the taxonomic group perspective: (1) no differences among all six greenspaces (i.e., sac fungi, copro-necrophagous beetles, bats), (2) higher richness in large outermost greenspaces and lower richness in smaller ones (i.e., basidiomycetous fungi, grasshoppers, amphibians, birds), and (3) highest richness of ants in the smallest greenspace and lower richness in the rest of studied greenspaces, except Macuiltépetl. Regarding plants and butterflies, we found no clear patterns, with the former having high richness in Tecajetes and Natura—two contrastingly different greenspaces—and similar values in the remaining greenspaces, and the latter showing highest richness in Natura, lowest richness in San Roque, and intermediate values in the remaining greenspaces (Table 1). In sum, we recorded an overall increase of species richness in larger peri-urban greenspaces (i.e., Natura, Santuario), intermediate values in Macuiltépetl and Tecajetes, and the lowest values in San Roque and Berros.

Table 1. Estimated species richness (Sm ± 84% CI) for the studied taxonomic groups in all studied greenspaces.

	Natura	Santuario	Macuiltépetl	San Roque	Tecajetes	Berros
Plants	50.3 ± 5.5 (AB)	43.1 ± 3.4 (B)	36.5 ± 2.9 (C)	33.1 ± 4.0 (C)	53.0 ± 5.3 (A)	19.8 ± 8.2 (D)
Sac fungi	18.9 ± 3.3 (A)	17.7 ± 2.9 (A)	19.1 ± 2.4 (A)	20.1± 3.0 (A)	18.6 ± 2.7 (A)	12.9 ± 7.6 (A)
Basidiomycetous fungi	21.9 ± 1.2 (B)	40.4 ± 1.7 (A)	20.2 ± 1.5 (B)	12.7 ± 2.4 (C)	5.0 ± 1.8 (D)	13.9 ± 1.6 (C)
Ants	6.4 ± 1.1 (B)	5.4 ± 0.8 (B)	8.2 ± 2.2 (AB)	5.0 ± 1.2 (B)	6.0 ± 0.0 (B)	12.1 ± 4 (A)
Grasshoppers	14.5 ± 1.9 (A)	12.1 ± 2.0 (A)	10.6 ± 2.0 (A)	12.8 ± 8.9 (A)	3	5
Copro-necrophagous beetles	4	5	4	5	5	3
Butterflies	30.2 ± 2.4 (A)	22.3 ± 6.2 (AB)	26.5 ± 6.3 (AB)	22.4 ± 4.3 (B)	27.9 ± 11.4 (AB)	26.2 ± 12.3 (AB)
Amphibians	7	5	2	1	1	3
Birds	18.0 ± 2.9 (A)	14.5 ± 4.7 (AB)	13.1 ± 2.9 (AB)	14.8 ± 3.9 (AB)	8.8 ± 7.1 (AB)	11.3 ± 2.2 (B)
Bats	5.6 ± 0.4 (AB)	4.6 ± 1.8 (AB)	6.6 ± 3.8 (AB)	3.7 ± 2.1 (B)	5.7 ± 0.7 (AB)	6 ± 0 (A)

Letters below values for each greenspace indicate statistically significant differences.

3.2. Composition Similarity

Taxonomic composition similarity, assessed using β_{sim} matrices represented in clustering dendrograms, shows that the assemblage similarity of plants, basidiomycetous and sac fungi, and birds among greenspaces is less than half of the overall recorded species (<54%; Figure 2). Regarding assemblages with higher average assemblage similarities, four of the studied taxonomic groups responded either to size and location, as well as to management and human activities. We found three taxonomic groups with similarity values >75% (i.e., copro-necrophagous beetles, amphibians, bats), recorded in Natura and Santuario. Regarding management and human activities of the studied greenspaces, only one group (i.e., grasshoppers) responded to low management and human activities that occur in San Roque and Santuario, and two groups (i.e., copro-necrophagous beetles, amphibians) responded to the high degree of greenspace management and human activities in Berros and Tecajetes (Figure 2). It is noteworthy that the two remaining groups (i.e., ants, butterflies) that showed no clear pattern regarding the size, location, management, or human activities of the studied greenspaces did not show high average similarity clustering values (~45%–70% similarity).

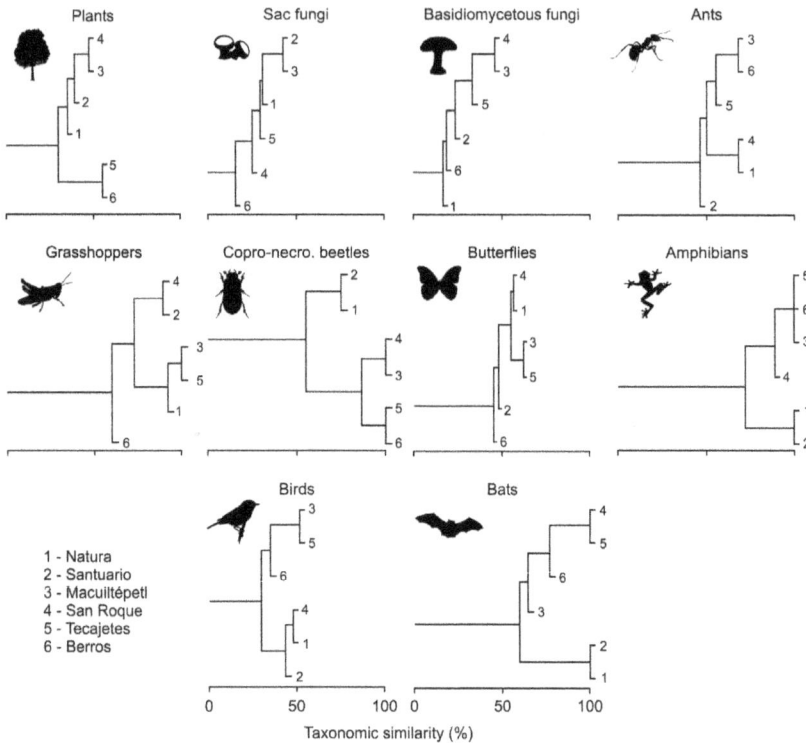

Figure 2. Hierarchical clustering dendrograms showing the taxonomic composition similarity of the studied taxonomic groups among greenspaces.

3.3. Greenspace Richness Contribution

Greenspace species richness contributions varied significantly among sites (GLM: $F(5, 54) = 3.42$, $p = 0.009$). Post-hoc general linear hypothesis tests show a significantly higher contribution of Natura when compared to Berros, San Roque, and Tecajetes (all p-values <0.016). However, no other statistical differences were recorded (all p-values >0.33), indicating that Santuario and Macuiltépetl have intermediate richness contributions in relation to the rest of studied greenspaces (Figure 3).

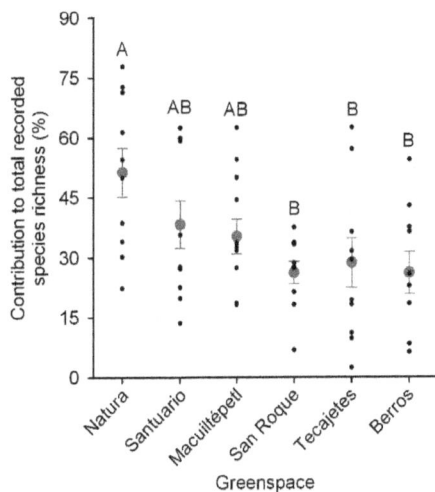

Figure 3. Average greenspace contribution to total species richness of all studied taxonomic groups (error bars represent SE). Black dots indicate the contribution of each taxonomic group per greenspace; letters above values for each greenspace indicate statistically significant differences.

4. Discussion

Using multi-taxonomic approaches to assess biodiversity responses to urbanization will lead to a better understanding of the phenomenon [20,21,61]. Our multi-site and multi-taxonomic approach allowed us to unveil some interesting patterns within the complex scenario given by the greenness of Xalapa (as well as its location in the Mexican Transition Zone), regarding the relationships between size, location, management, and human activities of urban greenspaces on 10 taxonomical groups. In general, our results show that overall species richness rises in peri-urban large greenspaces, with management and human activities driving the species richness and composition of some of the studied taxonomic groups. Nevertheless, given the nature and distribution of large-sized greenspaces in Xalapa, as well as the high insecurity in peri-urban small greenspaces, we could not have a balanced design. Thus, some of our results should be taken cautiously, particularly regarding the role of management and location.

Greenspaces with highest species richness values were the larger ones, both located in the peri-urban area of Xalapa (i.e., Natura, Santuario). As seen in other studies, large, peri-urban greenspaces mitigate the negative role of urbanized areas (e.g., habitat disruption, reduced resource availability), potentially favoring species richness for certain taxa [9,62,63]. This suggests that larger areas that are connected with the non-urban matrix can offer an important array of resources. Thus, by providing supplementary or complementary resources that may compensate for limited resource availability in greenspaces, they could allow the influx of individuals from the non-urban matrix, and enable the temporal establishment of populations of species of moderate sensibility to urbanization [64–66]. It is noteworthy to state that both Natura and Santuario have native ecosystem remnants and/or elements, which could be also driving our species richness results. Regarding the species richness of the intra-urban medium- to small-sized greenspaces, management and human activities seemed to play a more important role than size for some taxa, highlighting the importance of multi-taxonomic studies. For instance, San Roque, a medium-sized greenspace with low management and human activities had less richness for most groups that showed statistical differences among greenspaces (except sac fungi and birds), similar to that of smaller greenspaces with higher management and human activities (i.e., Tecajetes, Berros).

From a taxonomic group perspective, our species richness results show that there are assemblages with: (1) highest richness values in large greenspaces with low management and human activities and lowest values in small, heavily managed and visited greenspaces (i.e., basidiomycetous fungi, grasshoppers, amphibians, birds), (2) highest richness in the smallest heavily managed and visited greenspaces (i.e., ants), (3) no differences among the studied greenspaces (i.e., sac fungi, copro-necrophagous beetles), (4) mild differences (i.e., bats), and (5) inconsistent patterns (i.e., butterflies, plants). This shows how the species richness of some of the studied groups can be used as indicator groups for different scenarios if the requirements to be considered as bioindicators are met [67]. It is noticeable that inconsistent patterns for plant richness could be related to the nature of urban vegetation in the greenspaces of Xalapa, many of which are planted and exotic, which in turn could be affecting the butterfly and other groups' responses [68–70].

Setting our species richness results into context, assemblage composition similarity results revealed that some groups, such as plants, birds, basidiomycetous, and sac fungi, in the six studied greenspaces show a <54% similarity across the six studied greenspaces. As recorded in a previous multi-taxonomic study performed in Xalapa contrasting heavily versus lightly urbanized areas, fungi are, in general, site-specific [21]. Both studied fungi groups showed a nested pattern, highly differing in the ordering of greenspaces, which could be given as a response to habitat disruption [71,72]. However, our fungi results do not consider all active dormant mycelia [73], which is a reason why they should be interpreted cautiously. Plant assemblage composition in urban greenspaces responded primarily to the presence of native ecosystem remnants, with a mixture of Nearctic-Netropical elements, and also to the presence of planted species, making plant assemblages site-specific, responding to human decision-making [74]. The case for birds is quite interesting; their similarity values are intermediate and show a clear pattern, with two clusters: (1) Natura, Santuario, San Roque and (2) Macuiltépetl, Tecajetes, Berros. The first cluster represents the two largest peri-urban studied greenspaces, while San Roque is the intra-urban greenspace that has less management and human activities and includes cloud forest vegetation components. The second group includes more managed and visited intra-urban greenspaces. This shows that, although different bird assemblages are being sheltered in the studied greenspaces, size and location can be overridden by site-specific variables, as previously suggested by Evans et al. [75].

Assemblage composition similarity dendrograms for grasshoppers, ants, copro-necrophagous beetles, amphibians, and bats showed high average similarity values among the studied greenspaces (>75%), with many greenspace assemblages showing 100% shared species (i.e., copro-necrophagous beetles, amphibians, bats). This result could be given by the low species richness recorded in the studied greenspaces (average of about three to nine species per greenspace), which could be a result of the semi-permeable filtering of species from the non-urban matrix [9,65]. This was particularly true for copro-necrophagous beetles, amphibians, and bats, with contrastingly different assemblages in large peri-urban greenspaces. Also, management and human activities showed to play an important role in molding grasshoppers, copro-necrophagous beetles, and amphibians. In this sense, we recorded highly similar assemblages in Santuario and San Roque, two greenspaces with contrasting sizes and location, making them potential bioindicators of contrasting greenspace management and human activities, especially low ones. This agrees with previous studies that have shown grasshopper diversity to shift with habitat traits and management, mainly with those that mold their herbivorous diet [76,77]. In the case of copro-necrophagous beetles and amphibians, results are driven by the nestedness of three copro-necrophagous species in Berros also present in Tecajetes, and of one amphibian species recorded in Tecajetes and shared with Berros. It its noteworthy that we did not find any composition pattern for butterflies, suggesting that their assemblages are driven by variables that were not measured in this study (e.g., resource availability).

Finally, our greenspace richness contribution results show that, as Hostetler et al. [78] and Beninde et al. [79] highlight, creating green infrastructure, which increases connectivity, is only the first step in conserving urban biodiversity. As supported by our species richness results, Natura, the largest peri-urban greenspace, is the greenspace with the highest contribution, regardless of being, by comparison, the peri-urban greenspace with greater management and human activities. Interestingly, Santuario and Macuiltépetl, two contrastingly different greenspaces in terms of size, location, management, and human activities, contributed similarly in sheltering overall recorded species richness. It is surprising that Macuiltépetl, a small-sized inactive volcano (26.5 ha) located near the centroid of the city, can shelter similar biodiversity when compared to a lightly managed peri-urban greenspace, highlighting the importance of this unique site (as recorded previously [80,81]). Besides Macuiltépetl, the rest of studied intra-urban greenspaces sheltered significantly less overall species in average, as would be expected due to the isolation effects of urbanization [82]. Interestingly, Tecajetes sheltered a high proportion of copro-necrophagous beetles (62.5%) and bats (57.1%), while we recorded 54% of the total ant species in Berros. This underlines the value of small greenspaces for certain groups that benefit from managed, disturbed, and open areas, where resources are more abundant for them [83–85]. It is noticeable that among Berros, Tecajates, and San Roque (the three smallest intra-urban greenspaces), the latter was the one with lowest values for most studied taxonomic groups, noticing that although this is a lightly managed and unvisited greenspace with cloud forest elements, its contribution to local biodiversity is relatively low. Since our study design keeps us from separating the effects of management, human activities, and location, we hypothesize that these differences could be given by the biases of sampling some sites in highly heterogeneous urban greenspaces; however, further studies are needed to untangle this counterintuitive finding.

Many aspects of the biology and ecology of the studied taxonomic groups—which are most probably driving our results—are understudied, even more so within tropical and subtropical urban systems [21]. Although there is a rising interest among ecologists to study urban systems in developing countries, the current dearth of knowledge leaves us incipient evidence-based foundations to propose and/or demand different urban settlement, management, and planning strategies and practices [86]. Undoubtedly, the more we understand wildlife communities within complex urban areas, together with other physical and social components, the closer we will be to balancing the quality of human life quality and our detrimental effects on biodiversity and human welfare [87–90].

Acknowledgments: We are most thankful to Angelina Ruiz-Sánchez for her useful comments that enhanced the clarity and quality of our paper, as well as Jorge Valenzuela for his support in ant identification, Ina Falfán and Julian Avila-Campos for drawing figures, Ángeles Arenas Cruz for her support in amphibian surveys, and Cara Joos for proofreading the manuscript. Research seed funds were granted to Ian MacGregor-Fors and Federico Escobar through the "Proyectos de Investigación de Alto Valor Estratégico para la Sociedad" of the Dirección General del Instituto de Ecología, A.C. (project: "Patrones ecológicos y percepción social de la diversidad biológica que habita en la ciudad de Xalapa: Un enfoque multidisciplinario").

Author Contributions: I.M.-F. and F.E. conceived the experimental design; I.M.-F., F.E., S.A.-R., M.L.B., V.M.B., S.C.-Z., A.G.-S., F.G.-G., F.L.H., E.M.O., L.M., E.P., L.R.-R., E.R.-G., E.U.-B. performed field work; I.M.-F., R.R.-H., and F.E. analyzed the data; I.M.-F., R.R.-H., and F.E. led the writing of the paper.

Conflicts of Interest: The authors declare no conflict of interest. The founding sponsors had no role in the design of the study; in the collection, analyses, or interpretation of data; in the writing of the manuscript, and in the decision to publish the results.

References

1. Grimm, N.B.; Faeth, S.H.; Golubiewski, N.E.; Redman, C.L.; Wu, J.; Bai, X.; Brigs, J.M. Global change and the ecology of cities. *Science* **2008**, *319*, 756–760. [CrossRef] [PubMed]
2. Montgomery, M.R. The urban transformation of the developing world. *Science* **2008**, *319*, 761–764. [CrossRef] [PubMed]
3. United Nations, Department of Economic and Social Affairs, Population Division. *World Urbanization Prospects: The 2014 Revision*; United Nations: New York, NY, USA, 2015.
4. Czech, B.; Krausman, P.R. Distribution and causation of species endangerment in the United States. *Science* **1997**, *277*, 1116–1117. [CrossRef]
5. Czech, B.; Krausman, P.R.; Devers, P.K. Economic associations among causes of species endangerment in the United States. *Bioscience* **2001**, *50*, 593–601. [CrossRef]

6. Pickett, S.T.; Cadenasso, M.L.; Grobe, J.M.; Nilon, C.H.; Pouyat, R.B.; Zipperer, W.C.; Costanza, R. Urban ecological systems: Scientific foundations and a decade of progress. *J. Environ. Manag.* **2011**, *92*, 331–362. [CrossRef] [PubMed]

7. McKinney, M.L. Urbanization, biodiversity, and conservation. *Bioscience* **2002**, *52*, 883–890. [CrossRef]

8. McKinney, M.L. Urbanization as a major cause of biotic homogenization. *Biol. Conserv.* **2006**, *127*, 247–260. [CrossRef]

9. McKinney, M.L. Effects of urbanisation on species richness: A review of plants and animals. *Urban Ecosyst.* **2008**, *11*, 161–176. [CrossRef]

10. Radeloff, V.C.; Hammer, R.B.; Stewart, S.I.; Fried, J.S.; Holcomb, S.S.; McKeefry, J.F. The wildland-urban interface in the United States. *Ecol. Appl.* **2005**, *15*, 799–805. [CrossRef]

11. Lawson, D.M.; Lamar, C.K.; Schwartz, M.W. Quantifying plant population persistence in human-dominated landscapes. *Conserv. Biol.* **2008**, *22*, 922–928. [CrossRef] [PubMed]

12. Hahs, A.K.; McDonnell, M.J.; McCarthy, M.A.; Vesk, P.A.; Corlett, R.T.; Norton, B.A.; Clemants, S.E.; Duncan, R.P.; Thompson, K.; Schwarts, M.W.; et al. A global synthesis of plant extinction rates in urban areas. *Ecol. Lett.* **2009**, *12*, 1165–1173. [CrossRef] [PubMed]

13. MacGregor-Fors, I.; Morales-Pérez, L.; Schondube, J.E. Does size really matter? Species-area relationships in human settlements. *Divers. Distrib.* **2011**, *17*, 112–121. [CrossRef]

14. Bazzaz, F.A. Plant species diversity in old-field successional ecosystems in southern Illinois. *Ecol. Durham* **1975**, *56*, 485–488. [CrossRef]

15. Hobbs, E.R. Species richness of urban forest patches and implications for urban landscape diversity. *Landsc. Ecol.* **1988**, *1*, 141–152. [CrossRef]

16. Ortega-Álvarez, R.; MacGregor-Fors, I. Living in the big city: Effects of urban land-use on bird community structure, diversity, and composition. *Landsc. Urban Plan.* **2009**, *90*, 189–195. [CrossRef]

17. Nielse, A.B.; van den Bosch, M.; Maruthaveeran, S.; van den Bosch, C.K. Species richness in urban parks and its drivers: A review of empirical evidence. *Urban Ecosyst.* **2014**, *14*, 305–327. [CrossRef]

18. Myers, N.; Mittermeier, R.A.; Mittermeier, C.G.; da Fonseca, G.A.; Kent, J. Biodiversity hotspots for conservation priorities. *Nature* **2000**, *403*, 853–858. [CrossRef] [PubMed]

19. Marzluff, J.M.; Bowman, J.M.; Donnelly, R.E. A historical perspective on urban bird research: Trends, terms, and approaches. In *Avian Ecology and Conservation in an Urbanizing World*; Marzluff, J.M., Bowman, J.M., Donnelly, R.E., Eds.; Springer: Berlin, Germany, 2001; pp. 1–17.

20. Sattler, T.; Pezzatti, G.B.; Nobis, M.P.; Obrist, M.K.; Roth, T.; Moretti, M. Selection of multiple umbrella species for functional and taxonomic diversity to represent urban biodiversity. *Conserv. Biol.* **2014**, *28*, 414–426. [CrossRef] [PubMed]

21. MacGregor-Fors, I.; Avendaño-Reyes, S.; Bandala, V.M.; Chacón-Zapata, S.; Díaz-Toribio, M.H.; González-García, F.; Lorea-Hernández, F.; Martínez-Gómez, J.; Montes de Oca, E.; Montoya, L.; et al. Multi-taxonomic diversity patterns in a neotropical green city: A rapid biological assessment. *Urban Ecosyst.* **2015**, *18*, 633–647. [CrossRef]

22. Fontana, S.; Sattler, T.; Bontadina, F.; Moretti, M. How to manage the urban green to improve bird diversity and community structure. *Landsc. Urban Plan.* **2011**, *101*, 278–285. [CrossRef]

23. Avila-Campos, J.; Instituto de Ecología, A.C. Personal observation, 2016.

24. Instituto Nacional de Estadística, Geografía e Informática (INEGI). *Censo Nacional de Población y Vivienda 2010*; INEGI: Aguascalientes, Mexico, 2010.

25. Lemoine Rodríguez, R. Cambios en la Cobertura Vegetal de la Ciudad de Xalapa-Enríquez, Veracruz y Zonas Circundantes Entre 1950 y 2010. Master's Thesis, Universidad Veracruzana, Xalapa, Mexico, 31 October 2012.

26. Lot, A.; Chiang, F. *Manual de Herbario*; Consejo Nacional de la Flora de Mexico: Mexico City, Mexico, 1986.

27. Mueller, G.M.; Bills, G.F.; Foster, S.M. *Biodiversity of Fungi: Inventory and Monitoring Methods*; Elsevier Academic Press: San Diego, CA, USA, 2004.

28. Dennis, R.W.G. *British Ascomycetes*; J. Cramer: Vaduz, Liechtenstein, 1978.

29. Breitenbach, J.; Kränzlin, F. *Fungi of Switzerland. Vol. 1. Ascomycetes*; VerlagMykologia: Lucerne, Switzerland, 1984.

30. O'Dell, T.E.; Lodge, D.J.; Mueller, G.M. Approaches to Sampling Macrofungi. In *Biodiversity of Fungi—An Inventory and Monitoring Methods*; Elsevier Academic Press: San Diego, CA, USA, 2004.

31. Largent, D. *How to Identify Mushrooms to Genus I—Macroscopic Features*; Mad River Press: Eureka, CA, USA, 1973.
32. Largent, D.; Johnson, D.; Watling, R. *How to Identify Mushrooms to Genus III—Microscopic Features*; Mad River Press: Eureka, CA, USA, 1977.
33. Lodge, J.D.; Ammirati, J.F.; O'Dell, T.E.; Mueller, G.M. Collecting and Describing Macrofungi. In *Biodiversity of Fungi—An Inventory and Monitoring Methods*; Elsevier Academic Press: San Diego, CA, USA, 2004.
34. Moser, M. *Keys to Agarics and Boleti*; Roger Phillips: London, UK, 1983.
35. Pegler, D.N. *Agaric flora of the Lesser Antilles*; HMSO: London, UK, 1983.
36. Bessette, A.E.; Bessette, A.R.; Fischer, S.W. *Mushrooms of Northeastern North America*; Syracuse University Press: Syracuse, NY, USA, 1977.
37. Horak, E. *Röhrlinge und Blätterpilze in Europa*; Elsevier Spektrum Akademischer Verlag: Heidelberg, Germany, 2005.
38. Bestelmeyer, B.T.; Agosti, D.; Alonso, L.E.; Brandao, C.R.F.; Brown, W.L.; Delabie, J.H.C.; Silvestre, R. Field techniques for the study of ground-dwelling ants: An overview, description, and evaluation. In *Ants—Standard Methods for Measuring and Monitoring Biodiversity*; Agosti, D., Majer, J.D., Alonso, L.E., Shultz, T.R., Eds.; Smithsonian Institution Press: Washington, DC, USA, 2000; pp. 122–144.
39. MacKay, W.P.; MacKay, E.A. Guide to the species identification of the New World ants. *Sociobiology* **1989**, *16*, 3–47.
40. Bolton, B. *Identification Guide to the Ant Genera of the World*; Harvard University Press: Cambridge, MA, USA, 1994.
41. Southwood, T.R.E. *Ecological Methods with Particular Reference to the Study of Insect Populations*; Springer: London, UK, 1978.
42. Rivera-Garcia, E. Factores que Determinan la Estructura de las Comunidades de Chapulines (Orthoptera: Acridoidea) en el Desierto Chihuahuense. Ph.D. Thesis, Facultad de Ciencias, UNAM, Mexico City, Mexico, 18 January 2009.
43. Borror, D.J.; Tripplehorn, C.A.; Johnson, N.F. *An Introduction to the Study of Insects*; Saunders College Publishing: Philadelphia, PA, USA, 2009.
44. Morón, M.A.; Terrón, R.A. *Entomología Práctica*; Instituto de Ecología, A.C.: Mexico City, Mexico, 1988.
45. Arnett, R.H.; Thomas, M.C. *American Beetles, Volume I: Archostemata, Myxophaga, Adephaga, Polyphaga, Staphyliniformia*; CRC Press: Boca Raton, FL, USA, 2001.
46. Arnett, R.H.; Thomas, M.C.; Skelley, P.E.; Frank, J.H. *American Beetles, Volume II: Polyphaga: Scarabaeoidea through Curculionoidea*; CRC Press: Boca Raton, FL, USA, 2002.
47. Pollard, E.; Yates, T.J. *Monitoring Butterflies for Ecology and Conservation*; Chapman and Hall: London, UK, 1993.
48. Glassberg, J.A. *Swift Guide to the Butterflies of Mexico and Central America*; Sunstreak: Morristown, NJ, USA, 2007.
49. Hernández Baz, F.; Llorente Bousquets, J.E.; Luis Martínez, A.; Vargas Fernández, I. *Las Mariposas de Veracruz—Guía Ilustrada*; Consejo Veracruzano de Investigación Ciencia y Desarrollo Tecnológico: Veracruz, Mexico, 2010.
50. Heyer, R.; Donnelly, M.A.; Foster, M.; Mcdiarmid, R. *Measuring and Monitoring Biological Diversity: Standard Methods for Amphibians*; Smithsonian Institution: Washington, DC, USA, 2014.
51. Ralph, C.J.; Droege, S.; Sauer, J.R. *Managing and Monitoring Birds Using Point Counts: Standards and Applications*; USDA: Albany, NY, USA, 1995.
52. Ellison, L.E.; Everette, A.L.; Bogan, M.A. Examining patterns of bat activity in Bandelier National Monument, New Mexico, by using walking point transects. *Southwest. Nat.* **2005**, *50*, 197–208. [CrossRef]
53. Fenton, M.B. Reporting: Essential information and analysis. In *Bat Echolocation Research: Tools, Techniques and Analysis*; Brigham, R.M., Kalko, E.K.V., Jones, G., Parsons, S., Limpens, H.J.A., Eds.; Bat Conservation International: Austin, TX, USA, 2004; pp. 133–140.
54. Hsieh, T.C.; Ma, K.H.; Chao, A. iNEXT Online: Interpolation and Extrapolation (Version 1.0). Available online: https://chao.shinyapps.io/iNEXT (accessed on 27 October 2015).
55. Colwell, R.K.; Chao, A.; Gotelli, N.J.; Lin, S.Y.; Mao, C.X.; Chazdon, R.L.; Longino, J.T. Models and estimators linking individual-based and sample-based rarefaction, extrapolation, and comparison of assemblages. *J. Plant Ecol.* **2012**, *5*, 3–21. [CrossRef]

56. MacGregor-Fors, I.; Payton, M. Contrasting diversity values: Statistical inferences based on overlapping confidence intervals. *PLoS ONE* **2013**, *8*, e56794. [CrossRef]
57. Lennon, J.J.; Koleff, P.; Greenwood, J.J.D.; Gaston, K.J. The geographical structure of British bird distributions: Diversity, spatial turnover and scale. *J. Anim. Ecol.* **2001**, *70*, 966–979. [CrossRef]
58. Koleff, P.K.; Gaston, K.J.; Lennon, J.J. Measuring beta diversity for presence–absence data. *J. Anim. Ecol.* **2003**, *72*, 367–382. [CrossRef]
59. Gaston, K.J.; Davies, R.G.; Orme, C.D.L.; Olson, V.A.; Thomas, G.H.; Ding, T.S.; Rasmussen, P.C.; Lennon, J.J.; Benneti, P.M.; Owens, I.P.F.; et al. Spatial turnover in the global avifauna. *Proc. R. Soc. Lond. B Biol.* **2007**, *274*, 1567–1574. [CrossRef] [PubMed]
60. R Core Team. *R: A Language and Environment for Statistical Computing*; R Foundation for Statistical Computing: Vienna, Austria, 2015.
61. Soykan, C.U.; Brand, L.A.; Ries, L.; Stromberg, J.C.; Hass, C.; Simmons, D.A.; Patterson, W.J.D.; Sabo, J.L. Multitaxonomic diversity patterns along a desert riparian–upland gradient. *PLoS ONE* **2012**, *7*, e28235. [CrossRef] [PubMed]
62. Gavareski, C.A. Relation of park size and vegetation to urban bird populations in Seattle, Washington. *Condor* **1976**, *78*, 375–382. [CrossRef]
63. Puga-Caballero, A.; MacGregor-Fors, I.; Ortega-Alvarez, R. Birds at the urban fringe: Avian community shifts in different peri-urban ecotones of a megacity. *Ecol. Res.* **2014**, *29*, 619–628. [CrossRef]
64. Shwartz, A.; Shirley, S.; Kark, S. How do habitat variability and management regime shape the spatial heterogeneity of birds within a large Mediterranean urban park? *Landsc. Urban Plan.* **2008**, *84*, 219–229. [CrossRef]
65. MacGregor-Fors, I. How to measure the urban-wildland ecotone: Redefining 'peri-urban' areas. *Ecol. Res.* **2010**, *25*, 883–887. [CrossRef]
66. Fischer, J.D.; Schneider, S.C.; Ahler, A.A.; Miller, J.R. Categorizing wildlife responses to urbanization and conservation implications of terminology. *Conserv. Biol.* **2015**, *29*, 1246–1248. [CrossRef] [PubMed]
67. Moreno, C.E.; Pineda, E.; Escobar, F.; Sánchez-Rojas, G. Shortcuts for biodiversity evaluation: A review of terminology and recommendations for the use of target groups, bioindicators and surrogates. *Int. J. Environ. Heal.* **2007**, *1*, 71–86. [CrossRef]
68. Arias, H.P.F. Los Árboles de la Zona Urbana y Suburbana de Xalapa. Master's Thesis, Universidad Veracruzana, Xalapa, Mexico, 1983.
69. García-Campos, H.M. Las áreas verdes públicas de Xalapa. In *Ecología Urbana Aplicada a la Ciudad de Xalapa*; López-Moreno, I.R., Ed.; Instituto de Ecología, A.C., MAB UNESCO, H. Ayuntamiento de Xalapa: Xalapa, Mexico, 1993; pp. 99–132.
70. Ruiz-Montiel, C.; Vázquez-Torres, V.; Martínez-Hernández, M.J.; Murrieta-Pérez, L.; Perez-Hernández, M.S. Árboles y arbustos registrados en el Parque Ecológico Molino de San Roque. *Madera Bosques* **2014**, *20*, 143–152.
71. Martínez-Morales, M.A. Nested species assemblages as a tool to detect sensitivity to forest fragmentation: The case of cloud forest birds. *Oikos* **2005**, *108*, 634–642. [CrossRef]
72. Ganzhorn, J.U.; Eisenbeiß, B. The concept of nested species assemblages and its utility for understanding effects of habitat fragmentation. *Basic Appl. Ecol.* **2001**, *2*, 87–99. [CrossRef]
73. O'Dell, T.E.; Lodge, D.J.; Mueller, G.M. Approaches to sampling macrofungi. In *Biodiversity of Fungi: Inventory and Monitoring Methods*; Mueller, G.M., Bills, G.F., Foster, M.S., Eds.; Elsevier: San Diego, CA, USA, 2004.
74. Konijnendijk, C.C. *The Forest and the City—The Cultural Landscape of Urban Woodland*; Springer: Berlin, Germany, 2008.
75. Evans, K.L.; Newson, S.E.; Gaston, K.J. Habitat influences on urban avian assemblages. *IBIS* **2009**, *151*, 19–39. [CrossRef]
76. Wettstein, W.; Schmid, B. Conservation of arthropod diversity in montane wetlands: Effect of altitude, habitat quality and habitat fragmentation on butterflies and grasshoppers. *J. Appl. Ecol.* **1999**, *36*, 363–373. [CrossRef]
77. Nufio, C.R.; McClenahan, J.L.; Bowers, M.D. Grasshopper response to reductions in habitat area as mediated by subfamily classification and life history traits. *J. Insect Conserv.* **2011**, *15*, 409–419. [CrossRef]
78. Hostetler, M.; Allen, W.; Meurk, C. Conserving urban biodiversity? Creating green infrastructure is only the first step. *Landsc. Urban Plan.* **2011**, *100*, 369–371. [CrossRef]

79. Beninde, J.I.; Veith, M.; Hochkirch, A. Biodiversity in cities needs space: A meta-analysis of factors determining intra-urban biodiversity variation. *Ecol. Lett.* **2015**, *18*, 581–592. [CrossRef] [PubMed]

80. Ruelas, I.E.; Aguilar Rodríguez, S.H. La avifauna urbana del Parque Ecológico Macuiltépetl en Xalapa, Veracruz, Mexico. *Ornitol. Neotrop.* **2010**, *20*, 87–103.

81. González-García, F.; MacGregor-Fors, I.; Straub, R.; Lobato García, J.A. Birds of a neotropical green city: An up-to-date review of the avifauna of the city of Xalapa with additional unpublished records. *Urban Ecosyst.* **2014**, *17*, 991–1012. [CrossRef]

82. MacGregor-Fors, I.; Ortega-Álvarez, R. Fading from the forest: Bird community shifts related to urban park site-specific and landscape traits. *Urban For. Urban Green.* **2011**, *10*, 239–246. [CrossRef]

83. Ávila-Flores, R.; Fenton, M.B. Use of spatial features by foraging insectivorous bats in a large urban landscape. *J. Mammal.* **2005**, *86*, 1193–1204. [CrossRef]

84. Carpaneto, G.M.; Mazziota, A.; Piattella, E. Changes in food resources and conservation of scarab beetles: From sheep to dog dung in a green urban area of Rome (Coleoptera, Scarabaeoidea). *Biol. Conserv.* **2005**, *123*, 547–556. [CrossRef]

85. Guénard, B.; Cardinal-De Casas, A.; Dunn, R.R. High diversity in an urban habitat: Are some animal assemblages resilient to long-term anthropogenic change? *Urban Ecosyst.* **2015**, *18*, 449–463. [CrossRef]

86. MacGregor-Fors, I.; Ortega-Álvarez, R. Ecología urbana: Experiencias en América Latina. Available online: http://www1.inecol.edu.mx/libro_ecologia_urbana (accessed on 30 June 2016).

87. Rosenzweig, M.L. *Win-Win Ecology: How the Earth's Species Can Survive in the Midst of Human Enterprise*; Oxford University Press: London, UK, 2003.

88. McDonnell, M.J. Linking and promoting research and practice in the evolving discipline of urban ecology. *J. Urban Ecol.* **2015**, *1*, 1–6. [CrossRef]

89. Ramírez-Restrepo, L.; Cultid-Medina, C.A.; MacGregor-Fors, I. How many butterflies are there in a city of circa half a million people? *Sustainability* **2015**, *7*, 8587–8597. [CrossRef]

90. McDonnell, M.J.; MacGregor-Fors, I. The ecological future of cities. *Science* **2016**, *352*, 936–938. [CrossRef] [PubMed]

Section 2:
Socioecological Systems; Non-Market Valuation: Perceptions and Attitudes

forests

MDPI

Article

Exploring Relationships between Socioeconomic Background and Urban Greenery in Portland, OR

Lorien Nesbitt and Michael J. Meitner *

Department of Forest Resources Management, Faculty of Forestry, University of British Columbia, 2424 Main Mall, Vancouver, BC V6T 1Z4, Canada; lorien.nesbitt@ubc.ca
* Correspondence: mike.meitner@ubc.ca; Tel.: +1-604-822-0029

Academic Editors: Francisco Escobedo, Stephen John Livesley and Justin Morgenroth
Received: 1 May 2016; Accepted: 13 July 2016; Published: 29 July 2016

Abstract: Do urban residents experience societal benefits derived from urban forests equitably? We conducted a broad-scale spatial analysis of the relationship between urban greenery and socioeconomic factors in the Portland metropolitan area. The Normalized Difference Vegetation Index was derived from National Agriculture Imagery Program images to map urban vegetation cover, and Outdoor Recreation and Conservation Area data were used to identify green spaces. These measures of urban greenery were correlated with census data to identify socioeconomic factors associated with high levels of green inequity. Population density, house age, income, and race were strongly correlated with vegetation cover. However, the distribution of green spaces showed a much weaker relationship with socioeconomic factors. These results highlight the importance of different measures of access to urban greenery and suggest potential solutions to the problem of urban green inequity. Cities can use our methods to conduct targeted urban forest management to maximize urban forest benefits received by residents.

Keywords: climate change; equity; Portland; socioeconomics; urban forests; well-being

1. Introduction

Trees and green spaces are important to human well-being [1]. In cities, it is well established that trees and green spaces provide important environmental, economic, social and psychological benefits [2–9]. Trees and green spaces influence urban microclimates to reduce the "urban heat island effect" [10], improve air quality [5], and reduce surface water runoff [8]. They help us recover from stress [7], improve public health outcomes [11–13], increase social cohesion [14,15], may reduce violent crime [16], can increase property values [6], and support urban biodiversity [17,18]. As more and more people make cities their home, cities and societies need to consider how best to maximize the benefits of urban greenery and ensure that urban residents are all able to experience these benefits.

Although urban greenery has important, positive impacts on various aspects of urban quality of life, the distribution of urban greenery appears to be inequitable in many cities around the world [19–24]. Urban green spaces are most often located in wealthier neighbourhoods [6] and trees on private property are often larger and more abundant in high income neighbourhoods [25]. The canopy cover of public street trees can also vary widely within cities, and lower levels of canopy cover are more often found in lower-income neighbourhoods [20,23].

This disparity in access to urban greenery is central to the idea of *urban green equity*. We define urban green equity as equitable access to urban greenery, regardless of differentiating factors such as socioeconomic status, race, cultural background, or age. Equitable greenery access helps ensure that all residents have equitable access to the services and benefits provided by urban greenery that may be associated with higher levels of societal well-being [12,13,26]. Equitable access to urban greenery may be most important for lower socioeconomic groups. For example, the positive effects of urban greenery on health seem to be strongest among lower income groups who are more likely to suffer poor health outcomes and higher levels of mortality [11,13,27]. The availability of urban greenery also appears to have a stronger positive effect on residents' perceived general health in lower income groups [12].

While most cities likely experience some form of urban green inequity, it is not yet clear how this inequity should be defined and measured and what causes it. A diversity of study sites and measures of green cover and green access (or equity) has demonstrated general relationships between measures of privilege and urban green access but has yet to clarify the relative roles that these factors play [19–22,24,28–31]. As a first step in examining the concept of urban green equity and exploring its potential relationships to socioeconomic patterns, we undertook an analysis of the availability of urban greenery in the Portland–Vancouver metro area, United States of America (US). This first-step analysis does not seek to uncover causal relationships. Rather, our analysis examines correlations between the distribution of urban greenery and socioeconomic factors that may be related to this distribution. Further analysis would be needed to develop a theory of urban green equity that clarifies causal relationships.

Portland is considered an urban forestry leader in the US and was named one of the "10 Best Cities for Urban Forests" in 2013 by American Forests [32]. The City of Portland has a public tree inventory and a tree species diversity plan [33]. Urban forestry appears to be integrated with other municipal planning initiatives, including the city's climate change and green infrastructure plans [34,35]. From a regional perspective, while there is variation in local urban forestry policies and programs, as of 2010, 25 out of 30 jurisdictions (municipalities, counties, or regions in the Portland–Vancouver metro area) had ordinances regulating tree removal or preservation on private land, and 22 out of 30 jurisdictions regulated street tree removal on public land [36]. Five jurisdictions had adopted urban forestry management plans and four jurisdictions had an urban tree inventory [36].

A review of local plans and policies suggests that the City of Portland has some understanding of the importance of urban green equity. The Portland Urban Forestry Management Plan 2004 includes a "commitment to equity so environmental impacts and the costs of protecting the environment do not unfairly burden any one geographic or socioeconomic sector of the City" [37] (p. 105). While the 2007 Urban Forest Action Plan has no mention of or strategies to address urban green equity or socioeconomic inequalities through urban forest management [38], by 2015, the Implementation Update discusses an equity scan completed by the Environmental Services Tree Program in the 2013/2014 fiscal year and efforts to address inequities through programming [39]. The Update includes efforts to ensure equitable service delivery in low-income communities and communities of colour, and optimize tree planting to "support the equitable distribution of street trees throughout the city" [39] (p. 16). The City's new tree code, which took effect at the beginning of 2015, includes a "Tree Planting and Preservation Fund", one of the purposes of which is to advance the equitable distribution of tree-related benefits [40].

2. Materials and Methods

2.1. Data Collection

Our study site was the Portland–Vancouver metro area. Our analysis included the portion of the Portland–Vancouver metro area that falls within the state of Oregon in order to capture a range of development types and to better reflect the experiences of urban residents who may live and work in different parts of the metro area. We excluded the portion of the metro area that falls within the state of Washington due to data limitations. The metro area used in our analysis was the "Portland OR–WA" Urbanized Area (UA) as defined by the US Census Bureau for the most recent census year (2010) [41]. The Census Bureau defines urban areas as " . . . a densely settled core of census tracts and/or census blocks that meet minimum population density requirements, along with contiguous territory containing non-residential urban land uses as well as territory with low population density included to link outlying densely settled territory with the densely settled core." [42]. Those urban areas with populations of 50,000 or more are designated as UAs by the US Census Bureau [42].

To better understand how different segments of the Portland population may access urban greenery in different parts of the metro area, we measured urban greenery cover and park access by Block Group, the smallest unit for which US decennial census data and American Community Survey data are publicly-available across a wide range of topics [43]. Only Block Groups located entirely within the state of Oregon in the Portland OR–WA UA were included in the analysis. Block Group boundaries were obtained from the 2013 TIGER/Line Shapefiles produced by the US Census Bureau [44].

2.1.1. Greenery Cover

We estimated greenery cover using aerial imagery produced by the US National Agriculture Imagery Program (NAIP). The imagery was acquired by the NAIP program during the growing season ("leaf on") in the year 2010 and has a resolution of 1 m^2. The images were captured with four bands of data: red, green, blue, and near infrared. The four-band imagery allows for the calculation of the Normalized Difference Vegetation Index (NDVI) using the formula (NIR − Red)/(NIR + Red), where NIR is the near infrared wavelength band and Red is the red wavelength band. NDVI is a measure of the visible and near-infrared light reflected by green vegetation and is commonly used to estimate the density and type of vegetation in remotely-sensed images [45,46]. The benefits of using the NDVI to measure vegetation are that (1) it provides data that are comparable across large areas and many different contexts; and (2) it allows for differentiation between broad greenery types (e.g., woody vegetation and grassy vegetation).

We calculated NDVI values for each 1 m^2 pixel and reclassified each pixel as either "mixed vegetation" or "predominantly woody vegetation". This distinction reflects the different benefits urban residents can derive from urban vegetation generally and woody vegetation specifically [7,47,48]. "Mixed vegetation" was defined as all pixels with values of 0.12 or higher and represents all urban greenery, including grass, garden and crop plants, shrubs, hedges, and trees [49]. "Predominantly woody vegetation" was defined as all pixels with values of 0.3 or higher and includes trees of all sizes, large shrubs, and hedges [49]. These reclassifications had accuracies of 93% for "mixed vegetation" and 81% for "predominantly woody vegetation", as determined by a 100-point random sample comparing the reclassified pixel values with aerial photos.

To estimate the availability of greenery in each Block Group, we calculated the mean "mixed vegetation" pixels (Figure 1) and "primarily woody vegetation" pixels (Figure 2) per area of land for each block group.

2.1.2. Green Spaces

We used the Outdoor Recreation and Conservation Area (ORCA) feature class produced by the Portland Metro Regional Government to identify green spaces in the study area. This dataset includes a variety of green spaces including parks and/or natural areas, cemeteries, golf courses, areas owned by private home owners associations, school lands, and areas not specifically designated as any other type of green space, such as privately-owned plots, transportation corridor buffers, and trails that are not large enough to qualify as parks [50]. The ORCA dataset is thus a wide definition of green spaces and represents both public and private land.

To estimate access to green spaces in each Block Group, we calculated the Euclidean distance between the Block Group centroid and the nearest park edge (Figure 3). This simple method provides a reliable estimate of the average distance to the nearest green space edge from any point in the Block Group. The method treats all green spaces equally, regardless of size and ownership.

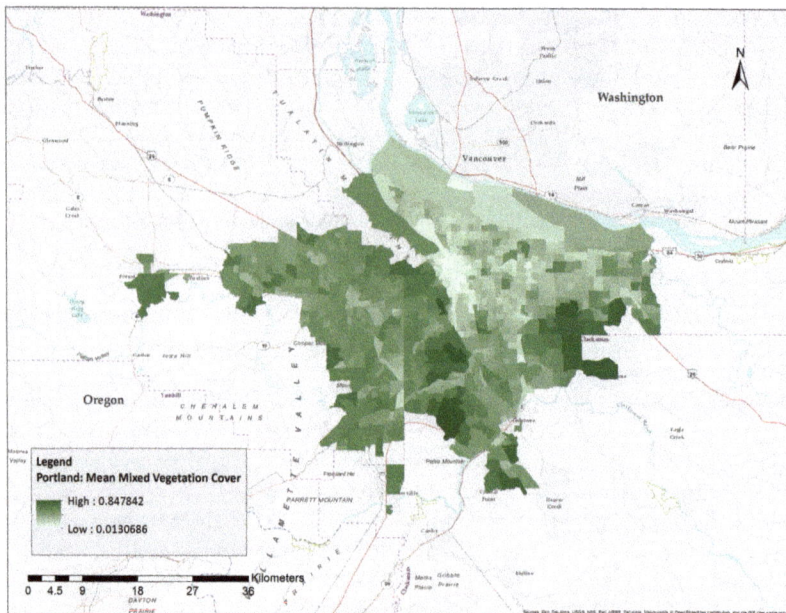

Figure 1. Mean number of pixels with Normalized Difference Vegetation Index (NDVI) values of 0.12 or greater per square metre of land in each Block Group.

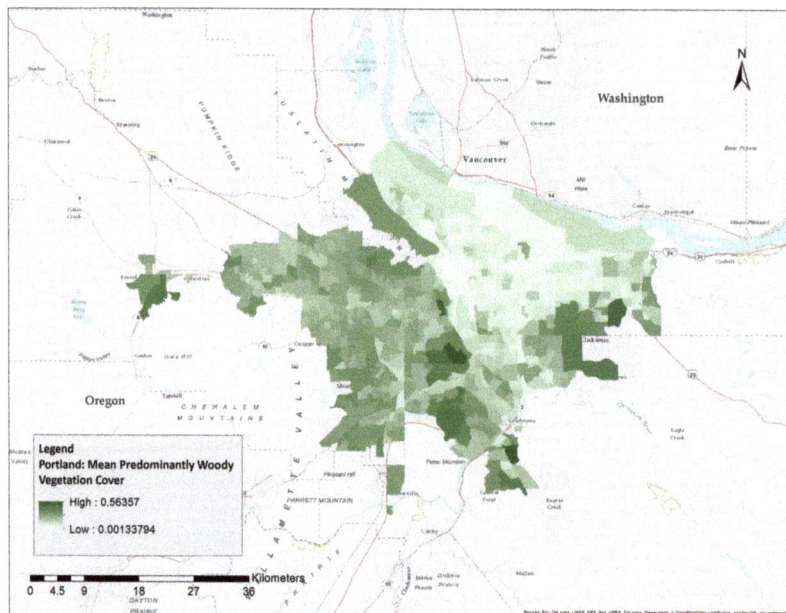

Figure 2. Mean number of pixels with Normalized Difference Vegetation Index (NDVI) values of 0.3 or greater per square metre of land in each Block Group.

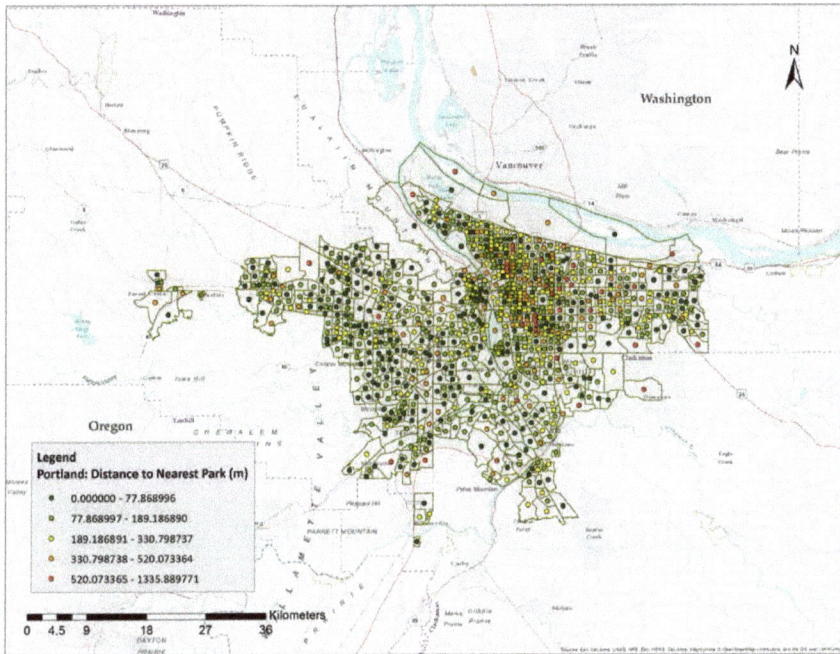

Figure 3. Euclidean distance from the Block Group centroid to the nearest park edge as defined by the Outdoor Recreation and Conservation Area (ORCA) feature class.

2.1.3. Socioeconomic and Land Use Data

We obtained socioeconomic data from the 2009–2013 American Community Survey 5-Year Estimates, produced by the US Census Bureau. The 2009–2013 American Community Survey 5-year estimates include results from the American Community Survey and describe the entire data collection period, from 1 January 2009 to 31 December 2013 [51]. We collected socioeconomic data on several topics, including:

- Age;
- Educational attainment;
- Employment status;
- Hispanic or Latino origin;
- House age (median);
- Income (annual per capita);
- Population density;
- Race.

We obtained zoning data for the Oregon counties and municipalities within the study area from the Portland Metro Regional Government. The dataset includes generalized zoning designation boundaries for the metro region in Oregon, including the City of Portland, Multnomah County, Washington County, and Clackamas County, Oregon. The generalized zoning categories are: Future Urban Development, Industrial, Commercial, Multi-family Residential, Mixed-Use Residential, Single-family Residential, Public Facility, Parks and Open Spaces, and Rural (Figure 4). Zoning data for Washington State were unavailable. Block Groups in Washington State were thus excluded from the analysis.

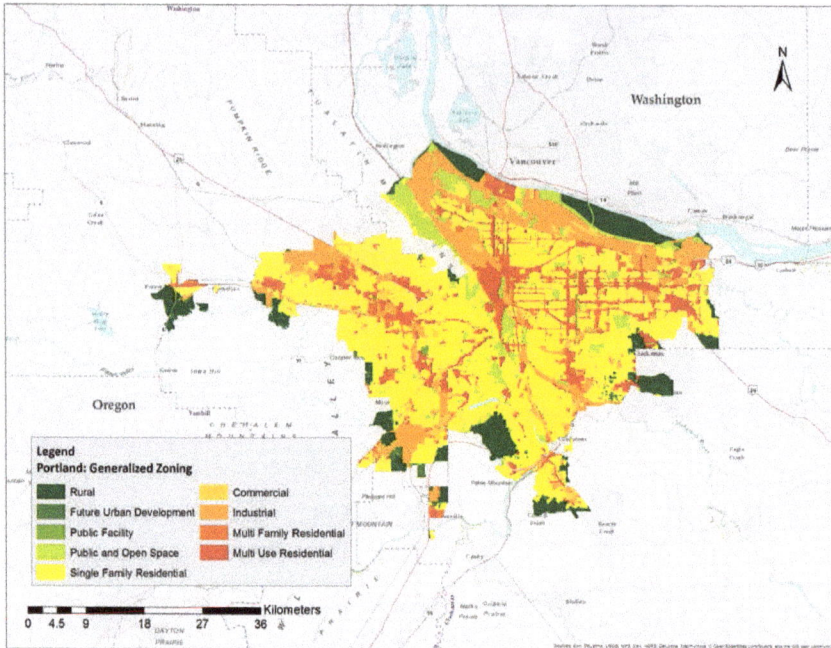

Figure 4. Zoning boundaries for nine generalized zoning categories.

2.2. Analysis

To better understand how different segments of the Portland population experience urban greenery in different areas of the metro region, we used step-wise multiple linear regression to examine correlations between socioeconomic factors and (1) mean "mixed vegetation" pixels; (2) mean "predominantly woody vegetation" pixels; and (3) distance to nearest green space by Block Group.

We investigated the relationship between zoning and greenery cover ("mixed vegetation" and "predominantly woody vegetation") using univariate Analysis of Variance (ANOVA).

3. Results

3.1. Greenery Cover

3.1.1. Mixed Vegetation

A multiple linear regression was calculated to predict mixed vegetation cover based on population density, median year built, income per capita, percent white, percent other race, percent professional degree, percent Master's degree and percent Asian. A significant regression equation was found ($F(8, 890) = 99.746$, $p < 0.01$) with an R^2 of 0.473 (Table 1). Predicted mixed vegetation cover is equal to $-6.643 - 30.532$ (population density) + 0.003 (median year built) + 1.089×10^{-6} (income per capita) + 0.312 (percent white) + 0.337 (percent other race) + 0.365 (percent professional degree) + 0.171 (percent masters degree) + 0.150 (percent Asian). Population density, house age, and per capita income showed the strongest relationships with mixed vegetation cover (Table 1). Block Groups with higher population densities, older houses, and lower per capita incomes had lower levels of mixed vegetation cover.

Table 1. Regression results for "mixed vegetation" cover and socioeconomic variables.

Variable	Mean "Mixed Vegetation" Pixels per Area Land		
	B	SE B	β
Population Density	−30.532	2.175	−0.362 **
Median Year Built	0.003	0.000	0.424 **
Income Per Capita	1.089×10^{-6}	0.000	0.014 *
Percent White	0.312	0.046	0.259 **
Percent Other Race	0.337	0.072	0.150 **
Percent Professional Degree	0.365	0.131	0.095 **
Percent Masters Degree	0.171	0.073	0.081 *
Percent Asian	0.150	0.069	0.071 *
R^2		0.473	
F		99.746 **	

B, unstandardized regression coefficient; SE, standard error; β, standardized regression coefficient; *p*, calculated probability. * $p < 0.05$; ** $p < 0.01$.

The ANOVA showed significant differences in mixed vegetation cover among generalized zoning boundaries ($F(1, 9) = 480.54$, $p < 0.01$) and zoning accounted for over 22% of the variation in the mixed vegetation cover data ($R^2 = 0.225$). Zoning types with the lowest levels of mixed vegetation cover were mixed-use residential, industrial, multi-family residential, and commercial (Figure 5). Zoning types with the highest levels of mixed vegetation cover were parks and open spaces, public facility, rural, and future urban development (Figure 5). Among residential zoning types, single-family residential areas had much higher levels of mixed vegetation cover than any other residential zoning types (Figure 5).

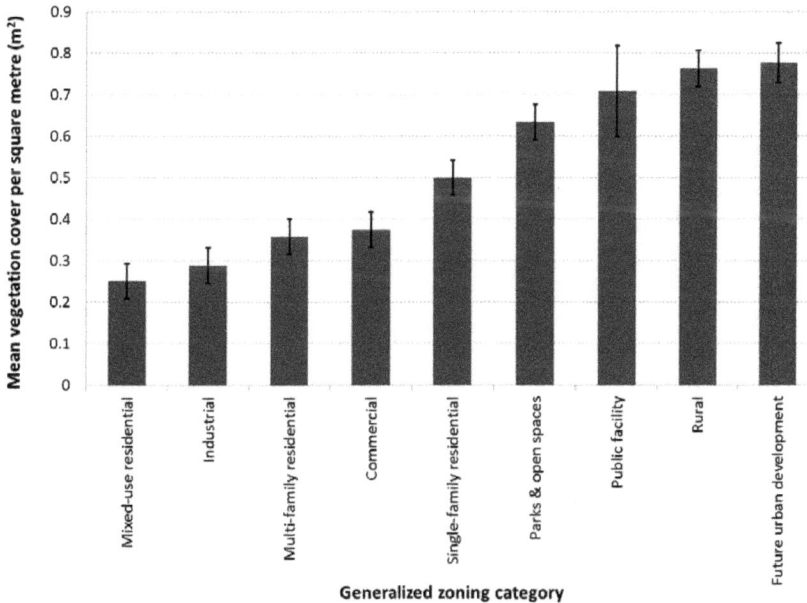

Figure 5. Mean mixed vegetation cover per square metre by generalized zoning category.

3.1.2. Predominantly Woody Vegetation

A multiple linear regression was calculated to predict predominantly woody vegetation cover based on median year built, population density, income per capita, percent other race, percent white, percent some school completed, percent high school diploma, and percent some college completed.

A significant regression equation was found ($F(8, 890) = 103.730$, $p < 0.01$) with an R^2 of 0.483 (Table 2). Predicted predominantly woody vegetation cover is equal to $-5.110 + 0.003$ (median year built) $- 17.728$ (population density) $+ 5.700 \times 10^{-7}$ (income per capita) $+ 0.296$ (percent other race) $+ 0.159$ (percent white) $- 0.175$ (percent some school) $- 0.134$ (percent high school diploma) $- 0.059$ (percent some college). House age, population density, and per capita income showed the strongest relationships with predominantly woody vegetation cover (Table 2). Block Groups with older houses, higher population densities, and lower per capita incomes had lower levels of predominantly woody vegetation cover.

Table 2. Regression results for "predominantly woody vegetation" cover and socioeconomic variables.

Variable	Mean "Predominantly Woody Vegetation" Pixels per Area Land		
	B	**SE B**	**β**
Median Year Built	0.003	0.000	0.493 **
Population Density	−17.728	1.451	−0.318 **
Income Per Capita	5.700×10^{-7}	0.000	0.079 *
Percent Other Race	0.296	0.046	0.199 **
Percent White	0.159	0.024	0.200 **
Percent Some School Completed	−0.175	0.042	−0.137 **
Percent High School Diploma	−0.134	0.035	−0.116 **
Percent Some College Completed	−0.059	0.029	−0.062 *
R^2		0.483	
F		103.730 **	

B, unstandardized regression coefficient; SE, standard error; β, standardized regression coefficient; *p*, calculated probability. * $p < 0.05$; ** $p < 0.01$.

The ANOVA showed significant differences in predominantly woody vegetation cover among generalized zoning boundaries ($F(1, 9) = 313.67$, $p < 0.01$) and zoning accounted for just over 15% of the variation in the predominantly woody vegetation cover data ($R^2 = 0.159$). Zoning types with the lowest levels of predominantly woody vegetation cover were mixed-use residential, industrial, multi-family residential, and commercial (Figure 6). Zoning types with the highest levels of predominantly woody vegetation cover were parks and open spaces, public facility, rural, and future urban development (Figure 6). Among residential zoning types, single-family residential areas had much higher levels of predominantly woody vegetation cover than any other residential zoning types (Figure 6).

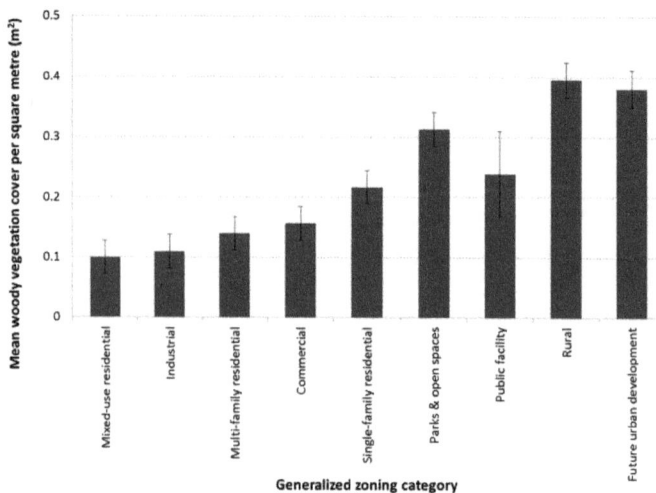

Figure 6. Mean cover of primarily woody vegetation per square metre by generalized zoning category.

3.2. Green Spaces

A multiple linear regression was calculated to predict distance to nearest park based on median year built, percent bachelors degree, and percent masters degree. A significant regression equation was found ($F(3, 895) = 24.089$, $p < 0.01$) with an R^2 of 0.075 (Table 3). Predicted distance to the nearest park is equal to 4725.338 $-$ 2.289 (median year built) $-$ 118.080 (percent bachelors degree) $-$ 183.887 (percent masters degree). Parks and green spaces thus appear to be more evenly distributed, showing a weak but significant relationship to socioeconomic factors. Block Groups with older houses, lower percentages of residents with bachelor's degrees, and lower percentages of residents with master's degrees had the largest distances to parks and green spaces.

Table 3. Regression results for park distance and socioeconomic variables.

	Distance to Nearest Park		
Variable	B	SE B	β
Median Year Built	−2.289	0.282	−0.271 **
Percent Bachelors Degree	−118.08	55.610	−0.840 *
Percent Masters Degree	−183.887	87.956	−0.083 *
R^2			0.075
F			60.238 **

B, unstandardized regression coefficient; SE, standard error; β, standardized regression coefficient; *p*, calculated probability. * $p < 0.05$; ** $p < 0.01$.

4. Discussion

The strong relationship between socioeconomic factors and greenery cover (Tables 1 and 2) suggests a somewhat uneven distribution of greenery in the Portland–Vancouver metro area. Population density and house age are the primary factors in this relationship, showing strong correlations with both mixed vegetation cover and predominantly woody vegetation cover. The correlation with population density is to be expected, given that higher population densities require higher building densities and thus reduced space for greenery near dwellings. This result is supported by the zoning analysis (Figures 5 and 6): mixed-use residential and multi-family residential zones, which are designed to accommodate higher residential population densities, showed some of the lowest levels of greenery cover among the zoning types analyzed. More surprisingly, the analysis indicates that greenery cover is more abundant in areas with newer houses, in contradiction to some previous research [19]. While this observation may appear to contradict the notion that older neighbourhoods generally contain older trees, it may reflect the fact that newer housing developments occur in peri-urban areas close to forested rural areas [52,53]. The zoning analysis supports this conclusion, showing the highest levels of urban greenery in rural areas and those zoned for future urban development (Figures 5 and 6).

The majority of the socioeconomic variables showing significant correlations with urban greenery cover reflect the income, race and education of residents, and indicate that Portland's residents may, in fact, suffer from green inequity (Tables 1 and 2). According to our analysis, residents who have higher incomes or those who are white, Asian, or who self-identify as "other race" (some race other than white, black, Aboriginal, Hawaiian, and Asian), or who hold a professional or masters degree, are more likely to live in areas with high levels of mixed vegetation cover (Table 1). Those with higher incomes, those who are white, or those who self-identify as "other race" are more likely to live in areas with high predominantly woody vegetation cover while those with limited secondary and post-secondary education are less likely to live in those areas (Table 2). Urban greenery cover is thus somewhat distributed according to typical socioeconomic divisions. The City of Portland appears to be aware of the potential for these divisions in greenery cover, as evidenced by programming to prioritize street tree planting in low-income communities and communities of colour described in the 2014 Urban Forest Action Plan Implementation Update [39]. While street tree planting is an important part of increasing greenery abundance in low-canopy neighbourhoods, our analysis does not provide guidance as to whether greenery on public or private land is primarily responsible for these inequities. Private land often provides a large portion of urban greenery [17,54] and is thus an important consideration in urban greening programs. Further analysis to differentiate between greenery on public and private land may help explain the relative contributions of public and private land to the observed green inequities and provide guidance for greening programs in low-canopy neighbourhoods.

While there is strong evidence of the environmental, psychological, physical health, and economic benefits of trees [2–8], and evidence of aesthetic preference for urban greenery [48], it is important to note that some urban residents may prefer not to live near urban greenery. Preferences for urban greenery appear to be influenced by culture [55–57] and may also be influenced by socioeconomic background [25]. Low levels of urban greenery and large distances to urban parks would not be considered to be problematic by such residents. Our current analysis cannot comment on such preferences and this may be an area for further research.

Interestingly, the distribution of green spaces in Portland appears to be highly equitable, with only weak relationships between the distance to the closest green space and socioeconomic variables (Table 3). While the analysis did not examine green space size or land tenure, it suggests that most Portland residents live within similar distances of green spaces of some size. Residents living in newer houses and with post-secondary degrees are more likely to live nearer to a green space, suggesting the presence of green inequity, but this does not appear to be a strong correlation. The relatively equitable distribution of green spaces in Portland highlights a policy tool that municipalities in the metro area may use to maintain or improve current levels of green equity. Although it may be difficult to influence greenery abundance on private property, maintaining or improving residents' access to high-quality green spaces would ensure that all residents are able to access greenery near their homes, regardless of socioeconomic status. Further analysis of green space access that considers size and land tenure would help clarify how green space development could be used to reduce green inequity.

The green inequities illustrated by our analysis become more serious when viewed through the lens of public health and well-being. Those residents of Portland who have the lowest levels of greenery access are also those who are most likely to suffer poor public health outcomes and who may experience the greatest benefits from higher greenery access [11–13]. In addition, these potential health inequities may be exacerbated by climate change. As climate change progresses and increases in intensity, cities may experience a range of changes and pressures that can have negative impacts on urban residents' health, such as rising sea levels and increased storm surges, heat stress, extreme precipitation events, flooding caused by increased precipitation and snow melt, landslides, drought, increased aridity, water scarcity, and air pollution [58]. Residents and communities with lower socioeconomic status may be more vulnerable to climate change impacts than other segments of the population [58,59], suggesting another way in which increased urban greenery may have stronger physical health benefits for lower socioeconomic groups. Although the specific impacts of climate change will vary by location, urban greenery can play a key role in moderating the impacts of climate change and helping urban communities adapt to its effects. The microclimatic regulation provided by urban greenery can work in concert with other types of infrastructure and adaptation programming in comprehensive climate change adaptation strategies that ensure adequate greenery cover in vulnerable areas [5,8,10,60–62].

Portland is clearly aware of the importance of urban green equity and some of its implications for urban forest management, although urban forest policy documents suggest that this awareness is relatively recent and it is thus unlikely that policy instruments will have had time to achieve any real reductions in urban green inequity [39]. Portland also appears to have begun integrating urban forest management with climate change and green infrastructure planning in the city [34,35]. Given the public health and climate change adaptation implications of urban green inequity, Portland would benefit from further integrating urban forest management into broader municipal planning initiatives, thereby attempting to ameliorate urban green inequities using a variety of policy mechanisms. Urban greenery is part of the larger urban "ecosystem", working together with other infrastructure and programming to improve residents' well-being.

5. Conclusions

Our results suggest that urban green equity is an issue affecting disadvantaged urban residents. However, our analysis examines the issue in only one metro area. Urban forest management is clearly influenced by local environment, development history, and municipal and regional policies, and will thus vary among cities and metro areas [63]. To truly begin to understand patterns of urban green inequity and associated factors, this analysis should be extended to other cities that represent a diversity of geographies, development histories, and policy approaches to urban forest management. This broader analysis would help unpack general and locally specific factors associated with urban green inequity.

In addition to a broader analysis of the topic, a deeper, case study analysis of urban green inequity would help explain important aspects of the topic not examined in our analysis. Factors such as land tenure and access, tree species and condition, aesthetic and recreational quality, and cultural preferences will likely affect how urban residents experience and benefit from urban greenery [54,55,64,65]. While our analysis demonstrated broad patterns of access to urban greenery, these finer scale factors influence whether urban greenery is truly accessible and the magnitude of benefits urban residents derive from it.

Both of the broad and fine scale analyses suggested above require high quality data to accomplish. Broad-scale spatial analyses require high-quality aerial imagery and land use and land tenure data that are not always available across cities and metro regions. When they are available, differences in local definitions and data collection methodologies can impede comparisons. Large data collection programs, such as the NAIP, or online clearinghouses would facilitate analyses across several cities or metro areas.

On the other end of the spectrum, fine scale data on tree species, condition, and local cultural values and preferences generally require field research to obtain it. Some municipalities have begun sharing local, fine scale data such as tree inventories, online, representing an important source of urban forest data. As government, organizations, academics, and citizens continue to collect such data, collaboration and open data access have the potential to facilitate continued in-depth research. As we improve our understanding of urban green equity and the factors associated with it, cities and societies will be able to craft more effective approaches to reducing urban green inequity, thereby improving residents' quality of life.

Acknowledgments: The authors would like to acknowledge and thank Stephen R. J. Sheppard and Cynthia Girling for their guidance on experimental methods and study design.

Author Contributions: Lorien Nesbitt and Michael J. Meitner conceived and designed the experiments; Lorien Nesbitt performed the experiments; Lorien Nesbitt and Michael J. Meitner analyzed the data; Michael J. Meitner contributed analysis tools; Lorien Nesbitt wrote the paper with guidance from Michael J. Meitner.

Conflicts of Interest: The authors declare no conflict of interest.

Abbreviations

The following abbreviations are used in this manuscript:

US	United States of America
UA	Urbanized Area
NAIP	National Agriculture Imagery Program
NDVI	Normalized Difference Vegetation Index
ORCA	Outdoor Recreation and Conservation Area
ANOVA	Analysis of Variance

References

1. Ward Thompson, C.; Aspinall, P.A. Natural environments and their impact on activity, health, and quality of life. *Appl. Psychol. Health Well Being* **2011**, *3*, 230–260. [CrossRef]
2. Kuchelmeister, G. Trees for the urban millennium: Urban forestry update. *Unasylva* **2000**, *51*, 49–55.
3. Thompson, C.W. Urban open space in the 21st century. *Landsc. Urban Plan.* **2002**, *60*, 59–72. [CrossRef]
4. Kenney, W.A.; Van Wassenaer, P.J.E.; Satel, A.L. Criteria and indicators for strategic urban forest planning and management. *Arboric. Urban For.* **2011**, *37*, 108–117.
5. Heidt, V.; Neef, M. Benefits of urban green space for improving urban climate. In *Ecology, Planning, and Management of Urban Forests*; Carreiro, M., Song, Y., Wu, J., Eds.; Springer: New York, NY, USA, 2008; pp. 84–96.
6. Poudyal, N.C.; Hodges, D.G.; Merrett, C.D. A hedonic analysis of the demand for and benefits of urban recreation parks. *Land Use Policy* **2009**, *26*, 975–983. [CrossRef]
7. Ulrich, R.S.; Simons, R.F.; Losito, B.D.; Fiorito, E.; Miles, M.A.; Zelson, M. Stress recovery during exposure to natural and urban environments. *J. Environ. Psychol.* **1991**, *11*, 201–230. [CrossRef]
8. Konijnendijk, C.; Nilsson, K.; Randrup, T.; Schipperijn, J. *Urban Forests and Trees*; Konijnendijk, C.C., Nilsson, K., Randrup, T.B., Schipperijn, J., Eds.; Springer-Verlag: Berlin/Heidelberg, Germany, 2005.
9. Gatrell, J.D.; Jensen, R.R. Growth through greening: Developing and assessing alternative economic development programmes. *Appl. Geogr.* **2002**, *22*, 331–350. [CrossRef]
10. McPherson, G.; Simpson, J.R.; Peper, P.J.; Xiao, Q. Benefit-cost analysis of Modesto's municipal urban forest. *J. Arboric.* **1999**, *25*, 235–248.
11. De Vries, S.; Verheij, R.A.; Groenewegen, P.P.; Spreeuwenberg, P. Natural environments—Healthy environments? An exploratory analysis of the relationship between greenspace and health. *Environ. Plan. A* **2003**, *35*, 1717–1731. [CrossRef]
12. Maas, J.; Verheij, R.A.; Groenewegen, P.P.; de Vries, S.; Spreeuwenberg, P. Green space, urbanity, and health: How strong is the relation? *J. Epidemiol. Community Health* **2006**, *60*, 587–592. [CrossRef] [PubMed]
13. Mitchell, R.; Popham, F. Effect of exposure to natural environment on health inequalities: An observational population study. *Lancet* **2008**, *372*, 1655–1660. [CrossRef]
14. Kweon, B.-S.; Sullivan, W.C.; Wiley, A.R. Green common spaces and the social integration of inner-city older adults. *Environ. Behav.* **1998**, *30*, 832–858. [CrossRef]
15. Sullivan, W.C. The Fruit of Urban Nature: Vital Neighborhood Spaces. *Environ. Behav.* **2004**, *36*, 678–700. [CrossRef]
16. Kuo, F.E.; Sullivan, W.C. Environment and crime in the inner city: Does vegetation reduce crime? *Environ. Behav.* **2001**, *33*, 343–367. [CrossRef]
17. Goddard, M.A.; Dougill, A.J.; Benton, T.G. Scaling up from gardens: Biodiversity conservation in urban environments. *Trends Ecol. Evol.* **2009**, *25*, 90–98. [CrossRef] [PubMed]
18. Rudd, H.; Vala, J.; Schaefer, V. Importance of backyard habitat in a comprehensive biodiversity conservation strategy: A connectivity analysis of urban green spaces. *Restor. Ecol.* **2002**, *10*, 368–375. [CrossRef]
19. Heynen, N.C.; Lindsey, G. Correlates of urban forest canopy cover: Implications for local public works. *Public Works Manag. Policy* **2003**, *8*, 33–47. [CrossRef]
20. Landry, S.M.; Chakraborty, J. Street trees and equity: Evaluating the spatial distribution of an urban amenity. *Environ. Plan. A* **2009**, *41*, 2651–2670. [CrossRef]
21. McConnachie, M.M.; Shackleton, C.M. Public green space inequality in small towns in South Africa. *Habitat Int.* **2010**, *34*, 244–248. [CrossRef]
22. Ogneva-Himmelberger, Y.; Pearsall, H.; Rakshit, R. Concrete evidence & geographically weighted regression: A regional analysis of wealth and the land cover in Massachusetts. *Appl. Geogr.* **2009**, *29*, 478–487.
23. *City of Vancouver Urban Forest Strategy*; Vancouver Board of Parks and Recreation: Vancouver, BC, Canada, 2014.
24. Talarchek, G. The urban forests of New Orleans: An exploratory analysis of relationships. *Urban Geogr.* **1990**, *11*, 65–86. [CrossRef]
25. Kirkpatrick, J.B.; Daniels, G.D.; Davison, A. Temporal and spatial variation in garden and street trees in six eastern Australian cities. *Landsc. Urban Plan.* **2011**, *101*, 244–252. [CrossRef]

26. Sanesi, G.; Gallis, C.; Kasperidus, H.D. *Forests, Trees and Human Health*; Nilsson, K., Sangster, M., Gallis, C., Hartig, T., de Vries, S., Seeland, K., Schipperijn, J., Eds.; Springer Netherlands: Dordrecht, The Netherlands, 2011.

27. Mitchell, R.; Popham, F. Greenspace, urbanity and health: Relationships in England. *J. Epidemiol. Commun. Health* **2007**, *61*, 681–683. [CrossRef] [PubMed]

28. Barbosa, O.; Tratalos, J.A.; Armsworth, P.R.; Davies, R.G.; Fuller, R.A.; Johnson, P.; Gaston, K.J. Who benefits from access to green space? A case study from Sheffield, UK. *Landsc. Urban Plan.* **2007**, *83*, 187–195. [CrossRef]

29. Lafary, E.W.; Gatrell, J.D.; Jensen, R.R. People, pixels and weights in Vanderburgh County, Indiana: Toward a new urban geography of human–environment interactions. *Geocarto Int.* **2008**, *23*, 53–66. [CrossRef]

30. Pearsall, H.; Christman, Z. Tree-lined lanes or vacant lots? Evaluating non-stationarity between urban greenness and socio-economic conditions in Philadelphia, Pennsylvania, USA at multiple scales. *Appl. Geogr.* **2012**, *35*, 257–264. [CrossRef]

31. Li, G.; Weng, Q. Measuring the quality of life in city of Indianapolis by integration of remote sensing and census data. *Int. J. Remote Sens.* **2007**, *28*, 249–267. [CrossRef]

32. American Forests American Forests Names the 10 Best U.S. Cities for Urban Forests. Available online: http://www.americanforests.org/newsroom/american-forests-names-the-10-best-u-s-citiesfor-urban-forests/ (accessed 21 April 2016).

33. American Forests. *Portland Urban Forest Fact Sheet*; American Forests: Washington, DC, USA, 2013.

34. *Climate Action Plan, Local Strategies to Address Climate Change*; City of Portland: Portland, OR, USA, 2015.

35. ENTRIX Inc. *Portland's Green Infrastructure: Quantifying the Health, Energy and Community Livability Benefits*; City of Portland: Portland, OR, USA, 2010.

36. Audubon Society of Portland; Portland State University. *Regional Urban Forestry Assessment and Evaluation for the Portland–Vancouver Metro Area*; Metro: Portland, OR, USA, 2010.

37. Portland Parks and Recreation; Urban Forestry Management Plan Technical Advisory Committee. *Portland Urban Forestry Management Plan 2004*; City of Portland: Portland, OR, USA, 2004.

38. Portland Parks and Recreation. *Urban Forest Action Plan*; City of Portland: Portland, OR, USA, 2007.

39. Portland Parks and Recreation. *Urban Forest Action Plan, 2014 Implementation Update*; City of Portland: Portland, OR, USA, 2015.

40. Bureau of Planning and Sustainability. *Title 11, Trees*; City of Portland: Portland, OR, USA, 2012.

41. US Census bureau. Qualifying urban areas for the 2010 census. *Fed. Regist.*; 2010; 77, pp. 18652–18669.

42. US Census bureau. Urban area criteria for the 2010 census. *Fed. Regist.*; 2010; 76, pp. 53030–53043.

43. US Census Bureau Geographic Terms and Concepts—Block Groups. Available online: https://www.census.gov/geo/reference/gtc/gtc_bg.html (accessed on 13 April 2016).

44. US Census Bureau. *2013 TIGER/Line Shapefiles (Machine-Readable Data Files)*; US Census Bureau: Washington, DC, USA, 2013.

45. Sellers, P.J. Relations between canopy reflectance, photosynthesis and transpiration: Links between optics, biophysics and canopy architecture. *Adv. Sp. Res.* **1987**, *7*, 27–44. [CrossRef]

46. Tucker, C.J. Red and photographic infrared linear combinations for monitoring vegetation. *Remote Sens. Environ.* **1979**, *8*, 127–150. [CrossRef]

47. Nowak, D.J.; Crane, D.E.; Stevens, J.C. Air pollution removal by urban trees and shrubs in the United States. *Urban For. Urban Green.* **2006**, *4*, 115–123. [CrossRef]

48. Chiesura, A. The role of urban parks for the sustainable city. *Landsc. Urban Plan.* **2004**, *68*, 129–138. [CrossRef]

49. McBride, J. Mapping Chicago area urban tree canopy using color infrared imagery. In *LUMA-GIS Thesis*; Lund University: Lund, Sweden, 2011.

50. Metro Regional Government RLIS Discovery: ORCA Sites. Available online: http://rlisdiscovery.oregonmetro.gov/metadataviewer/display.cfm?meta_layer_id=3332 (accessed on 20 April 2016).

51. US Census Bureau American FactFinder—About the Data. Available online: http://factfinder.census.gov/faces/affhelp/jsf/pages/metadata.xhtml?lang=en&type=dataset&id=dataset.en.ACS_13_5YR (accessed on 20 April 2016).

52. Sieghardt, M.; Mursch-Radlgruber, E.; Paoletti, E.; Couenberg, E.; Dimitrakopoulus, A.; Rego, F.; Hatzistathis, A.; Randrup, T.B. The abiotic urban environment: Impact of urban growing conditons on urban vegetation. In *Urban Forests and Trees*; Konijnendijk, C., Nilsson, K., Randrup, T., Schipperijn, J., Eds.; Springer-Verlag: Berlin/Heidelberg, Germany, 2005; pp. 281–323.

53. Jim, C.Y. Green-space preservation and allocation for sustainable greening of compact cities. *Cities* **2004**, *21*, 311–320. [CrossRef]

54. Greene, C.S.; Millward, A.A.; Ceh, B. Who is likely to plant a tree? The use of public socio-demographic data to characterize client participants in a private urban forestation program. *Urban For. Urban Green.* **2011**, *10*, 29–38. [CrossRef]

55. Fraser, E.D.G.; Kenney, W.A. Cultural background and landscape history as factors affecting perceptions of the urban forest. *J. Arboric.* **2000**, *26*, 106–113.

56. Buijs, A.E.; Elands, B.H.M.; Langers, F. No wilderness for immigrants: Cultural differences in images of nature and landscape preferences. *Landsc. Urban Plan.* **2009**, *91*, 113–123. [CrossRef]

57. Braverman, I. Everybody loves trees: Policing American cities through street trees. *Duke Environ. Law Policy Forum* **2008**, *19*, 81–118.

58. Revi, A.; Satterthwaite, D.E.; Aragón-Durand, F.; Corfee-Morlot, J.; Kiunsi, R.B.R.; Pelling, M.; Roberts, D.C.; Solecki, W. Urban areas. In *Climate Change 2014: Impacts, Adaptation, and Vulnerability. Part A: Global and Sectoral Aspects. Contribution of Working Group II to the Fifth Assessment Report of the Intergovernmental Panel on Climate Change*; Field, C.B., Barros, V.R., Dokken, D.J., Mach, K.J., Mastrandrea, M.D., Bilir, T.E., Chatterjee, M., Ebi, K.L., Estrada, Y.O., Genova, R.C., et al., Eds.; Cambridge University Press: New York, NY, USA, 2014; pp. 535–612.

59. Paavola, J.; Adger, W.N. Fair adaptation to climate change. *Ecol. Econ.* **2006**, *56*, 594–609. [CrossRef]

60. Mathey, J.; Rößler, S.; Lehmann, I.; Bräuer, A. Urban Green Spaces: Potentials and Constraints for Urban Adaptation for Climate Change. In *Resilient Cities: Cities and Adaptation to Climate Change Proceedings of the Global Forum 2010*; Otto-Zimmerman, K., Ed.; Springer: New York, NY, USA, 2011; pp. 479–485.

61. Fields, B. From green dots to greenways: Planning in the age of climate change in post-Katrina New Orleans. *J. Urban Des.* **2009**, *14*, 325–344. [CrossRef]

62. Tyler, S.; Moench, M. A framework for urban climate resilience. *Clim. Dev.* **2012**, *4*, 311–326. [CrossRef]

63. Hoalst-Pullen, N.; Patterson, M.W.; Gatrell, J. Empty spaces: Neighbourhood change and the greening of Detroit, 1975–2005. *Geocarto Int.* **2011**, *26*, 417–434. [CrossRef]

64. Gobster, P.H.; Westphal, L.M. The human dimensions of urban greenways: Planning for recreation and related experiences. *Landsc. Urban Plan.* **2004**, *68*, 147–165. [CrossRef]

65. Zhang, H.; Jim, C.Y. Species diversity and performance assessment of trees in domestic gardens. *Landsc. Urban Plan.* **2014**, *128*, 23–34. [CrossRef]

forests

Article

Comparative Study of Local and National Media Reporting: Conflict around the TV Oak in Stockholm, Sweden

Johan Östberg [1,*] and Daniela Kleinschmit [2]

[1]   Department of Landscape Architecture, Planning and Management, SLU, Alnarp SE-230 53, Sweden
[2]   Forest and Environmental Policy, University of Freiburg, Freiburg D-79106, Germany;
      daniela.kleinschmit@ifp.unifreiburg.de
*   Correspondence: Johan.Ostberg@slu.se; Tel.: +46-40-41-51-26

Academic Editors: Francisco Escobedo, Stephen John Livesley and Justin Morgenroth
Received: 16 April 2016; Accepted: 30 September 2016; Published: 13 October 2016

Abstract: The TV oak (Television Oak) conflict concerned felling an old tree in a wealthy area of Stockholm. The case received great public attention in different media formats with different scopes (e.g., newspapers, television, internet). The TV Oak issue involved actors with different, partly conflicting perceptions. Assuming that the relevance of urban tree management issues in particular leads to increased interest among the local audience, this paper compared differences in reporting on the TV Oak case in local and national newspapers. The comparison comprised the actors "speaking" in the newspapers, the interest roles attributed to different actors and the frames used. The empirical materials used were articles concerning the TV Oak published between October 2011 and June 2012 in one local and two national Swedish newspapers. Quantitative analysis of statements in these articles showed that the geographical scope of the newspaper was not the major driving force framing the TV Oak conflict and that variety of framings, ranging from a humanised perception of the oak to a more analytical hazard perception, were used. Differences between the interest roles allocated to different actors (e.g., in terms of victim, causer, and helper in the oak conflict) showed that the framing of conflicts very much depended on single actors, in particular a high profile journalist in the national newspapers and private individuals writing letters to the editor in the local newspaper.

Keywords: media analysis; mediated conflicts; framing; newspaper; urban tree cut

---

1. Introduction

Trees contribute to the delivery of many ecosystem services in urban areas, such as moderation of local climate [1–4], storm water management [5,6], recreation and human well-being [7], and cultural value [8]. Long-term management and renewal of urban trees is crucial for sustainable urban development. However, there are conflicts concerning the management of urban trees as they are associated with multiple values (e.g., cultural, historical, and aesthetic) [8–11]. Such conflicts, with the perceptions and values of the stakeholders involved, are shaped in a certain way when they attract the attention of the media. These 'mediatized conflicts' can be understood as a "process, something that occurs through repeated and socially situated interaction between individuals and groups whose perceptions and actions are structured by and expressed through media" [12]. Hence, media reports and constitutes conflicts at the same time [13]. How media depicts a conflict has implications for the way the conflict proceeds and for the manner in which decisions are taken in the conflict [12].

The media is of high importance in contemporary democratic societies around the world and is able to reflect a diversity of possible opinions, providing information and serving as an indicator of public opinion for political decision makers [14]. However, the media also plays a powerful role in selecting news for publication, a process in which it follows its own logic [15,16]. Actors in various conflicts recognise the central position of the media in influencing decision-making and therefore adapt to this logic in order to compete for public attention [17,18].

The central role of the media is even more obvious in the situation of 'mediated conflicts'. Those actors who have successfully placed their statement in the media have a 'standing' as "an actor with voice, not merely as an object being discussed by others" [19]. Actors with standing have substantial opportunities getting support from advocates and changing the minds of their opponents [20,21]. Hence, they are able to shape the political discussion, for example, on ecological management decisions in a certain direction.

Actors with standing use their power to define the problem by selecting and emphasising specific information [22]. This 'framing' gives meaning to complex situations. Media frames are defined as "a central organizing idea or story line that provides meaning to an unfolding strip of events .... The frame suggests what the controversy is about, the essence of the issue" [23].

One way to analyse the framings used is to differentiate the three different interest positions in the conflict: victim, causer, and helper [24,25]. Diagnostic framing implies the two major roles of actors in a conflict, the causers and the victim. Both are preconditions in order to initiate political action [26]. The former is important, as identifying the source of the problem is necessary if the problem is to be remedied, while the latter attributes a particular situation as a problem from which people suffer and which therefore needs remedying. Prognostic framing contains the problem solving dimension of a problem identifying actors as helpers, in other words, those who can contribute to the solution of a problem.

The interest position attributed to actors can either weaken or strengthen their position in the mediated conflict. According to von Prittwitz [27], and further developed by Krumland [28], the interests of the causer are assumed to benefit from a destroying action and causers only rarely gain acceptance in the public discussion. The interests of the victim are twofold, in that they are often attributed a moral advantage, while they can also be perceived as powerless, unable to resolve problematic situations. The interests of the helper are assumed to solve, or help to solve, the problem. The actors attributed this role gain acceptance and legitimacy.

Selecting news for publishing is a central instrument in mediated conflicts. Media studies have identified different factors affecting this selection [29,30]. These news factors are elements that should safeguard the attention of a lay audience in particular. A major assumption in the theory is increasing the number of news factors that can be attributed to a news item increases the probability of it being published and being placed prominently in the media [29]. One of the factors adding value to news is the geographical nearness of the audience to the event reported, with increased interest among an audience in events that are occurring locally. This raises the question of whether there are differences between local and national newspapers in reporting certain events, for example, events with a narrow scope could be assumed to be mainly relevant for local newspaper reporting.

On 24 October 2011 a large oak in Stockholm that was due to be cut down was rescued by a group of demonstrators. The tree was first called the 'Radio Oak' and later on the 'TV Oak' because of its close proximity to the Swedish radio and TV building. This example of urban tree management gathering media attention can contribute to understanding whether resources (e.g., financial or human resources) are the most relevant factor to explain the chance of actors to shape media reporting and whether this influence changes with the scope of the media. Hence, the aim of this paper is to assess how different actors shaped the media debate about the TV Oak conflict in local and national newspaper reporting. This topic is of high relevance not only for the specific situation in Stockholm, but as a phenomenon appearing in many cities in Europe (e.g., 2015 in Sheffield, United Kingdome, where there were protest against the felling of urban trees) but also around the globe (e.g., 2016 in Perth, Australia, where protesters tried to save an old tree from being felled). Hence, the results of this study give implications beyond the specific case of the TV Oak, shedding light on the possible influence of different actors in shaping the media debate and thus influencing the management of (urban) green spaces.

Based on the aim of this paper the following research questions were examined in a comparative analysis of media with different scope (local and national):

(a) Who are the main actors with standing in the mediated conflict of the TV Oak?
(b) Which actors have been attributed the role of causer, victim, and helper in the problem?
(c) Which additional frames were used in the mediated conflict of the TV Oak?

2. Experimental Section

The media analysis was limited to newspapers, although the TV Oak conflict has also been reported in magazines and on TV and discussed on the internet (e.g., in the comments field for the different newspaper articles and on Facebook). Despite the current digital trend, newspapers can still be considered relevant as they are generally recognised as having a major influence on public opinion. In particular, political decision makers rely on newspapers when searching for indicators of public opinion [14]. A total of 165 articles were included in this media analysis (60 from *Dagens Nyheter*, 17 from *Svenska Dagbladet*, and 88 from *Östermalmsnytt*). The articles were thereby rather evenly distributed between national (*Dagens Nyheter* and *Svenska Dagbladet*) and local neighbourhood (*Östermalmsnytt*) newspapers, with 77 and 88 articles, respectively.

The newspaper *Dagens Nyheter* (hereafter DN), which is based in Stockholm, is Sweden's largest quality newspaper (with a 'quality newspaper' referring to a publication which is well-perceived by other journalists and political decision makers), with a readership of around 824,500 people each day. DN is published every day of the week and is politically independent. The newspaper readership is spread over the whole of Sweden, and thus DN is regarded as a national newspaper.

The newspaper *Svenska Dagbladet* (hereafter SvD), which is also based in Stockholm, was selected because it is another national quality newspaper with broad scope (daily readership 500,000), but also has a specific section dedicated to Stockholm news. Like DN, SvD is published every day of the week and is politically independent.

The newspaper *Östermalmsnytt* (hereafter Ösn) is a local newspaper and was selected because it covers the city area of Östermalm in Stockholm, where the TV Oak is located. Ösn is published one day a week (40,900 copies) and is distributed free of charge to all households within the city area of Östermalm. The newspaper is of unknown political opinion.

A media search was conducted on 1 June 2012 on the websites of these three newspapers, using the single search term *"TV-eken"* (the TV Oak). The majority of the articles found were within the period of 15 October 2011 to 1 June 2012. One article dated June 2010 was excluded from the analysis because it did not relate to the TV Oak conflict.

A quantitative content analysis was carried out on the 165 selected articles. Bryman [31] defines content analysis as an "approach to the analysis of documents and texts … that seeks to quantify content in terms of predetermined categories and in a systematic and replicable manner". Content analysis employs a category system, which allows material relevant to the problem dimensions to be coded, thereby reflecting the theoretical considerations. In the present study, two units of analysis were considered: the entire newspaper article and the statements of speakers in the article. With respect to the entire article, the formal categories used included "date of publishing", "author", "section in which the article appeared", and "style of article".

The statements of different speakers were used as the propositional units of analysis. Statements were taken as being all those claims presented as either direct or indirect quotes made by one person or entity in one article, with claims in statements by other people counted as different statements. Consequently, the number of statements was the same as the number of speakers.

The starting assumption of the analysis was that the media pays attention to conflicts. The subjects of the analysis were the actors gaining media representation and the frames attributing different roles to actors. For this reason, we recorded the following three attributes for each statement: (1) Speakers, defined as the person or entity appearing in the article and making the statement; (2) actors identified by the speakers as causer, victim, or helper; and (3) frames used for describing the central perspective of the conflict.

The different actor roles (speaker, causer, victim, and helper) fell within the categories: Scientists, experts, arborists, politicians, administration, police, judiciary, journalists in own media, journalists from other media, non-governmental organizations (NGOs), companies, private individuals, and others (Table 1).

Table 1. Definition of the different actors.

Actor	Examples and/or Definitions
Scientists	Scientists, persons from universities or academic degrees (e.g., researcher and professor).
Experts	People referred to as experts (e.g., tree expert and municipalities tree expert).
Arborists	All persons called arborist.
Politicians	Political speakers (e.g., persons referred to as coming from a political party).
Administration	The municipality's administration and local government.
Police	The police.
Judiciary	Lawyers and attorneys.
Journalists in own media	Journalists from the newspapers.
Journalists from other media	Journalists from the other media (e.g., other papers, TV, and radio).
NGO (Non-governmental organization)	Different types of NGOs (Non-governmental organizations) (e.g., friends of different parks and associations of arborists).
Companies	Different types of companies (e.g., smaller arborist companies and consultant companies).
Private individuals	Letter to the editor and persons being interviewed by the media.
Others	All other speakers.

The causer, victim, and helper roles contained two additional categories concerning those who could not appear as speaker. If they did not have the ability to speak they were categorised as "nature", while if they did not form an actor consistently speaking with one voice they were categorised as "society". However, it was possible to identify them as causer, victim, or helper.

The framing categories were identified inductively from the material [32], by observing particular storylines in the text and combining them into larger categories. The framing categories were: Cultural-historical, art, anthropomorphism (humanisation of the oak), technology, science, hazard, power play (measure of power), and not recognisable.

A coding manual was developed in which all categories and sub-categories were defined, operationalized, and illustrated with examples. Prior testing was carried out to test and improve the coding manual, and two coders were trained to code the articles using the manual. The data obtained were then entered into an Excel spreadsheet (Microsoft Corporation, Washington, DC, USA). Intercoder reliability was continuously confirmed during coding by contact between the two intercoders and between the intercoders and the second author where definitions and interpretations was discussed. The Excel data documents allowed further processing and interpretation of the data.

3. Results

The prominence of the newspaper articles in the different newspapers generally followed a similar trend (see Figure 1). In the following, the major peaks in attention and the reasons for these are described. The timeline of the conflict on the TV Oak is summarised in Table 2. The media attention started with a first peak in week 42 in all three newspapers. That was the first time cutting of the tree was stopped by protesters. In the following period, the peaks of attention varied between the newspapers. Ösn had a second peak in week 44, when a large number of letters to the editor were published. In weeks 46 and 47, there was another peak in attention in all three newspapers after a second attempt to cut down the tree was stopped, an independent Norwegian arborist was contacted, and the tree was finally cut down during a night-time intervention. DN and Ösn had their respective third and fourth peak in attention in week 52, in articles summarising the main events of the year.

DN and Ösn continued reporting, although on a lower level, until the end of our search for articles referring to the TV Oak on 1 June 2012 (Table 2, Figure 1).

Table 2. Calendar week and date of important events surrounding media attention to the TV Oak.

Week and Date	Date and Event
	2011
Week 41, 15 October	News about planned felling of the TV Oak reaches the media.
Week 42, 17 October	Protesters occupy the site around the TV Oak.
Week 43, 24 October	The first attempt to fell the TV Oak is stopped by protesters.
Week 43, 27 October	The second felling is stopped by protesters and the city authorities decide to bring in an arborist from Norway for a second opinion.
Week 45, 9 November	The Norwegian arborist arrives and examines the tree.
Week 46, 17 November	After further investigation by the city authorities, it is decided to cut the tree down anyway, without having received the report from the Norwegian arborist.
Week 47, 21 November	The independent arborist reports that the tree is in no immediate risk of falling.
Week 47, 25 November	The tree is cut down in the middle of the night by the city authorities, accompanied by a large group of police officers.
	2012
Week 9, 28 February	Three experts are flown in from Germany, Denmark, and England. Together they carry out further investigations of the tree trunk and the cut remains.
Week 11, 14 March	The city authorities reveal two proposals for a new development at the site of the TV Oak. Both propositions include a tram track straight over the site of the TV Oak.
Week 16, 20 April	Closing day for responses in public consultation on the city's development plan.
Week 18, 4 May	The planning proposals are handed over to the city board for a final decision.

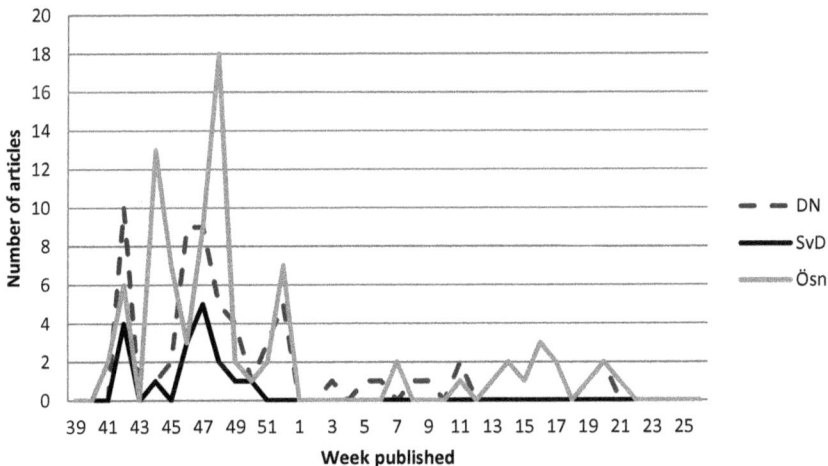

Figure 1. Number of articles per week including the term *"TV-eken"* (the TV Oak) in the newspapers *Dagens Nyheter* (DN), *Svenska Dagbladet* (SvD), and *Östermalmsnytt* (Ösn) during the search period.

3.1. Speakers

The 165 articles found contained 470 statements, 204 in DN, 65 in SvD, and 201 in Ösn. Major actors appearing as speakers in the media discussion were journalists, private individuals, city administration officials, experts, and NGOs. In both the national newspapers, the journalists themselves appeared as the major representatives of the 'speaker' category. In the case of DN, the category 'journalist' was mainly represented by one specific person (41% of DN statements). In contrast, private individuals dominated the discussion on the TV Oak conflict in Ösn and journalists were in second place. This was mainly the result of a large number of letters to the editor. In the national newspapers, administrations were seen as representing the speaker category, with the second largest number of statements. A major actor within this group giving an opinion in the media was Stockholm Parks and Streets Department, which is responsible for urban trees and safety issues relating to traffic. Other speakers given a high degree of attention in the national newspapers were experts and NGOs. Those appearing as experts were assigned this title by the journalist in question and they were often individuals specifically dealing with trees in the urban area. The NGOs comprised a variety of groups, from those working more generally on environmental and aesthetic issues to those formed only for the sake of the TV Oak (e.g., Friends of the TV Oak). In the local newspaper, the main speakers in TV Oak reporting were administration and experts, followed by private individuals and journalists. NGOs did not appear often as speaker in this forum (Table 3).

Table 3. Total number of speakers in the newspapers DN, SvD, and Ösn and percentage distribution in different speaker groups.

Speakers	DN	SvD	Ösn	Mean
Total Number of Speakers	*204*	*65*	*208*	*159*
Administration	16%	18%	13%	15%
Arborist	8%	5%	4%	6%
Experts	10%	9%	9%	9%
Journalists in own media	26%	29%	21%	24%
Journalists from other media	2%	8%	1%	2%
NGO (Non-governmental organization)	11%	8%	5%	8%
Politicians	8%	3%	6%	6%
Scientists	2%	1%	1%	2%
Private individuals	4%	8%	31%	17%
Other	13%	11%	9%	11%
Total	**100%**	**100%**	**100%**	

3.2. Interest Positions

3.2.1. Victim

A victim of the TV Oak conflict was mentioned in 187 statements. The most frequently mentioned victim in all newspapers was the TV Oak itself (141/187, 75%): "Ancient oak threatened" (Kristoffer Törnmalm, SvD journalist, SvD 23 October 2011). In DN, the notion of the oak as the victim prevailed (86%), in comparison with 77% in SvD and 63% in Ösn. In all newspapers, private individuals were also named as victims in the case. However, SvD most often referred to private individuals as potential victims, taking into account that falling tree branches could injure bypassers and citing: "The tree is rotten, completely dead and may fall at any time and it is thus lethal" (Mats Frej, spokesman at Stockholm Parks and Streets Department, SvD 18 November 2011).

In contrast, Ösn presented administration as victims (in 15%), referring to a case in which city staff had been threatened (e.g., in the article "Tree hugger under attack" (Hanna Svensson, Ösn 10 December 2011).

3.2.2. Causer

Causers were mentioned less often (110 times) than victims in the reporting about the TV Oak, perhaps because the allocation of this role was less obvious than in the case of victim. The actors most commonly mentioned as causers of the problem were the city authorities (38 times), nature (25 times), and the oak (13). DN (43%) and Ösn (30%) mainly referred to administrations as those causing the problem as they were the ones who planned to cut down the oak: "The Traffic office takes the easy way out and levels the tree to the ground. I think this is euthanasia, instead of offering the help needed." (Maria Antinsson, private individual, interviewed in Ösn 29 October 2011). SvD mentioned administrations as causer in 19% of statements. In contrast, SvD cited nature as the main causer (50%), drawing attention to fungi attacking the oak by including the statement: "According to measurements on the amount of decay in the tree, it is in danger of collapsing" (Britt-Marie Alvem, Stockholm Parks and Streets Department, SvD 23 October 2011). However, DN and Ösn also mentioned nature as causer (20% and 16% of statements, respectively). DN and SvD directly depicted the oak as the causer of the problem too (16% and 19%, respectively), referring to problems with traffic, as the roads were closed down because branches or the whole tree could fall down.

3.2.3. Helper

Most statements did not pinpoint a helper in the TV Oak problem. A helper was named in only 29 of the 470 statements, and the most frequently mentioned helper in these were politicians (31%) and experts or private individuals (10% each). Most statements mentioning a helper were published in DN, with 17 statements, where the most frequently named helper was from the politicians category (29% of relevant statements). Politicians were most frequently mentioned as the helper in Ösn too, in 33% of the total of nine statements mentioning helpers. Administrations were the second most mentioned helper, with 22% of the 470 statements. SvD contained only three statements mentioning a helper, the dominant helper mentioned being private individuals (67%) and politicians (33%).

3.3. Framing the TV Oak

In the following, we only present the results from the three frames most often used in the discussion about the TV Oak. The main framing was hazards, which accounted for 40% of statements. The cultural-historical perspective was the second most common frame, with 36%, and anthropomorphism third, with 17%.

The hazard framing was type most commonly used in DN (46% of statements) and SvD (52%), but only ranked third in Ösn, with 28%. In DN, cultural history was the second most common framing, with 3%, while anthropomorphismaccounted for 8%. In SvD, cultural history was ranked second, with 28% and anthropomorphism was third, with 12%. In Ösn, the most common framing was cultural history (40%) followed by anthropomorphism (31%) and hazards (28%). In the following we describe these frames in more detail (Table 4, also see Appendix A for the number of statements for the different article styles).

Table 4. Number of statements in each newspaper and framing of the statements, shown as a percentage for each framing.

Framing	DN (*Dagens Nyheter*)	SvD (*Svenska Dagbladet*)	Ösn (*Östermalmsnytt*)	Mean
Number of statements	*83*	*25*	*65*	*58*
Art	1%	8%	4%	3%
Cultural-historical	24%	24%	29%	26%
Anthropomorphism	8%	12%	31%	17%
Power play (measure of power)	10%	4%	0%	5%
Hazards	46%	52%	28%	40%
Science	3%	0%	0%	1%
Technology	8%	0%	8%	7%
Total	**100%**	**100%**	**100%**	

3.4. Hazard Frame

The hazard frame used mainly referred to the risk of injuring people and sometimes human-made objects (cars). In comparison, loss of culture was not described as a risk. One of the main arguments put forward by administrations to justify felling the TV Oak was that hazards were associated with not felling it. The administrations, and more specifically the Parks and Streets Department, argued that the amount of dead wood and major fungal infection had made the oak a hazard and this was the reason why it was finally cut down.

When analysing the statements in the newspapers, 77 (40%) of all statements on framing mentioned the presence of a risk. Of these, 61% stated that there was a risk (DN 41%; SvD 80%; Ösn 68%), for example: "According to the Parks and Streets Department, the oak poses a dangerous risk and must therefore be cut down" (SvD 17 November 2011) and "The traffic department's tree expert Björn Embrén says that the TV-oaks status is worse than he first thought and that it needs to be cut down immediately since it poses a threat to the public" (DN 17 November 2011). The remaining 39% of the statements referring to a hazard stated that there was no risk (DN 59%; SvD 20%; Ösn 32%), for example: "Experienced arborists: The TV Oak will not collapse" (header, DN 20 November 2011). In DN this resulted mainly from a series of articles (28% of all relevant articles published in DN) published by one specific journalist.

3.5. Cultural-Historical Perspective

As Table 4 shows, 26% of the framings used concerned the cultural-historical background of the TV Oak (DN 24%; SvD 24%; Ösn 29%). According to these statements, the TV Oak spanned more than three centuries of history and was therefore a contemporary witness of important historical events: "The experiences of the TV Oak", SvD 24 November 2011. Beyond this historical perspective, the oak was also presented as a rare object in the city environment, not least because of its shape: "The oak is a piece of art" (Åke Askensten, politician, SvD 3 November 2011). This framing resulted in the claim "Leave it in place and declare it a culture heritage!" (Rolf Lindell, politician, DN 21 October 2011). The cultural-historical frame also included references to similar past events, where trees had been at the centre of urban management conflicts, for example, in the past demonstrations concerning the elms in Kungsträdgården: "The Elm Battle in Kungsträdgården showed the ruling government in 1968 that there was a civil movement to be reckoned with. In 2011, the parallel is of course the somewhat pathetic TV Oak" (Joakim Högström, The Year's Events, Ösn 17 December 2011).

3.6. Personification of the Oak

The TV Oak was framed in some statements as a human being, in particular in Ösn (31% of all framings). This framing was mainly used by those who opposed cutting down the tree. However, even those in favour partly framed the tree in a humanised way, for example, by referring to sickness in the tree: "But it (*oak*) feels really sick" (Björn Embrén, Stockholm Parks and Streets Department, DN 25 October 2011), and asking for it to be cut down in a respectful way: "The oak feels better where it belongs" (Louise Treschow, private individual, Ösn 5 November 2011). The anthropomorphism (humanisation) framing of the oak was even more prominent after it had been cut down. The newspaper reporting referred to the felled tree as if it were a dead human being, for example, by discussing what to do with the wood "The TV Oak's last resting place" (Susanna Baltscheffsky, SvD 02 December 2011). This perspective was supported by an event at the site of the oak after it had been cut down: "When the action group on Saturday held a "memorial" with light (!), placement of flowers (!!) and a minute's silence (!!!), we experienced Stockholm's most bizarre event in a long time" (editorial column, DN 02 December 2011). As the oak was described as a dead person, the administrations that had it cut down were referred to as murderers. Furthermore, the oak was described as a person with family: "Born: A Radio Oak, March 2012. The little baby is the child of the 800-year-old Radio Oak, also called the TV Oak or Vädlaeken, which the Parks and Streets Department killed in November" (Kerold Klang, opinion column, Ösn 7 April 2012).

During the conflict threats were aimed at different city officials. One example is "The Parks and Streets Department have charged an environmental politician for threating a city official" (DN 17 December 2011).

4. Discussion

The TV Oak conflict raised public attention in the media, perhaps partly as a result of the appeal of conflict issues in reporting [17]. However, the broad attention it raised beyond the local level is surprising for a local conflict such as the TV Oak. There are several possible reasons for this high media attention. First, the oak, with its close proximity to the national radio and TV house, was highly visible to journalists. Although the DN and SvD offices are not located in the direct neighbourhood of the oak, the media attention drawn to this issue by radio and television journalists might have led to a multiplier effect, resulting in increased attention among other media sources too. Furthermore, even the name "TV Oak" hints at a specific relationship between the media and the tree. Reporting the conflict about this issue allowed journalists to refer to "their own" tree, in this way being self-referential [28]. This argument is supported by the fact that in radio reporting the tree was referred to as the Radio Oak. Secondly, the media interest in the TV Oak conflict might have originated from one specific journalist with a particular interest in the issue. Evidence of this is provided by the minor number of articles published in SvD (total of 17 articles), in comparison with the 60 articles published in DN, 17 of which were written by one particular journalist.

The attention of the media changed with events connected to the TV Oak and there was no difference between the national and local neighbourhood newspaper in this respect. According to literature on media attention, after an event the discussion can be expected to abate [33]. However, in the present case the discussion about the TV Oak continued, although on a lower level. Even now, reference is made to the TV Oak conflict in the media (e.g., in an article on demolishing an old staircase in Stockholm (DN, 15 February 2013)). This shows that conflicts on urban tree management, though no longer active, are reproduced in media reporting for a long time, as also seen with the Elm Battle in Stockholm, and thereby kept in people's minds. This is important to acknowledge from an urban tree management point of view since the media will reproduce previous conflict on urban tree management when a new conflict arises, and this thereby, to some extent, might shape the framing of the new conflict.

In the national media, journalists were the main speakers in the discussion on the TV Oak. This result is supported by other studies of conflict issues in quality newspapers (e.g., [24]). The high number of journalists among the speakers resulted from them introducing an issue, leading from one argument to another and then giving final statements and their own opinions in reports and commentaries. This dominant position indicates that if there is a specific journalist in charge for a specific issue, a certain bias might be created. Private individuals were the main speakers in the local newspaper, as a result of letters to the editor. The amount of direct experience increases with proximity to the site of an event and allows private individuals to share their experiences and values. Furthermore, free local newspapers such as Ösn usually do not have the resources to contract journalists for all reporting issues and instead they depend on contributions from other sources. Because of the dominance of private individuals, the strict differentiation between journalists and letter writers is not reflected in local reporting any more [34]. In such cases, journalists become the managers and coordinators of information rather than seeking the information themselves.

The TV Oak was most often mentioned as a victim of the conflict. This is not surprising, as the oak was the main subject of the conflict. Attributing the role of victim includes two perspectives: first, a problem has mobilisation potential if someone is already suffering; and second, the role as a victim can lead to an actor being characterised as weak and lacking the ability to help itself. If the tree is attributed the role of victim according to the first interpretation the mobilisation potential is retained, but in the second it becomes a factor that can be disregarded.

The major difference between the national and local newspapers was that the latter more often named the authorities, and in particular the Parks and Streets Department, as a victim of the problem. This resulted from stressing the topic of attacks on officials from the Parks and Streets Department in reporting. This issue was the focus of three major articles, including interviews with those threatened. In comparison, the national newspapers just reported once about the threatened attacks. It can be assumed that the reason for weighting the threats differently lay with the relationship between the media and their informants. Given the detailed information about the threat and the fact that it was the first newspaper to report about it, despite being a local weekly publication, Ösn appears to have had good contact with the Parks and Streets Department, which offered the latter an opportunity to gain acceptance for its position. These contacts can therefore be of importance when urban tree managers want to gain acceptance for specific management actions. Contact between the media and the political administration system, in this case the Parks and Streets Department, is generally a phenomenon that benefits the media too, as there is plenty of information that can be assumed to be trustworthy and might become relevant for publication [35].

In regards to causers, the newspapers referred to different sources. DN and Ösn cited the authorities as causer, while SvD referred to fungi. It can be assumed that the first increases the mobilisation potential of the conflict more than the latter, as it identifies the source of the problem to remedy. However, depicting the authorities as causer does not pinpoint a specific authority, but puts the blame on them in general.

Helpers of the problem were only mentioned in 6% of all statements, while all other statements did not mention any solutions to the problem. This means that prognostic framing was not used to any great extent in the reporting, as those statements referring to helpers mainly named politicians as those who could help solve the problem. The finding that problems described in the media are meant to be solved by politicians has also been made in previous media studies [24].

The most often used frame was that associating the oak with hazards, with some reported as claiming that there was a risk and others denying the risk. However, our analysis revealed that DN was the only newspaper in which the majority of the statements claimed that there was no risk. This perception resulted mainly from one journalist advocating the no risk case. This is further evidence of the fact that the media highly influences processes by the way in which it describes them. According to media theory, this means the media is not an independent but an intervening variable, providing an additional dynamic and can influence the reporting in a certain direction [21]. The finding in the present case that a frame can depend on one journalist further highlights the powerful position of the media. The dominant perspective of the other two newspapers in the hazard framing context was that the TV Oak presented a hazard. Thus, the local-national newspaper dichotomy did not explain the difference in reporting, but rather the approach taken by one specific journalist.

Speakers use different strategies to increase acceptance of their position. In particular, collective symbols and metaphors are used in the media, as they help simplify the content. Attributing importance to the cultural-historical aspect of the TV Oak assisted in adaptation to existing patterns of interpretation in the culture of the media and the audience [28]. This meant that the TV Oak was linked to existing cultural and historical ideas, such as the Elm Battle. Based on this interpretation, the cultural-historical aspects cannot be assumed to be the reason for the reporting, but were used as an instrument to gain attention and acceptance. The cultural or historical aspects highlighted depend on the position, subculture, and values of the respective speaker [36].

The anthropomorphism of the tree by those who were against cutting it down appeared as a dominant rhetoric in the reporting, especially by those articles published in Ösn. Whether this has been a strategy to gain acceptance for a specific position in the TV Oak conflict or if this is an expression of the relatedness to the nonhuman environment remains an open question. In general, the perception of an object as humanlike allows for the potential to evoke feelings of empathy and by doing so, allows for moralization [37]. The fact that one of the DN editorials made fun of the personification of the oak shows that the anthropomorphism of a tree is not undisputed in a society which is mainly dominated by the program of modern science [37].

5. Conclusions about Local-National Reporting

The local newspaper, in this study Ösn, was characterised by the dominance of letters to the editor. In general, this allowed a more open discussion and showed the diversity of possible opinions in the local community. In contrast, the national newspapers depended very much on a single journalist for the story, framing the reporting in a specific direction. However, despite the huge range of letters to the editor in the local newspaper, the specific interest roles allocated to different actors involved in the TV Oak conflict were similar in the national and local newspapers when it came to causers and helpers of the problem. One exception was statements naming private individuals as causers and complaining about protesters being responsible for traffic problems in the area. In regards to allocating the role of the victim, the empirical results revealed that the local newspaper represented the city authority in question (Parks and Streets Department) as being threatened. The results of this study can be used to gain a deeper understanding of conflicts surrounding urban trees around the world. The results indicated good contact between the local newspaper and the authorities, resulting in a specific direction of the reporting. The results of the framing analysis identified different priorities of perspectives in the local and national newspapers. The national newspaper devoted more attention to the hazards associated with the TV Oak, whereas the local newspaper tended more towards an anthropomorphism of the oak, owing to the tone taken in many letters to the editor.

Author Contributions: For research articles with several authors, a short paragraph specifying their individual contributions must be provided. The following statements should be used "Johan Östberg and Daniela Kleinschmit conceived and designed the experiments; Johan Östberg performed the experiments; Johan Östberg and Daniela Kleinschmit analyzed the data; Johan Östberg and Daniela Kleinschmit wrote the paper." Authorship must be limited to those who have contributed substantially to the work reported.

Conflicts of Interest: The authors declare no conflict of interest.

Appendix A

Table A1. Number of statements for the different article styles and framing of the statements, shown as a percentage for each framing.

Style of Article	Front Page	Report	Announcement	Interview	Reportage	Letters to the Editor	Editorial or Comment	Other
Number of Statements	*2*	*101*	*11*	*14*	*1*	*26*	*8*	*10*
Art	0%	4%	0%	14%	0%	0%	0%	0%
Cultural-historical	0%	23%	18%	21%	0%	38%	50%	30%
Anthropomorphism	50%	11%	18%	21%	0%	31%	25%	30%
Power play (measure of power)	0%	7%	9%	0%	0%	0%	13%	0%
Hazards	50%	48%	45%	29%	100%	19%	13%	40%
Science	0%	1%	0%	0%	0%	4%	0%	0%
Technology	0%	7%	9%	14%	0%	8%	0%	0%
Total	**100%**	**100%**	**100%**	**100%**	**100%**	**100%**	**100%**	**100%**

References

1. Dimoudi, A.; Nikolopoulou, M. Vegetation in the urban environment: Microclimatic analysis and benefits. *Energy Build.* **2003**, *35*, 69–76. [CrossRef]
2. Nowak, D.J.; Noble, M.H.; Sisinni, S.M.; Dwyer, J.F. People and trees: Assessing the US urban forest resource. *J. For.* **2001**, *99*, 37–42.
3. Nowak, D.J.; Crane, D.E.; Stevens, J.C. Air pollution removal by urban trees and shrubs in the United States. *Urban For. Urban Green.* **2006**, *4*, 115–123. [CrossRef]
4. Yokohari, M.; Brown, R.D.; Kato, Y.; Yamamoto, S. The cooling effect of paddy fields on summertime air temperature in residential Tokyo, Japan. *Landsc. Urban Plan.* **2001**, *53*, 17–27. [CrossRef]
5. Bolund, P.; Hunhammar, S. Ecosystem services in urban areas. *Ecol. Econ.* **1999**, *29*, 293–301. [CrossRef]
6. Xiao, Q.; McPherson, E.G. Rainfall interception by Santa Monica's municipal urban forest. *Urban Ecosyst.* **2002**, *6*, 291–302. [CrossRef]
7. Todorova, A.; Asakawa, S.; Aikoh, T. Preferences for and attitudes towards street flowers and trees in Sapporo, Japan. *Landsc. Urban Plan.* **2004**, *69*, 403–416. [CrossRef]

8. Alcamo, J.; Hassan, R.; Bennett, E. *Ecosystems and Human Well-Being: A Framework for Assessment/Millennium Ecosystem Assessment*; Alcamo, J., Ed.; Island Press: Washington, DC, USA, 2003.
9. Morgenroth, J.; Östberg, J.; van den Bosch, C.K.; Nielsen, A.B.; Hauer, R.; Sjöman, H.; Chen, W.; Jansson, M. Urban Tree Diversity—Taking Stock and Looking Ahead. *Urban For. Urban Green.* **2016**, *15*, 1–5. [CrossRef]
10. Östberg, J.; Delshammar, T.; Wiström, B.; Nielsen, A.B. Grading of Parameters for Urban Tree Inventories by City Officials, Arborists, and Academics Using the Delphi Method. *Environ. Manag.* **2013**, *51*, 694–708. [CrossRef] [PubMed]
11. Randrup, T.B. Development of a Danish model for plant appraisal. *J. Arboricult.* **2005**, *31*, 114–123.
12. Hutchins, B.; Lester, L. Theorizing the enactment of mediatized environmnetal conflicts. *Int. Commun. Gaz.* **2015**, *77*, 337–377. [CrossRef]
13. Cottle, S. *Mediatized Conflict: Developments in Media and Conflict Studies*; Open University Press: Maidenhead, UK, 2006.
14. Kleinschmit, D.; Krott, M. The media in forestry: Government, governance and social visibility. In *Public and Private in Natural Resource Governance: A False Dichotomy?*; Sikor, T., Ed.; Earthscan: London, UK, 2008.
15. Altheide, D.L.; Snow, R.P. *Media Logic*; Sage: London, UK, 1979.
16. Esser, F.; Matthes, J. Mediatization Effects on Political News, Political Actors, Political Decisions, and Political Audiences. In *Democracy in Teh Age of Globalization and Mediatization*; Kriesi, H., Bochsler, D., Matthes, J., Lavenex, S., Bühlmann, M., Esser, F., Eds.; Palgrave Macmillan: Basingstoke, UK, 2013; pp. 177–201.
17. Schulz, W. Reconstructing Mediatization as an Analytical Concept. *Eur. J. Commun.* **2004**, *1*, 87–101. [CrossRef]
18. Frederiksson, M.; Schillemans, T.; Pallas, J. Determination of organizational mediatization: An analysis of the adaptation of Swedish government agencies to news media. *Public Adm.* **2015**, *93*, 1049–1067. [CrossRef]
19. Ferree, M.M.; Gamson, W.A. *Shaping Abortion Discourse. Democracy and the Public Sphere in Germany and the United States (Communication, Society and Politics)*; Cambridge University Press: Cambridge, UK, 2002.
20. Kepplinger, H.M.; Brosius, H.-B.; Staab, J.F. Instrumental Actualization: A Theory of Mediated Conflicts. *Eur. J. Commun.* **1991**, *6*, 263–290. [CrossRef]
21. Gerhards, J.; Schäfer, M. Is the Internet a better Public Sphere? Comparing old and new media in the US and Germany. *New Media Soc.* **2010**, *12*, 143–160.
22. Schön, D.A.; Rein, M. *Frame Reflection: Toward the Resolution of Intractable Policy Controversies*; Basic Books: New York, NY, USA, 1995.
23. Gamson, W.A.; Modigliani, A. The Changing Culture of Affirmative Action. In *Research in Political Sociology*; Braungart, R.G., Braungart, M.M., Eds.; JAI Press: Greenwich, UK, 1987.
24. Feindt, P.H.; Kleinschmit, D. The BSE crisis in German newspapers: Reframing responsibility. *Sci. Cult.* **2011**, *20*, 183–208. [CrossRef]
25. Kleinschmit, D.; Sjöstedt, V. Between science and politics: The Swedish newspaper reporting on forests in a changing climate. *Environ. Sci. Policy* **2014**, *35*, 117–127. [CrossRef]
26. Benford, R.; Snow, D. Framing Processes and Social Movements: An Overview and Assessment. *Annu. Rev. Sociol.* **2006**, *26*, 611–639. [CrossRef]
27. Von Prittwitz, V. *Das Katastrophenparadox: Elemente Einer Theorie der Umweltpolitik*; Leske + Budrich: Opladen, Germany, 1990.
28. Krumland, D. *Beitrag der Medien Zum Politischen Erfolg*; PeterLang: Frankfurt, Germany, 2004.
29. Galtung, J.; Ruge, M. The structure of foreign news: The presentation of the Congo, Cuba and Cyprus crises in four Norwegian newspapers. *J. Peace Res.* **1965**, *2*, 64–91. [CrossRef]
30. Staab, J.F. The Role of News Factors in News Selection: A Theoretical Reconsideration. *Eur. J. Commun.* **1990**, *4*, 423–443. [CrossRef]
31. Bryman, A. *Social Research Methods*, 3rd ed.; Oxford University Press: Oxford, UK, 2008.
32. Chinn, P.L.; Kramer, M.K. *Theory and Nursing: A Systematic Approach*; Mosby Year Book: St. Louis, MO, USA, 1999.
33. Downs, A. The Issue-Attention Cycle and The Political Economy of Improving Our Environment. In *The Political Economy of Environmental Control*; Downs, A., Bain, J., Ilchman, W., Eds.; University of California Press: Berkley, CA, USA, 1972; pp. 9–34.
34. Grimme, E. *Zwischen Routine und Recherche: Eine Studie über Lokaljournalisten und Ihre Informanten*; Westdeutscher Verlag: Opladen, Germany, 1990.

35. Herman, E.S.; Chomsky, N. *Manufacturing Consent: The Political Economy of the Mass Media*; Pantheon Books: New York, NY, USA, 2002.
36. Lakoff, G.; Johnson, M. *Metaphors We Live by*; The University of Chicago Press: Chicago, IL, USA, 1980.
37. Gebhard, U.; Nevers, P.; Billmann-Mahecha, E. Moarlizing Trees: Anthromorphism and Identity in Children's Relationshsips to Nature. In *Identity and the Natural Environment: The Psychological Significance of Nature*; Clayton, S., Opotow, S., Eds.; MIT Press: Cambridge, UK, 2003; pp. 91–112.

forests

Editorial

The Biodiversity of Urban and Peri-Urban Forests and the Diverse Ecosystem Services They Provide as Socio-Ecological Systems

Stephen J. Livesley [1,*], Francisco J. Escobedo [2] and Justin Morgenroth [3]

[1] School of Ecosystem and Forest Science, Faculty of Science, Burnley campus, The University of Melbourne, Victoria 3121, Australia

[2] Functional and Ecosystem Ecology Unit, Faculty of Natural Sciences and Mathematics, Universidad del Rosario, Kr 26 No. 63B-48, Bogotá D.C. 111221492, Colombia; franciscoj.escobedo@urosario.edu.co

[3] School of Forestry, College of Engineering, The University of Canterbury, Christchurch, New Zealand; justin.morgenroth@canterbury.ac.nz

* Correspondence: sjlive@unimelb.edu.au

Received: 21 October 2016; Accepted: 6 November 2016; Published: 24 November 2016

Abstract: Urban and peri-urban forests provide a variety of ecosystem service benefits for urban society. Recognising and understanding the many human–tree interactions that urban forests provide may be more complex but probably just as important to our urbanised society. This paper introduces four themes that link the studies from across the globe presented in this Special Issue: (1) human–tree interactions; (2) urban tree inequity; (3) carbon sequestration in our own neighbourhoods; and (4) biodiversity of urban forests themselves and the fauna they support. Urban forests can help tackle many of the "wicked problems" that confront our towns and cities and the people that live in them. For urban forests to be accepted as an effective element of any urban adaptation strategy, we need to improve the communication of these ecosystem services and disservices and provide evidence of the benefits provided to urban society and individuals, as well as the biodiversity with which we share our town and cities.

Keywords: urban ecology; urban landscape; climate change adaptation; climate change mitigation; tree canopy cover; urban planning

1. Introduction

The term "ecosystem service" is used prodigiously with respect to current urban ecosystems and in urban forest research [1]. These studies often regard ecosystem functions and benefits in an abstract fashion, without connecting with, or considering fully, the human–environment interactions that pervade our urban landscapes. Furthermore, many of these studies frequently refer to ecosystem services without attempting to quantify those services or qualify what enhances—or detracts—from the level of benefits to society. These two issues may be holding back the use of urban trees as a strategy that can help tackle many of the "wicked problems" that urban society faces, such as climate change, community welfare and wellbeing, and biodiversity conservation and management. The aim of this Special Issue is to help fill this void in the current research by focusing on the diversity of urban forests and the benefits (e.g., cultural, environmental, personal, and economic) that different societies across the world gain through the biodiversity and ecological functions that urban and peri-urban forests provide.

This Special Issue was conceptualised by Professor Francisco Escobedo as a means to progress our global discussion of urban forest function using a more social-ecological approach. His proposal for this Special Issue coincided with preparations for the 2nd International Conference on Urban Tree Diversity, held in Melbourne in February 2016, and co-organised by Stephen Livesley and Justin Morgenroth amongst others. Despite the simple title, this conference aimed to provide a research and management platform to discuss the many "diverse" services and functions that urban trees provide us and our urban landscape. As such, this Special Issue was promoted at that Conference, through our combined research networks and by MDPI itself. We have been able to bring together a large number of international studies covering a wide spectrum of ecosystem services, ecological functions that urban trees and urban forests can provide—from supporting faunal biodiversity to the diversity of urban forests themselves; from urban forests for carbon sequestration to air quality improvements through particulate deposition; from indicators of resilience and health in urban forest planning to socio-economic drivers and inequity in urban forest cover. This Special Issue includes research performed in every continent except Antarctica. These studies originate from the USA, Germany, Canada, Colombia, Sweden, South Korea, Chile, South Africa, Mexico, Italy, and China. We could not have hoped to create a more internationally inclusive and relevant Special Issue, and are very proud to present as Guest Editors this collection of urban forest studies.

2. Human–Tree Interactions in an Urbanised Society

Östberg and Kleinschmit [2] describe the role of the media in reporting and maintaining local and national interest in the removal of a significant urban tree in Stockholm, Sweden. This case study highlights the important role of "champions", be they from the media or private individuals making shrewd use of the media. Changes to urban forests can lead to passionate protest and demand, and this can come from any one of the many stakeholders concerned with the vegetation environment of their local street, neighbourhood, or city. A good way to minimise confusion, anger, and protest is to provide information in advance of tree-related changes, to educate communities and stakeholders on the issues at hand, and to consult and involve them in the high-level planning and decision-making for that change. This has been successfully demonstrated by several cities across the world including the City of Melbourne, through their exhaustive round of neighbourhood workshops to communicate, educate, consult, and co-plan the future of their urban forests [3].

The changing face of urban forest management and consultation is further investigated by Barron et al. [4] in their contribution to the Special Issue that looks at the disconnect between what we as urban forest managers or researchers measure and monitor and what we actually expect or want urban forests to deliver. With greater management and public demands from our urban forests comes the need for clear indicators of performance that can track progress and the success or failure of initiatives and interventions. Barron et al. [4] tackle this issue using the Delphi method to rank issues and indicators that international urban forest managers or researchers regarded as important, followed by targeted interviews with Canadian urban forest professionals. The study noted that many indicators regarded as being of "high importance" are not being measured in many municipal urban forestry programs, particularly social indicators of human health and well-being. This is a real concern for managers seeking (or being asked) to track the efficacy of funded urban forest programs to deliver the socio-ecological and ecosystem service benefits they claim and expect.

3. Urban Tree Inequity

It is now widely recognised that there can be considerable inequity within urban society as to access to green space or urban forests and tree cover itself [5]. In this Special Issue, Nesbitt and Meitner [6] assess the correlation between urban vegetation cover in Portland, Oregon, and socio-economic variables collected in the census of the United States. Neighbourhoods of higher population density, lower average household income, and fewer residents identifying as white or Asian had less vegetation cover. This study provides more evidence that green inequity is a very real phenomenon, and future research needs to tackle what impacts this inequity may be having upon physical and mental health and well-being. The approach presented in this study provides a guide that can be used to identify and target areas that need urban forest intervention to address the stark inequity in urban greening. Escobedo et al. [7] add a temporal level of understanding to this issue of green inequity through a study of spatial and temporal dynamics in Santiago's urban forest over 12 years. Average tree mortality and overall tree basal area remained stable across the city, whereas tree canopy cover and basal area increased in the more affluent suburbs, whilst decreasing in the intermediate- and low-income suburbs. The study further reinforces the observation that green inequity is a universal issue, and a contemporary issue as tree canopy cover is changing now and progressing towards increasingly negative outcomes.

Tackling green inequity will mean conserving the trees and green space that exist, whilst adding new tree plantings and hopefully new green spaces in the areas of our cities that need it most. Widney et al. [8] examine the growth and survival of urban tree planting initiatives in three US cities (Detroit, Indianapolis, and Philadelphia) to model the expected ecosystem service benefits 5 and 10 years in the future. The news is not good, because the current (and accepted) levels of planted tree mortality in these three cities means that these new tree planting initiatives cannot keep up with concurrent mortality and the loss of the larger "legacy" trees already in the urban landscape. Widney et al. [8] make a plea for improved and early intervention measures to raise tree survival rates in those crucial establishment years, so that the social, ecological, and ecosystem service benefits these trees were planted to maintain, if not increase, can be realised.

4. Climate Change Mitigation through Carbon Sequestration in Our Own Neighbourhoods

Mitigation of global climate change may not be the most recognised function that urban forests can provide for society at a global scale, but there is great regional and local interest in the carbon sequestration potential of urban vegetation systems both above- and below-ground. This is probably because society needs more information so as to become more pro-active and empowered as to how green space and vegetation in "their" landscape can help in some way. In Colombia, Clerici et al. [9] developed a cost-effective method combining high-resolution, remotely sensed imagery classification with ground-truthed plot data to estimate and monitor the above-ground tree biomass and carbon stocks in peri-urban Andean forests. In China, Lv et al. [10] studied above- and below-ground carbon stocks in more than 200 plots and surmised that soil carbon increases in urban green spaces have sequestered an additional 25% on top of that stored above-ground in the existing or planted urban forests of the Harbon City region. In a similar study of South Korean cities that have developed rapidly in recent decades, Yoon et al. [11] were able to estimate soil carbon density in a range of urban green space and forest types and then scale up to male whole-of-city estimates for Seoul, Daegu, and Daejeon.

5. Urban Biodiversity: The Trees Themselves and the Fauna Habitat They Provide

Interest in maintaining and even enhancing biodiversity within urban landscapes is increasing, not only for the inherent value of biodiversity conservation itself but also because of the tangible societal benefits (e.g., environmental awareness, and the mental health and well-being) realised from viewing and interacting with biodiversity. MacGregor-Fors et al. [12] report on an extensive city-wide study of fauna and flora biodiversity in Mexico covering ten taxonomic groups in a very little studied region of the world. They are able to relate species richness to key size and location traits of the urban green space and forests that they measured. A common and passionate debate that runs throughout urban biodiversity research relates to the use of exotic, native, or indigenous plant species as the cornerstone of faunal biodiversity habitats [13]. In this Special Issue, Shackleton [14] contributes to this debate with a simple but intriguing study of over 1200 street trees in Grahamstown, Eastern Cape, South Africa. Shackleton [14] is able to demonstrate the importance of native trees for bird species richness and abundance. However, at the same time, exotic trees are important for supporting parasitic mistletoes that provide interesting habitats for invertebrates in their own right, and as such, foraging resources to insectivorous animals. This study adds a layer of complexity to the debate of urban forests being "novel ecosystems" and reiterates that the native-and-exotic tree debate in novel urban landscapes is far from black and white. Nitoslawski and Duinker [15] look at the diversity of urban forests themselves, and again through a native and exotic lens. By assessing the impacts of sub-division development on the tree species composition of urban forests in Halifax and London in Canada, they are able to determine whether the pre-urban landscape (woodlands or agriculture) lead to differences in urban tree diversity following urbanization. In both cities, regardless of the previous landscape, the newer neighbourhoods had greater tree species richness and evenness and are characterised by substantially more native tree species. This study provides hope that these newer suburbs will provide high quality, native tree habitats to support faunal biodiversity, albeit highly fragmented habitats on small building lots, interspersed with non-native trees.

6. Summary and Future Directions

We are pleased to present this Special Issue and believe that many of the studies from across the world will make a lasting contribution to raising the recognition of the ecological, environmental and socio-economic value of trees to our towns and cities. Urban trees play a vital role in maintaining and enhancing the resilience and integrity of many social, cultural, and ecosystem functions. There is a real need to recognise and tackle the issues of "green inequity" or "tree inequity" in the towns and cities of all countries. Without accepting the need for action, we cannot effectively use urban trees as a single mechanism to provide greater social, ecological, and ecosystem service benefits in urban landscapes while minimising the ecosystem disservices. Urban tree planting initiatives should not contribute to the growing divide between the haves and have-nots in modern urban society. For tree planting to provide the greatest and most cost-effective ecosystem service benefit, the first areas to be planted should be those with the least green space or tree canopy cover [16]. If this can be done, it will provide a great opportunity for urban forest researchers to concurrently monitor and measure the gradual, long-term delivery of benefits that increased urban tree cover, tree diversity, and tree health can provide. Measuring the relevant social and ecosystem indicators will of course be essential to evaluating this and the success of separate municipal urban forest programs.

Engaging the urban population with greenery and nature is a must and can indeed improve our awareness, appreciation, and willingness to tackle all our pressing environmental issues, be they urban, rural, local, national, or global. Soon, the majority of the world's population will be urban residents and they will have a profound relationship with the trees around them, providing cues for memories and a "sense of place", and invoking emotions that these trees are perceived as active participants in urban life [17]. As such, the role of a vibrant, diverse, and healthy urban forest cannot be underestimated. Several studies point to the important role of diversity in tree populations, as well as the positive role that urban forests can have for maintaining fauna biodiversity, creating opportunities for local communities to make a greater connection with nature. There is a real need for greater research on the human health and wellness benefits of urban biodiversity and urban forests themselves. These urban forests may not provide critical habitats for threatened or endangered animals in the same way that more remote or larger nature reserves might. Similarly, these urban forests may sequester only a small fraction of the carbon sequestered by managed plantations and natural forest systems. However, an increasing number of us live an "urban life", so it is this urban forest that provides the best, or most frequent, opportunity for society to interact with nature, to be environmentally aware in the truest sense, to directly observe the impacts of climate change, and to feel empowered that your urban landscape contributes in some small way to a better world.

Acknowledgments: We would like to acknowledge the contributions made by the authors and all reviewers of the 16 manuscripts in this Urban Forest and Peri-Urban Forest Special Issue, as well as the support provided by the editors and managers of the journal *Forests* of MDPI.

Author Contributions: Francisco Escobedo proposed the Special Issue. Francisco Escobedo, Stephen Livesley, and Justin Morgenroth guest-edited the Special Issue and wrote this editorial together.

Conflicts of Interest: The authors declare no conflict of interest.

References

1. Haase, D.; Larondelle, N.; Andersson, E.; Artmann, M.; Borgström, S.; Breuste, J.; Gomez-Baggethun, E.; Gren, Å.; Hamstead, Z.; Hansen, R. A quantitative review of urban ecosystem service assessments: Concepts, models, and implementation. *Ambio* **2014**, *43*, 413–433. [CrossRef] [PubMed]
2. Östberg, J.; Kleinschmit, D. Comparative study of local and national media reporting: Conflict around the tv oak in Stockholm, Sweden. *Forests* **2016**, *7*, 233. [CrossRef]
3. City of Melbourne. *Urban Forest Precinct Plans*; City of Melbourne: Melbourne, Australia, 2016.
4. Barron, S.; Sheppard, R.S.; Condon, M.P. Urban forest indicators for planning and designing future forests. *Forests* **2016**, *7*, 208. [CrossRef]
5. Schwarz, K.; Fragkias, M.; Boone, C.G.; Zhou, W.; McHale, M.; Grove, J.M.; O'Neil-Dunne, J.; McFadden, J.P.; Buckley, G.L.; Childers, D.; et al. Trees grow on money: Urban tree canopy cover and environmental justice. *PLoS ONE* **2015**, *10*. [CrossRef] [PubMed]
6. Nesbitt, L.; Meitner, J.M. Exploring relationships between socioeconomic background and urban greenery in Portland, Oregon. *Forests* **2016**, *7*, 162. [CrossRef]
7. Escobedo, J.F.; Palmas-Perez, S.; Dobbs, C.; Gezan, S.; Hernandez, J. Spatio-temporal changes in structure for a mediterranean urban forest: Santiago, Chile 2002 to 2014. *Forests* **2016**, *7*, 121. [CrossRef]
8. Widney, S.; Fischer, C.B.; Vogt, J.; Dahlhausen, J.; Biber, P.; Rötzer, T.; Uhl, E.; Pretzsch, H. Tree mortality undercuts ability of tree-planting programs to provide benefits: Results of a three-city studytree species and their space requirements in six urban environments worldwide. *Forests* **2016**, *7*, 65. [CrossRef]
9. Clerici, N.; Rubiano, K.; Abd-Elrahman, A.; Posada Hoestettler, M.J.; Escobedo, J.F. Estimating aboveground biomass and carbon stocks in periurban Andean secondary forests using very high resolution imagery. *Forests* **2016**, *7*, 138. [CrossRef]
10. Lv, H.; Wang, W.; He, X.; Xiao, L.; Zhou, W.; Zhang, B. Quantifying tree and soil carbon stocks in a temperate urban forest in northeast China. *Forests* **2016**, *7*, 200. [CrossRef]
11. Yoon, K.T.; Seo, W.K.; Park, S.G.; Son, M.Y.; Son, Y. Surface soil carbon storage in urban green spaces in three major South Korean cities. *Forests* **2016**, *7*, 115. [CrossRef]
12. MacGregor-Fors, I.; Escobar, F.; Rueda-Hernández, R.; Avendaño-Reyes, S.; Baena, L.M.; Bandala, M.V.; Chacón-Zapata, S.; Guillén-Servent, A.; González-García, F.; Lorea-Hernández, F.; et al. City "green" contributions: The role of urban greenspaces as reservoirs for biodiversity. *Forests* **2016**, *7*, 146. [CrossRef]
13. Sjöman, H.; Morgenroth, J.; Sjöman, J.D.; Sæbø, A.; Kowarik, I. Diversification of the urban forest—Can we afford to exclude exotic tree species? *Urban For. Urban Green.* **2016**, *18*, 237–241. [CrossRef]
14. Shackleton, C. Do indigenous street trees promote more biodiversity than alien ones? Evidence using mistletoes and birds in South Africa. *Forests* **2016**, *7*, 134. [CrossRef]
15. Nitoslawski, A.S.; Duinker, N.P. Managing tree diversity: A comparison of suburban development in two Canadian cities. *Forests* **2016**, *7*, 119. [CrossRef]
16. Norton, B.A.; Coutts, A.M.; Livesley, S.J.; Harris, R.J.; Hunter, A.M.; Williams, N.S.G. Planning for cooler cities: A framework to prioritise green infrastructure to mitigate high temperatures in urban landscapes. *Landsc. Urban Plan.* **2015**, *134*, 127–138. [CrossRef]
17. Pearce, L.M.; Davison, A.; Kirkpatrick, J.B. Personal encounters with trees: The lived significance of the private urban forest. *Urban For. Urban Green.* **2015**, *14*, 1–7. [CrossRef]

Section 3:
Ecosystem Service Tradeoffs;
Climate Change

forests

MDPI

Communication

Surface Soil Carbon Storage in Urban Green Spaces in Three Major South Korean Cities

Tae Kyung Yoon [1,†], Kyung Won Seo [2,†], Gwan Soo Park [3], Yeong Mo Son [4] and Yowhan Son [5,6,*]

[1] Environmental Planning Institute, Seoul National University, Seoul 08826, Korea; yoon.ecology@gmail.com
[2] Forest Practice Research Center, National Institute of Forest Science, Pocheon 11186, Korea; lorangetree@korea.kr
[3] Department of Environment and Forest Resources, Chungnam National University, Daejeon 34134, Korea; gspark@cnu.ac.kr
[4] Division of Forest Industry Research, National Institute of Forest Science, Seoul 02455, Korea; treelove@korea.kr
[5] Department of Environmental Science and Ecological Engineering, Korea University, Seoul 02841, Korea
[6] Department of Biological and Environmental Sciences, Qatar University, Doha 2713, Qatar
* Correspondence: yson@korea.ac.kr; Tel.: +82-2-3290-3015
† These authors contributed equally to this work.

Academic Editors: Francisco Escobedo, Stephen John Livesley and Justin Morgenroth
Received: 17 March 2016; Accepted: 23 May 2016; Published: 28 May 2016

Abstract: Quantifying and managing carbon (C) storage in urban green space (UGS) soils is associated with the ecosystem services necessary for human well-being and the national C inventory report of the Intergovernmental Panel on Climate Change (IPCC). Here, the soil C stocks at 30-cm depths in different types of UGS's (roadside, park, school forest, and riverside) were studied in three major South Korean cities that have experienced recent, rapid development. The total C of 666 soil samples was analyzed, and these results were combined with the available UGS inventory data. Overall, the mean soil bulk density, C concentration, and C density at 30-cm depths were 1.22 $g \cdot cm^{-3}$, 7.31 $g \cdot C \cdot kg^{-1}$, and 2.13 $kg \cdot C \cdot m^{-2}$, respectively. The UGS soil C stock ($Gg \cdot C$) at 30-cm depths was 105.6 for Seoul, 43.6 for Daegu, and 26.4 for Daejeon. The lower C storage of Korean UGS soils than those of other countries is due to the low soil C concentration and the smaller land area under UGS. Strategic management practices that augment the organic matter supply in soil are expected to enhance C storage in South Korean UGS soils.

Keywords: park; riverside; roadside; settlement; soil carbon sequestration

1. Introduction

Enhancing ecosystem services and human well-being is especially vital for urbanized lands (see the definition in the footnotes of Table 1) [1,2] where half of the global population and 90% of the South Korean population lives. Designing, establishing, and managing urban green spaces (UGS's), as well as preserving remnant ecosystems in a city, could be one way to secure or restore ecosystem services in urbanized lands [3]. Soils support various ecosystem services such as nutrient cycling, habitat provision, food production, storm water management, and carbon (C) storage [3–6]. Here, the emphasis on the conservation or sequestration of soil C, with a C stock greater than that of atmosphere and biomass combined [7–11], is extended to urban soil C, both organic and inorganic C [12–14]. The significance of UGS's for terrestrial C inventories and the need to treat them distinctly from other major land types (e.g., forest, agricultural land, grassland, and wetland) is underscored by its special treatment as a separate land category, *settlements*, which encompasses all developed land according to the Good Practice Guidance (GPG) and Guidelines (GL) of the

Intergovernmental Panel on Climate Change (IPCC) [15,16]. Moreover, organic and inorganic C in soils increases their quality, indirectly supporting other ecosystem services such as nutrient cycling and pollution mitigation [3]. Not surprisingly, there is an increasing concern to manage UGS soil C among various interested stakeholders, including scientists, local governments and NGOs, and international institutions [3,4,15,16]. To date, most studies on UGS soil C have focused on *old cities*, that is, cities with a relatively long history of urbanization and hence more potential to store abundant C in UGS soils [3,17,18]. Given the changing global trends of urbanization in the last 100 years, UGS soil C of *young cities* with a relatively recent history of urbanization has not been adequately studied. In South Korea, a few studies have reported on UGS soil C [19,20]; however, these studies investigated UGS soil C at a single site or in a single city.

Here, we sought to quantify the soil C stocks of the UGS's at 30-cm depths in three South Korean cities and thereby provide a current estimate of the soil C status. Because land use, land cover change, and/or urbanization history can all affect UGS soil C [17,20–22], soil C densities in different types of UGS's such as roadside, park, school forest, and riverside were quantified and then compared to vegetation C density and site history. In addition, we reviewed the management of soil C in UGS's to enhance the ecosystem services of C storage. South Korea is a populous urbanized country with 505 people·km^{-2} and has experienced rapid urbanization since the 1960s; for instance, since then, its urban population has increased from 10 million to 45 million (Korean Statistical Information Service; http://www.kosis.kr). By studying UGS soils in multiple South Korean cities, we can provide needed information on baselines of UGS soil C in recently developed cities.

2. Materials and Methods

The study was conducted in three major South Korean cities: Seoul, Daegu, and Daejeon (Table 1). Seoul, the capital city since the late 14th century, has the highest population, and its population density and urbanized land cover are also high compared to the other cities [23,24]. Daegu, a principal city in Southeastern Korea since the 15th century, has the fourth highest population in South Korea, while Daejeon, which was established early in the 20th century, now has the fifth highest population [25]. The population density and relative urbanized land cover is now similar between Daegu and Daejeon. In tandem with human population growth, the expansion of road networks and residential developments was mostly completed by the early 1990s in Seoul and Daegu [23–25]. Since then, their infrastructure has been renovated for the well-being of citizens in various ways, for example, by restoring streams and establishing UGS's [24]. Being the youngest city in this study, the population growth and infrastructure development in Daejeon has remained steady until 2014 [25]. The three cities have a climate best described as *Dfa* (hot summer continental climate) under the Köppen climate classification; however, Daegu is moderately warmer with less precipitation than the other two cities.

Table 1. Environmental, demographic, and land use characteristics of the three studied cities.

	Seoul	Daegu	Daejeon
Environment *			
Location	37°34′ N, 126°58′ E	35°52′ N, 128°36′ E	36°21′ N, 127°23′ E
Mean annual temperature	12.5 °C	14.1 °C	13.0 °C
Mean annual precipitation	1451 mm	1064 mm	1459 mm
Demography †			
Area	605.18 km^2	883.63 km^2	540.1 km^2
Population	10.02 million	2.52 million	1.54 million
Population density	16,659 people·km^{-2}	2857 people·km^{-2}	2873 people·km^{-2}
World urban areas rank by population density ‡	#238	#272	#181

Table 1. *Cont.*

	Seoul	Daegu	Daejeon
Land uses [§]			
Urbanized land area (% of total city area)	316.5 km^2 (52.1%)	154.1 km^2 (17.5%)	89.0 km^2 (16.5%)
Impervious area (% of total city area)	347.3 km^2 (57.2%)	172.2 km^2 (19.5%)	102.4 km^2 (19.0%)
Forest (% of total city area)	127.8 km^2 (21.1%)	274.1 km^2 (31.0%)	249.9 km^2 (46.3%)
Green space area (% of total city area)	44.1 km^2 (7.3%)	19.4 km^2 (2.2%)	18.5 km^2 (3.4%)
Roadside area	4.06 km^2	1.30 km^2	1.01 km^2
Park area	30.31 km^2	12.69 km^2	14.10 km^2
School forest area	0.98 km^2	0.85 km^2	0.36 km^2
Riverside area	6.19 km^2	0.15 km^2	0.54 km^2

* Korea Meteorological Administration (http://www.kma.go.kr). [†] Korean Resident Registration Demographics (http://rcps.egov.go.kr:8081). [‡] Demographia World Urban Areas [26]. Note that the population density rank of Seoul is downgraded because it includes other cities near Seoul (e.g., Incheon, Ansan, and Suwon) in the database. [§] The categories of land uses are based on the National Urban Forest Inventory [27] and the Environmental Geographic Information Service (http://egis.me.go.kr) and defined as follows: Urbanized land is the city district area excluding forest, agriculture land, grassland, wetland, bare land, and water cover. Forest is the forested land excluding green space. Green space is the aggregate of vegetated area close to the living space of inhabitants. Roadside is the vegetated area of sidewalk, roadway, and traffic island. Park is the artificially vegetated area, as defined by the Urban Park Act. School forest is the vegetated area on school grounds. Riverside is the vegetated area adjacent to flowing water, as defined by the River Act.

We investigated total soil C in the roadside, park, school forest, and riverside, following the legal classification of UGS's as defined by the National Urban Forest Inventory [27] and the Environmental Geographic Information Service (http://egis.me.go.kr) (full details in the footnotes of Table 1). Because the inorganic C produced by carbonation reactions in calcareous materials (e.g., concrete, cement, parent material) can form a fraction of soil C, in the present study, in addition to organic C [12–14,28], we focused on total C, covering both organic and inorganic C. The UGS's exclude remnant, non-urbanized, and mountainous forest areas, such as green belt and conservation areas in the city districts (as defined by the administrative system of Korea). The four UGS types represent approximately 90% of all the UGS areas in South Korea. Since mountainous forests dominate the land cover of Korea, urban areas are developed in the lowland basins of the city districts; together, the remnant, non-urbanized, and mountainous forests account for an appreciable or considerable portion of city land cover (27.5% for Seoul, 63.8% for Daegu, and 70.4% for Daejeon) and of ecosystem C sequestration [29], even in highly inhabited cities [26]. However, this study excluded measurement of soil C in these non-UGS types, which are accounted for in the *forest* category of IPCC [15,16] and the National Forest Inventory of Korea [30]. It should be noted that soils beneath impervious surfaces (e.g., paved roads), which may have a significant potential C storage capacity with C density equivalent to that of UGS's soils [31], were not investigated in this study.

In each city, we selected 3 to 10 replicate sites per each UGS type—except at riverside due to its lack of suitable sites for replication or limited accessibility—considering their distribution, history, area, location, vegetation cover, and accessibility after discussion with experts (see Table 2). Unfortunately, detailed records of soil management practices undertaken at each UGS site were unavailable to us; however, it is commonly agreed upon that this management had been poor and lacked specific guidelines, regulations, or acts. In general, UGS soils have an insufficient supply of organic matter from aboveground. For the park and school forest UGS's, the grass was regularly mowed, and grass clippings and leaf litter were removed to maintain surface cleanliness and prevent inconvenience for users. Park, school forest, and riverside soils were irrigated but generally not fertilized, whereas roadside soils were hardly managed. These aspects of UGS soil management have been listed in local government ordinances or guidelines since 2010. Site history (year of construction) was available for 34 of the 64 sites. Tree C density at each site (unpublished data)—a corollary of vegetation cover—was estimated by a biometric approach that used measurements of tree stem diameters at breast height and urban-specific allometric equations [32].

Table 2. Soil bulk density, C concentration, and area-based C density in urban green spaces (UGS) soils for 0–30 cm depth. Values in parentheses represent one standard deviation.

	N		Bulk Density	Soil C Concentration	Soil C Density
	Site	Soil	$(g \cdot cm^{-3})$ *	$(g \cdot C \cdot kg^{-1})$	$(kg \cdot C \cdot m^{-2})$ [†]
Roadside					
Seoul	12	120	1.24 (0.22)	9.97 (7.31)	3.03 (2.15)
Daegu	9	60	1.04 (0.34)	9.08 (6.97)	2.10 (1.33)
Daejeon	10	72	1.07 (0.23)	6.62 (5.62)	1.56 (1.23)
Total	31	252	1.14 (0.27)	8.92 (6.95)	2.41 (1.85)
Park					
Seoul	3	117	1.34 (0.19)	6.66 (3.94)	2.24 (1.34)
Daegu	3	36	1.07 (0.31)	8.37 (6.26)	2.34 (1.94)
Daejeon	3	18	1.11 (0.13)	4.96 (3.99)	1.28 (0.97)
Total	9	171	1.27 (0.23)	6.48 (4.32)	2.05 (1.40)
School forest					
Seoul	13	150	1.27 (0.22)	6.13 (4.38)	1.88 (1.39)
Daegu	3	30	1.18 (0.31)	5.38 (3.48)	1.49 (0.89)
Daejeon	4	24	1.28 (0.15)	4.62 (3.69)	1.36 (0.97)
Total	20	204	1.26 (0.23)	5.82 (4.21)	1.76 (1.30)
Riverside					
Seoul	1	18	1.27 (0.25)	9.61 (6.31)	2.96 (1.73)
Daegu	2	12	1.23 (0.34)	8.58 (3.34)	2.90 (1.24)
Daejeon	1	9	1.32 (0.14)	6.09 (2.57)	1.92 (0.83)
Total	4	39	1.28 (0.24)	8.29 (4.94)	2.63 (1.45)
All					
Total	64	666	1.22 (0.25)	7.31 (5.63)	2.13 (1.59)

* Bulk density was corrected by gravel content (see Equation (1)). [†] Soil C concentration was multiplied by a ratio of soil dry weight to total volume of soil and gravel (see Equation (2)).

Soil cores 30 cm deep were taken in random triplicate using a soil sampler (Ø 5.5 cm, 50-cm length; Shinill Science Inc., Seoul, Korea) at each UGS site in August 2009 for Seoul, August 2010 for Daejeon, and August 2011 for Daegu. The soil sampler using a manually driving hammer probe may result in a potential error in determining bulk density due to compaction or stretching of the soil core [33]; nonetheless, this method is considered best practice for sampling urban soils and has been documented to present minimal site disturbance, soil excavation, and error [33]. A total of 666 soil samples were taken: 252 from roadside, 171 from park, 204 from school, and 39 from riverside UGS's (see Table 2).

The soil cores were separated into 10-cm sections in the field. In laboratory, the samples were air-dried and filtered through 2-mm sieves to exclude roots and gravel (*i.e.*, >2 mm). The oven-dried (105 °C) weight of soils (<2 mm) and gravel were determined separately. The bulk density was determined by the ratio of the oven-dried weight to the volume of the sieved soils, which was calculated by extracting the volume of the gravel from the volume of the original soil core (Equation (1)). The volume of the gravel was determined from its weight, assuming a gravel density of 2.65 g· cm^{-3} [34,35]. This correction procedure is required to accurately report bulk density, which can be affected by the gravel content. The subsamples of air-dried soils were ball-milled before their C concentrations were determined by an elemental analyzer (Vario Macro CN analyzer, Elementar Analysensysteme GmbH, Hanau, Germany).

The soil C concentration (g·C·kg^{-1}) was converted to soil C density (mg·C·cm^{-3} or kg·C·m^{-2}) by multiplying the density of fine soil in the soil core (see Equations (2) and (3)) instead of multiplying the bulk density, to avoid the miscalculation of C density due to gravel content in the soil core. The total UGS's C stock (Gg·C) of each city was estimated as the product of the land area covered by each UGS type and their corresponding soil C density (Equation (4)).

$$\text{Bulk density } (g \cdot cm^{-3}) = DW_s/(V_t - V_g), \tag{1}$$

$$\text{Soil C density } (mg \cdot C \cdot cm^{-3}) = C_s \times DW_s/V_t, \tag{2}$$

$$\begin{aligned}\text{Area} - \text{based soil C density } (kg \cdot C \cdot m^{-2}; \ 30 \ cm \ deep) = \text{mean soil C density}\\ (mg \cdot C \cdot cm^{-3}) \text{ in each soil profile} \times 0.3 \ (m), \text{ and}\end{aligned} \tag{3}$$

$$\begin{aligned}\text{Soil C stock } (Gg \cdot C; \ 30 \ cm \ deep) = \text{area} - \text{based soil C density } (kg \cdot C \cdot m^{-2}; \ 30 \ cm \ deep)\\ \times \text{ land cover } (km^2),\end{aligned} \tag{4}$$

where DW_s is the dry weight of soil (g), V_t is the total volume of soil and gravel (cm^3), V_g is the volume of gravel (cm^3), and C_s is the soil C concentration (g·C·kg^{-1}).

All response variables satisfied data normality, according to the Shapiro–Wilk test. The effects of city, UGS type, and soil depth on the variables of soil bulk density, C concentration, and C density were tested using a factorial three-way ANOVA. When the ANOVA model was significant ($p < 0.05$), Duncan's multiple range test was performed to determine the differences between the means for each city, UGS type, and/or soil depth. Regression analysis was performed among soil bulk density, C concentration, and C density. In addition, the effects of site history and tree C density on soil C density were determined using regression analysis. All statistical analyses were conducted using *R* (version 3.2.3) [36].

3. Results

The main effects of city, UGS type, and soil depth all significantly affected soil bulk density, C concentration, and C density at 30-cm depths (Tables 2 and 3). The soil C density was greatest in Seoul (2.37 kg·C·m^{-2}), followed by Daegu (2.07 kg·C·m^{-2}), and then Daejeon (1.39 kg·C·m^{-2}); this order seemed largely due to the higher bulk density of Seoul (1.28 g·cm^{-3}) and higher soil C concentrations of Seoul (7.58 g·C·kg^{-1}) and Daegu (8.22 g·C·kg^{-1}) (Table 3, Figure 1). Considering the four UGS types, the soil C density values were significantly higher in soils at riverside (2.62 kg·C·m^{-2}) and roadside (2.41 kg·C·m^{-2}) than at park (2.05 kg·C·m^{-2}) and school (1.76 kg·C·m^{-2}), which corresponded to the order of their soil C concentrations (Table 3). Bulk density increased, while both soil C concentration and C density decreased significantly with soil depth, from surface (0–10 cm) to deeper soil depths (10–30 cm) (Table 3). The descriptive summary for mean values of soil bulk density, C concentration, and C density in each city and UGS type are presented in Table 2.

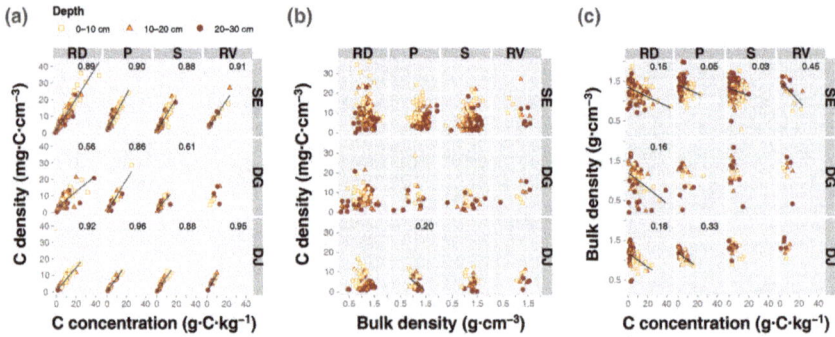

Figure 1. Pairwise relationships among soil bulk density, C concentration, and volume-based C density of soils in different urban green space types for three cities in South Korea (**a**): soil C concentration-C density, (**b**): soil bulk density-C density, (**c**): soil C concentration-bulk density). Where the linear regression was significant ($p < 0.05$), a solid line and an R^2 value number are drawn (numbers in the upper right corner of each sub-panel). Abbreviations are explained in Table 1.

Table 3. A three-way factorial ANOVA that tested for significant differences in soil bulk density, C concentration, and C density by city, UGS type, and soil depth. The table gives the *p*-values; those highlighted in bold are statistically significant ($p < 0.05$).

	df	Bulk Density	Soil C Concentration	Soil C Density
City (C)	2	**<0.001** SE > DJ > DG [†]	**<0.001** DG = SE > DJ	**<0.001** SE > DG > DJ
Type (T)	3	**<0.001** RV = P = S > RD [‡]	**<0.001** RD = RV > P = S	**<0.001** RV ⩾ RD ⩾ P > S
Depth (D)	2	**<0.001** 20–30 = 10–20 > 0–10	**<0.001** 0–10 > 10–20 = 20–30	**<0.001** 0–10 > 10–20 > 20–30
C × T *	6	**<0.001**	0.46	**<0.05**
C × D	4	**<0.001**	**<0.05**	0.33
T × D	6	0.09	0.44	0.07
C × T × D	12	0.18	0.80	0.81

* ×: interaction effect; [†] SE: Seoul, DG: Daegu, DJ: Daejeon; [‡] RD: roadside, P: park, S: school forest, RV: riverside.

Pairwise correlations among soil bulk density, C concentration, and C density suggested that soil C density was largely regulated by soil C concentration, rather than bulk density (Figure 1). In most cases, the soil C concentration showed a strong relationship with soil C density; while the bulk density did not. However, the soil C concentration and bulk density variables were independent of, or weakly dependent on, each other. The mean soil C density was not significant correlated with either tree C density ($p = 0.20$; data not shown) or year since construction ($p = 0.85$; data not shown) for all UGS types.

The total soil C stocks of each city at 30-cm depths were estimated by combining our empirical data with known UGS inventory data (Table 1). The total UGS soil C stocks in Seoul, Daegu, and Daejeon were estimated to be 105.6, 43.6, and 26.4 Gg· C, respectively (Figure 2). Parks contributed the most to the soil C stocks, accounting for more than two thirds of the total UGS soil C stock. Soil C stocks in other UGS types were consistent with their relative ranking in land cover in each city. For example, riverside had a more abundant soil C stock in Seoul than the other cities due to its relatively large coverage there. Similarly, the school contributed negligible soil C stocks (<2%) due to its limited cover in all cities. Nationwide, UGS soils (385.1 km^2) stored 812.8 Gg· C, as estimated by applying mean soil C densities for each UGS type to nationwide UGS areas.

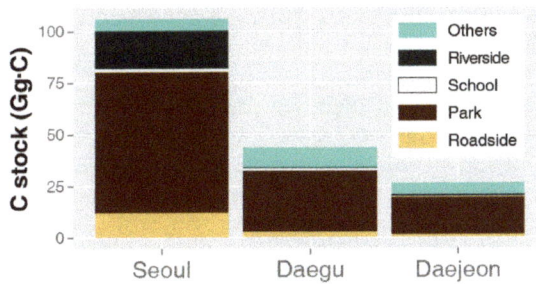

Figure 2. Soil C stock at 30-cm depths of urban green spaces in Seoul, Daegu, and Daejeon (South Korea).

4. Discussion

The soil C density of UGS's in Korea was lower than those in other countries. For instance, area-based total C densities of urban soils at 30-cm depths are reported to be 8.2 kg· C· m^{-2} for turf and 7.9 kg· C· m^{-2} for tree-planted areas in urban parks of Tokyo, Japan [21], 7.3 kg· C· m^{-2} for urban lawns in Fort Collins, USA [37], 5.0–17.2 kg· C· m^{-2} for golf courses in Melbourne, Australia [38], and 4.9–17.5 kg· C· m^{-2} for UGS's and urban forests in Auckland, New Zealand [22,39]. These values are much higher than the range of C densities, also at 30-cm depths, found in this study (1.55–3.02 kg· C· m^{-2}; Table 2). However, the bulk density in the present study (1.22 g· cm^{-3}) was comparable to that in other studies [31,40], apart from a few extraordinary cases of high bulk densities in compacted soils (>1.6 g· cm^{-3}) [41,42]. In contrast, the soil C concentrations found in this study (5.38–9.97 g· C· kg^{-1}) were much lower than not only other UGS soils, such as 20.9–32.0 g· C· kg^{-1} for golf courses in Melbourne, Australia [38], 39.2 g· C· kg^{-1} for UGS's in Auckland, New Zealand [39], and 12.3 g· C· kg^{-1} for lawns in Phoenix, USA [43], but also the 19.2 C· kg^{-1} for remnant urban forests in South Korea [44].

The factors driving the exceptionally low soil C concentration and density of UGS's in South Korea relative to other cities worldwide are not understood. Conceivably, the relatively short history of UGS (<~30 years) and the inappropriate management of UGS soils (e.g., loss of organic matter supply [3,45]) might explain the low soil C storage. That Daejeon had the lowest soil C density among the three cities (Tables 2 and 3) might reflect its relatively short history of urbanization and limited UGS development. Different management practices might also affect the C densities in UGS types. For instance, the riverside soils, which were less disturbed (*i.e.*, not subject to leaf litter and clipping removal) than other UGS soils, also stored the highest C content (Tables 2 and 3). On the other hand, human interference could reduce the input of organic matter and consequently the soil C concentration of park and school soils. In one of the park sites used in our study, Bae and Ryu [20] reported a noticeable increase in total soil C concentration (256%) over a decade as a result of sound management practices, such as proper irrigation and the application of a compost amendment made from removed site litter. It has been further suggested that management practices to enhance organic matter supply and soil environment should also be considered for enhancing C storage in UGS soils [46].

To better manage UGS soil C including organic and inorganic C, sufficient organic matter supply from litter, most of all, must first be designed [3]. The litter cycling of natural ecosystems may not be applicable to urban settings where social demands and services predominate, namely for maintaining surface cleanliness and users' convenience. Nonetheless, leaving a minimum amount of litter and grass clippings or alternative practices which offset the loss of organic matter supply to UGS soils, could be adopted. Compost amendment application and mulching, or both, can increase urban soil C storage [46,47]. Recently, in South Korea, collected litter has been reutilized as compost amendments for agriculture. Thus, these composts can be applied to UGS soils. Black C or biochar applications to urban soils may be effective not only for increasing soil organic C storage but also for enhancing soil fertility for vegetation growth [3,4]. On the other hand, inorganic C from carbonate reactions may be difficult to control through management practices, given the current dearth of knowledge in this field [3].

Bulk density estimates in the present study were within the normal range. Hence, soil compaction, which represents soil physical structures, might not be a critical concern for soil C density in this study. Nevertheless, we note that soil physical structures could be improved by having sufficient pore spaces, silt, and clay to support rhizosphere activity and soil organic matter accumulation [45,48]. Compost practices are also effective at reducing soil compaction and improving soil physical structure [49–51]. The resulting enhanced physical and chemical soil characteristics could support more vigorous vegetation growth, which in turn leads to more organic matter supply from aboveground and belowground [45,48], and consequently more soil organic C too. For instance, compost amendment practices during soil rehabilitation activities have successfully reduced bulk density and enhanced tree growth in highly urbanized soils [52]. Here, constructing physically fine soils at the beginning of the establishment phase is essential because, unlike agricultural soils, the practices that improve the physical properties of UGS soils are not easily undertaken later. In South Korea, specific guidelines on the construction and management of UGS soils, which were unavailable in the past, have been developed recently [53], and applying them can promote conditions for C sequestration. The citywide assessments in this study described the current status of UGS soil C; however, we could not analyze the full effects of management practices on UGS soil C. For this reason, further experimental studies and long-term monitoring on UGS soil C stocks are warranted.

A comparison of C storage across land uses would suggest that the C storage in UGS's is negligible in South Korea, due to its small land cover and low soil C density when compared to major land uses such as forests and agricultural lands. Soil C stock in South Korean forests covering approximately 60,000 km^2 is estimated to be 341.7 Tg·C at 30-cm depths, according to the Korean Forest Soil Carbon model [30]. The UGS's of South Korea have neither noteworthy land cover (0.4%) nor high soil C density, resulting in their low C storage (0.8 Tg·C; 30 cm deep) being an order of magnitude different from the nationwide values.

In summary, we report the current status of total C storage in UGS soils in three South Korean cities. These results may provide a baseline to help construct the inventory of settlements on the IPCC-GPG and GL, develop sound management guidelines for improving UGS soil C, and safeguard the ecosystem services of UGS soils in terms of C storage.

Acknowledgments: This study was supported by research grants from the National Research Foundation of Korea (2015R1A6A3A01058445) and the Korean Ministry of Environment (2016001300004). This study is a result of reworking the dissertation of Kyung Won Seo [54].

Author Contributions: Tae Kyung Yoon, Kyung Won Seo, and Gwan Soo Park were responsible for the fieldwork data from Daegu, Seoul, and Daejeon, respectively; Tae Kyung Yoon and Kyung Won Seo analyzed the data; Yowhan Son and Yeong Mo Son conceived and designed the experiments; Tae Kyung Yoon wrote the manuscript.

Conflicts of Interest: The authors declare no conflicts of interest. The funding sponsors had no role in the design of the study; in the collection, analyses, or interpretation of data; in the writing of the manuscript; or in the decision to publish the results.

References

1. Millennium Ecosystem Assessment. *Ecosystems and Human Well-being: Synthesis*; Island Press: Washington, DC, USA, 2005.
2. Secretariat of the Convention on Biological Diversity. *Cities and Biodiversity Outlook*; Secretariat of the Convention on Biological Diversity: Montreal, QC, Canada, 2012.
3. Lorenz, K.; Lal, R. Managing soil carbon stocks to enhance the resilience of urban ecosystems. *Carbon Manag.* **2015**, *6*, 35–50. [CrossRef]

4. Renforth, P.; Leake, J.R.; Edmondson, J.; Manning, D.A.; Gaston, K.J. Designing a carbon capture function into urban soils. *Proc. ICE-Urban Des. Plan.* **2011**, *164*, 121–128. [CrossRef]
5. Pavao-Zuckerman, M.A. Urbanization, soils, and ecosystem services. In *Soil Ecology and Ecosystem Services*; Wall, D.H., Bardgett, R.D., Behan-Pelletier, V., Herrick, J.E., Jones, T.H., Ritz, K., Six, J., Strong, D.R., van der Putten, W.H., Eds.; Oxford University Press: Oxford, UK, 2012; pp. 270–281.
6. Haase, D.; Larondelle, N.; Andersson, E.; Artmann, M.; Borgström, S.; Breuste, J.; Gomez-Baggethun, E.; Gren, Å.; Hamstead, Z.; Hansen, R.; *et al.* A quantitative review of urban ecosystem service assessments: Concepts, models, and implementation. *AMBIO* **2014**, *43*, 413–433. [CrossRef] [PubMed]
7. Houghton, R.A. Why are estimates of the terrestrial carbon balance so different? *Glob. Chang. Biol.* **2003**, *9*, 500–509. [CrossRef]
8. Lal, R. Soil carbon sequestration to mitigate climate change. *Geoderma* **2004**, *123*, 1–22. [CrossRef]
9. Scharlemann, J.P.W.; Tanner, E.V.J.; Hiederer, R.; Kapos, V. Global soil carbon: Understanding and managing the largest terrestrial carbon pool. *Carbon Manag.* **2014**, *5*, 81–91. [CrossRef]
10. Köchy, M.; Hiederer, R.; Freibauer, A. Global distribution of soil organic carbon—Part 1: Masses and frequency distributions of SOC stocks for the tropics, permafrost regions, wetlands, and the world. *SOIL* **2015**, *1*, 351–365. [CrossRef]
11. Batjes, N.H. Harmonized soil property values for broad-scale modelling (WISE30sec) with estimates of global soil carbon stocks. *Geoderma* **2016**, *269*, 61–68. [CrossRef]
12. Washbourne, C.L.; Renforth, P.; Manning, D.A.C. Investigating carbonate formation in urban soils as a method for capture and storage of atmospheric carbon. *Sci. Total Environ.* **2012**, *431*, 166–175. [CrossRef] [PubMed]
13. Washbourne, C.-L.; Lopez-Capel, E.; Renforth, P.; Ascough, P.L.; Manning, D.A.C. Rapid removal of atmospheric CO_2 by urban soils. *Environ. Sci. Technol.* **2015**, *49*, 5434–5440. [CrossRef] [PubMed]
14. Whitmore, A.P.; Kirk, G.J.D.; Rawlins, B.G. Technologies for increasing carbon storage in soil to mitigate climate change. *Soil Use Manag.* **2015**, *31*, 62–71. [CrossRef]
15. Intergovernmental Panel on Climate Change. *Good Practice Guidance for Land Use, Land-Use Change and Forestry*; Institute for Global Environmental Strategies: Kanagawa, Japan, 2003.
16. Intergovernmental Panel on Climate Change. *IPCC Guidelines for National Greenhouse Gas Inventories*; Institute for Global Environmental Strategies: Kanagawa, Japan, 2006.
17. Pouyat, R.V.; Yesilonis, I.D.; Nowak, D.J. Carbon storage by urban soils in the United States. *J. Environ. Qual.* **2006**, *35*, 1566–1575. [CrossRef] [PubMed]
18. Lorenz, K.; Lal, R. Biogeochemical C and N cycles in urban soils. *Environ. Int.* **2009**, *35*, 1–8. [CrossRef] [PubMed]
19. Jo, H.-K. Impacts of urban greenspace on offsetting carbon emissions for middle Korea. *J. Environ. Manag.* **2002**, *64*, 115–126. [CrossRef]
20. Bae, J.; Ryu, Y. Land use and land cover changes explain spatial and temporal variations of the soil organic carbon stocks in a constructed urban park. *Landsc. Urban Plan.* **2015**, *136*, 57–67. [CrossRef]
21. Takahashi, T.; Amano, Y.; Kuchimura, K.; Kobayashi, T. Carbon content of soil in urban parks in Tokyo, Japan. *Landsc. Ecol. Eng.* **2008**, *4*, 139–142. [CrossRef]
22. Weissert, L.F.; Salmond, J.A.; Schwendenmann, L. Variability of soil organic carbon stocks and soil CO_2 efflux across urban land use and soil cover types. *Geoderma* **2016**, *271*, 80–90. [CrossRef]
23. Kim, H.M.; Han, S.S. Seoul. *Cities* **2012**, *29*, 142–154. [CrossRef]
24. Kim, K.J.; Choe, S.C. In search of sustainable urban form for Seoul. In *Megacities: Urban Form, Governance, and Sustainability*; Sorensen, A., Okata, J., Eds.; Springer: Tokyo, Japan, 2011; pp. 43–65.
25. Nam, J.; Yun, B.-H.; Park, G.-W. The analysis on feasibility of urban regeneration through the evaluation of urban growth stage—The application of differential urbanization model and cyclic urbanization model. *J. Korea Plan. Assoc.* **2015**, *50*, 153–177. (In Korean with English Abstract) [CrossRef]
26. Demographia. *Demographia World Urban Areas: 11th Annual Edition*; Demographia, 2015. Available online: http://www.demographia.com/db-worldua.pdf (accessed on 14 April 2016).
27. Korea Forest Service. *National Urban Forest Inventory*; Korea Forest Service: Daejeon, Korea, 2014. (In Korean).
28. Xu, N.; Liu, H.; Wei, F.; Zhu, Y. Urban expanding pattern and soil organic, inorganic carbon distribution in Shanghai, China. *Environ. Earth Sci.* **2011**, *66*, 1233–1238.

29. Lee, J.; Lee, G.; Kim, J. Calculating total urban forest volume considering the carbon cycle in an urban area —Focusing on the city of Chuncheon in South Korea. *For. Sci. Technol.* **2014**, *10*, 80–88. [CrossRef]

30. Lee, J.; Yoon, T.K.; Han, S.; Kim, S.; Yi, M.J.; Park, G.S.; Kim, C.; Son, Y.M.; Kim, R.; Son, Y. Estimating the carbon dynamics of South Korean forests from 1954 to 2012. *Biogeosciences* **2014**, *11*, 4637–4650. [CrossRef]

31. Edmondson, J.L.; Davies, Z.G.; McHugh, N.; Gaston, K.J.; Leake, J.R. Organic carbon hidden in urban ecosystems. *Sci. Rep.* **2012**, *2*, 963. [CrossRef] [PubMed]

32. Yoon, T.K.; Park, C.-W.; Lee, S.J.; Ko, S.; Kim, K.N.; Son, Y.; Lee, K.H.; Oh, S.; Lee, W.-K.; Son, Y. Allometric equations for estimating the aboveground volume of five common urban street tree species in Daegu, Korea. *Urban For. Urban Green.* **2013**, *12*, 344–349. [CrossRef]

33. Walter, K.; Don, A.; Tiemeyer, B.; Freibauer, A. Determining soil bulk density for carbon stock calculations—A systematic method comparison. *Soil Sci. Soc. Am. J.* **2016**. [CrossRef]

34. Rowell, D.L. *Soil Science: Methods & Applications*; Routledge: New York, NY, USA, 2014.

35. Throop, H.L.; Archer, S.R.; Monger, H.C.; Waltman, S. When bulk density methods matter: Implications for estimating soil organic carbon pools in rocky soils. *J. Arid Environ.* **2012**, *77*, 66–71. [CrossRef]

36. R Development Core Team. *R: A Language and Environment for Statistical Computing*; R Foundation for Statistical Computing: Vienna, Austria, 2016; Available online: https://www.R-project.org (accessed on 14 April 2016).

37. Kaye, J.P.; McCulley, R.L.; Burke, I.C. Carbon fluxes, nitrogen cycling, and soil microbial communities in adjacent urban, native and agricultural ecosystems. *Glob. Chang. Biol.* **2005**, *11*, 575–587. [CrossRef]

38. Livesley, S.J.; Ossola, A.; Threlfall, C.G.; Hahs, A.K.; Williams, N.S.G. Soil carbon and carbon/nitrogen ratio change under tree canopy, tall grass, and turf grass areas of urban green space. *J. Environ. Qual.* **2016**, *45*, 215–223. [CrossRef] [PubMed]

39. Curran-Cournane, F.; Lear, G.; Schwendenmann, L.; Khin, J. Heavy metal soil pollution is influenced by the location of green spaces within urban settings. *Soil Res.* **2015**, *53*, 306–315. [CrossRef]

40. Pouyat, R.V.; Yesilonis, I.D.; Russell-Anelli, J.; Neerchal, N.K. Soil chemical and physical properties that differentiate urban land-use and cover types. *Soil Sci. Soc. Am. J.* **2007**, *71*, 1010–1019. [CrossRef]

41. Jim, C.Y. Urban soil characteristics and limitations for landscape planting in Hong Kong. *Landsc. Urban Plan.* **1998**, *40*, 235–249. [CrossRef]

42. Scharenbroch, B.C.; Lloyd, J.E.; Johnson-Maynard, J.L. Distinguishing urban soils with physical, chemical, and biological properties. *Pedobiologia* **2005**, *49*, 283–296. [CrossRef]

43. Green, D.M.; Oleksyszyn, M. Enzyme activities and carbon dioxide flux in a Sonoran desert urban ecosystem. *Soil Sci. Soc. Am. J.* **2002**, *66*, 2002–2008. [CrossRef]

44. Yoon, T.K.; Noh, N.J.; Han, S.; Kwak, H.; Lee, W.-K.; Son, Y. Small-scale spatial variability of soil properties in a Korean swamp. *Landsc. Ecol. Eng.* **2015**, *11*, 303–312. [CrossRef]

45. Craul, P.J. *Urban Soils: Applications and Practices*; John Wiley & Sons: New York, NY, USA, 1999.

46. Brown, S.; Miltner, E.; Cogger, C. Carbon sequestration potential in urban soils. In *Carbon Sequestration in Urban Ecosystems*; Lal, R., Augustin, B., Eds.; Springer: New York, NY, USA, 2012; pp. 173–196.

47. Beesley, L. Carbon storage and fluxes in existing and newly created urban soils. *J. Environ. Manag.* **2012**, *104*, 158–165. [CrossRef] [PubMed]

48. Day, S.D.; Eric Wiseman, P.; Dickinson, S.B.; Roger Harris, J. Tree root ecology in the urban environment and implications for a sustainable rhizosphere. *J Arboric.* **2010**, *36*, 193–205.

49. Sæbø, A.; Ferrini, F. The use of compost in urban green areas—A review for practical application. *Urban For. Urban Green.* **2006**, *4*, 159–169. [CrossRef]

50. Larney, F.J.; Angers, D.A. The role of organic amendments in soil reclamation: A review. *Can. J. Soil Sci.* **2012**, *92*, 19–38. [CrossRef]

51. Sloan, J.J.; Ampim, P.A.Y.; Basta, N.T.; Scott, R. Addressing the need for soil blends and amendments for the highly modified urban landscape. *Soil Sci. Soc. Am. J.* **2012**, *7*, 1133–1141. [CrossRef]

52. Layman, R.M.; Day, S.D.; Mitchell, D.K.; Chen, Y.; Harris, J.R.; Daniels, W.L. Below ground matters: Urban soil rehabilitation increases tree canopy and speeds establishment. *Urban For. Urban Green.* **2016**, *16*, 25–35. [CrossRef]

53. Chungnam National University. *Development of Soil Management Guideline and Monitoring Manual for Urban Forest in Living Area*; Chungnam National University: Daejeon, Korea, 2014. (In Korean)
54. Seo, K.W. Estimation of Soil Carbon Storage by Urban Green Spaces. Ph.D. Thesis, Korea University, Seoul, Korea, August 2015. (In Korean with English Abstract).

forests

MDPI

Article

Estimating Aboveground Biomass and Carbon Stocks in Periurban Andean Secondary Forests Using Very High Resolution Imagery

Nicola Clerici [1],*, Kristian Rubiano [1], Amr Abd-Elrahman [2], Juan Manuel Posada Hoestettler [1] and Francisco J. Escobedo [1]

[1] Programa de Biología, Facultad de Ciencias Naturales y Matemáticas, Universidad del Rosario,
     Cr. 24 No 63C-69, Bogotá 111221, Colombia; kristianru_30@hotmail.com (K.R.);
     juan.posada@urosario.edu.co (J.M.P.H.); franciscoj.escobedo@urosario.edu.co (F.J.E.)
[2] Geomatics Program, School of Forest Resources and Conservation, University of Florida, 1200 N. Park Road,
     Plant City, FL 33563, USA; aamr@ufl.edu
*   Correspondence: nicola.clerici@urosario.edu.co; Tel.: +571-2970200 (ext. 4024)

Academic Editor: Eric J. Jokela
Received: 30 April 2016; Accepted: 4 July 2016; Published: 9 July 2016

Abstract: Periurban forests are key to offsetting anthropogenic carbon emissions, but they are under constant threat from urbanization. In particular, secondary Neotropical forest types in Andean periurban areas have a high potential to store carbon, but are currently poorly characterized. To address this lack of information, we developed a method to estimate periurban aboveground biomass (AGB)—a proxy for multiple ecosystem services—of secondary Andean forests near Bogotá, Colombia, based on very high resolution (VHR) GeoEye-1, Pleiades-1A imagery and field-measured plot data. Specifically, we tested a series of different pre-processing workflows to derive six vegetation indices that were regressed against in situ estimates of AGB. Overall, the coupling of linear models and the Ratio Vegetation Index produced the most satisfactory results. Atmospheric and topographic correction proved to be key in improving model fit, especially in high aerosol and rugged terrain such as the Andes. Methods and findings provide baseline AGB and carbon stock information for little studied periurban Andean secondary forests. The methodological approach can also be used for integrating limited forest monitoring plot AGB data with very high resolution imagery for cost-effective modelling of ecosystem service provision from forests, monitoring reforestation and forest cover change, and for carbon offset assessments.

Keywords: periurban forests; carbon offsets; ecosystem service mapping; Colombia; remote sensing; vegetation indices

1. Introduction

Urban areas and forests play an important role in the carbon (C) cycle and the global climate because they are main drivers in the increase and regulation, respectively, of atmospheric carbon dioxide (CO_2), a major greenhouse gas [1,2]. Cities in the developing world are often documented as sources of high levels of carbon dioxide emissions [3]. It is also reported that without forests and their regulating ecosystem process and services, *sensu* [4], carbon dioxide concentration in the atmosphere would be about 43% higher than current concentrations [5]. Indeed, the *International Panel for Climate Change* (IPCC) has highlighted the importance of developing baseline estimates of carbon stocks in terrestrial ecosystems, which require mapping vegetation biomass distribution from the local to the global scale [6,7]. Precise carbon stock estimates in and around urban areas, therefore, represent key and necessary information to define carbon emission mitigation strategies and programs at the local and regional level (e.g., REDD+, [3,8]).

Latin America and the Caribbean regions have about 80% of their population residing in urban areas. Most periurban areas in Neotropical Latin America have historically undergone land use change [9]. Recently this urbanization trend has increased and led to the concern of deleterious effects to the hydrology and to increased carbon emissions [10,11]. In particular, periurban forests in Neotropical America are providers of key ecosystem services to the highly urbanized population in the region, but they are increasingly being lost [6,12,13]. Among terrestrial ecosystems, tropical forests are possibly the most important in regulating global carbon cycle. Lewis et al. [14] estimated that tropical forests are sinks for about 40% of terrestrial carbon on Earth, and responsible for almost half of the terrestrial net primary productivity [15]. Pan et al. [7] calculated that tropical forests store about 475×10^9 Mg of carbon, highlighting the paramount role these ecosystems have in regulating climate and mitigating climate change. Thus, information is needed on how to rapidly map, monitor, and estimate carbon stocks and other ecosystem services in periurban, Neotropical forests in a rapid, cost-effective manner.

In South America, estimations of aboveground biomass (AGB), a key proxy for carbon stocks [16], have been derived especially for lowland tropical rainforests [17–19] and for semi-arid savannas and dry forests [20]. One of the less studied Neotropical types of forest cover, but with a high potential contribution to the terrestrial C cycle, are early secondary forests whose development follows highly intervened forest land covers. These high Andean periurban secondary forest formations, typical of mountain areas, are present at elevation between ca. 2600 and 3200 m [21]. This vegetation type is characterized in general terms by a closed, dense canopy and low-to-medium woody plants that develop via ecological succession after land abandonment—typically cattle, pastures, or crops. In Colombia, it is estimated that in 1995 they covered about 7% of the country's terrestrial area [22]. Due to the dynamics of rural land abandonment and migration to urban areas and subsequent urbanization characterizing Colombia in recent decades [9,23], it is very likely that Colombian secondary forests are currently following an expansion trend. More importantly, Neotropical secondary forests are still not well-characterized in terms of composition, successional dynamics, carbon cycle, and even biodiversity [24]. In this context, the estimation of AGB and carbon in these ecosystems would represent important information to better characterize such a little studied successional forest type from the Neotropics.

Ground-based methods for forest AGB estimation are generally based on permanent plots, forest inventory, and monitoring methods that use species/family-specific or general allometric equations obtained from destructive methods and in situ dendrometric measurements (e.g., diameter at breast height (DBH), wood density, height, volume, and crown area (CA)), [19,25–27]. Although these approaches are common and can reach acceptable levels of accuracy, they are not adequate to map AGB distribution at regional scales in the Neotropics, because of the costs, access, and safety issues, and time necessary to perform extended field-based forest inventories.

Remote sensing, both active and passive, provide some of the most time-efficient and cost-effective approaches to derive AGB estimation at regional and national scale. Radar, optical, and LiDAR data have been extensively used to estimate AGB with a variety of methods (see [28,29] for comprehensive reviews). The use of optical data (visible, near-infrared) has been largely exploited for vegetation biomass estimation, due to the large amount and type of optical imagery available. Especially, sensors from the Landsat series (i.e., TM, ETM+, OLI) have been historically used to map biomass and carbon in a variety of ecosystems, for the relevance of their spectral bands, the continuity of the program, and the suitability of the 30 m spatial resolution for regional mapping (e.g., [30–33]). Several research discussed AGB modelling using Landsat imagery in tropical regions [34–37]. Most of these studies reported difficulties in modelling AGB in complex forest stands such as successional forest. Lu [38] demonstrated that the complexity of forest vegetation results in highly variable standing stocks of AGB and an even more variable rate of AGB accumulation following a deforestation event such as the case in successional forests.

Most available optical-based methods are established based on the relationships between field-derived AGB and different vegetation indices (VI). In doing so the aim is the identification of an analytical relationship between on-the-ground AGB estimates and spectral indices sensitive to photosynthetically active radiation and scaling with amount of biomass [39–42]. A vegetation index is generally derived from the spectral reflectance of two or more bands, and proportional to the value of biophysical parameters like leaf area index (LAI), net primary productivity (NPP), and absorbed photosynthetically active radiation (APAR) [43]. The choice of adequately performing indices (VIs) and satellite data type depends on the scale of analysis, type of ecosystem and environmental conditions, vegetation density, spectral information available, and the nature of the field information available.

In this work—due to the difficulty in AGB measurements in often inaccessible, dense Andean forests, as well as a limited spatial extent and availability of secondary forest plot data—we hypothesize that the use of very high resolution (VHR) data from commercial satellites, such as the GeoEye-1 and Pleiades-1A satellites, coupled with limited field data can be used to model AGB in complex secondary forests of the Colombian Andes. Previous work with VHR imagery includes the use of IKONOS by Thenkabail et al. [44] to estimate AGB and carbon stock of oil palm plantations in African savannas. Zhou and colleagues [45] exploited VIs and texture images derived from Quickbird and topographic information to derive AGB of black locust plantations in the Loess Plateau of China. Multispectral high-resolution Worldview-2 images were also used by Zhu et al. [46] to derive AGB for mangrove ecosystems using neural networks algorithms. Pereira et al. [47] used GeoEye-1 imagery to estimate biomass and carbon stock of coffee crops in Brazil exploiting a series of VIs.

The aim of this study is to develop an approach to model AGB and carbon stocks in complex secondary periurban Neotropical forests using very high resolution satellite imagery and limited field data from periurban early and late secondary forest monitoring plots. Specific objectives of this research are to:

(i)   identify an exploitable relationship between VIs derived from VHR satellite imagery and AGB estimated in the field in Andean periurban secondary forests;

(ii)   assess the effect of data processing on the performance of the biomass estimation models.

Such an approach should yield a low cost and rapid method for forest inventory and monitoring activities, and assessing the regulating ecosystem services and carbon dioxide offset potential of secondary periurban Neotropical forests in the Colombian Andes [8,19,24].

2. Experimental Section

2.1. Study Area

The study area is located in the Cundiboyacense high plain, in the Eastern Colombian Andes, at an average altitude of 2800 m (Figure 1). The annual temperature is 14 °C, and precipitation varies from 600 mm· year^{-1} in the center of the region and 1200 mm· year^{-1} in the western part. The high plain is characterized by two rainy periods, occurring from April to June and from September to November [48]. The high elevation plain environment is highly influenced by the adjacent city of Bogotá and contiguous urbanizations, with about 10 million inhabitants and a very high population density [49]. The region is one of the more important agro-industrial centers of the country, with agriculture, pastures, and mining as dominant activities in the rural non-urban areas [10,50].

Secondary periurban Andean forests are present throughout the eastern and northern peripheries of Bogotá [12]. The five most common vegetation families found in 20 permanents plots (20 m × 20 m)—established in early and late secondary forests in the study are—are Ericaceae, Melastomataceae, Cunoniaceae, Primulaceae, and Asteraceae, representing 56% of individuals with a basal diameter above 5 cm (Norden, Posada, et al., unpublished data). The five dominant genera are *Miconia*, *Weinmannia*, *Cavendisha*, *Myrsine*, and *Myrcianthes*, representing 51% of individuals. Eighty species of shrubs and trees have been identified in the area and the five dominant are

Weinmannia tomentosa Linnaeus filius 1782, *Cavendishia bracteata* Hoerold 1909, *Miconia ligustrin* Triana 1872, *Miconia squamulosa* Triana 1872, and *Myrcianthes leucoxyla* McVaugh 1963, with 44% of stems. Other studies have reported that below 2700 m common forest species are *Ilex kunthiana* Triana 1872, and *Vallea stipularis* Linnaeus filius 1782, while in drier areas shrubs of the genera *Myrcianthes* and *Morella* are more frequent. Other studies report that in the mountains between 2700 and 3000 m the genera that dominate forests are *Clusia, Miconia, Weinmania,* and *Cedrela* [21].

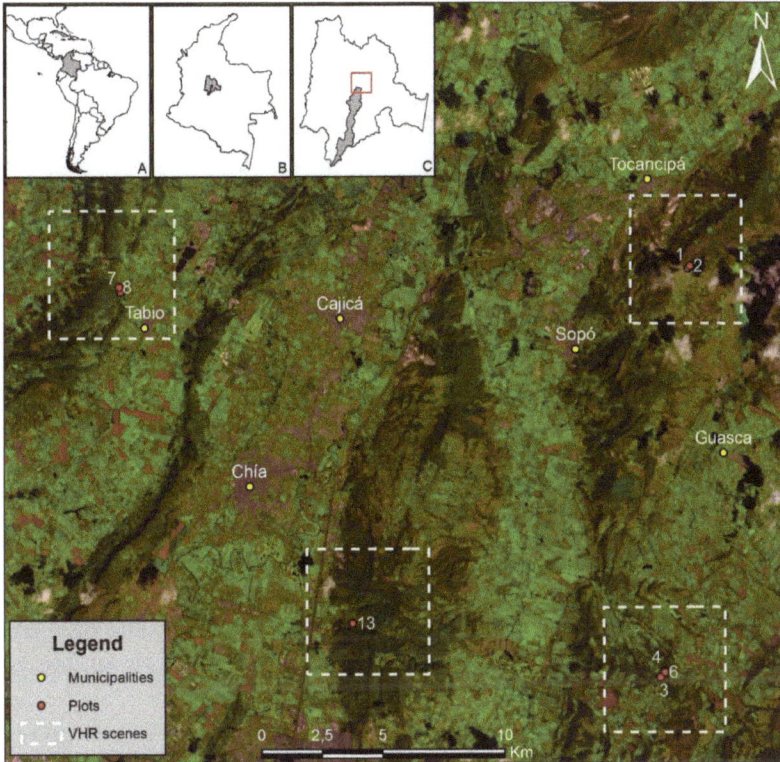

Figure 1. The Bogotá, Colombia, periurban study area, (A) Colombia (grey); (B) Department of Cundinamarca (grey); and (C) Bogotá D.C. city limits (grey) and study regional context (red polygon). Very High Resolution scenes boundaries (white dotted polygons) and plot locations (red points, numbered).

In this study, we used eight available 20 m × 20 m permanent plots of young and old secondary vegetation, established in Bogotá city´s northern limits in the suburban municipalities of Guatavita, Guasca, Torca and Tabio, Colombia (Figure 1), at altitudes varying between 2600 and 3100 m, in moderate slope hills. Most plots are dominated by species of low to medium height, as a result of an early stage of succession, with the exception of two plots of late secondary forests. We identified 57 species in these plots and the most common species were *Cavendishia bracteata, Miconia squamulosa, Myrcianthes leucoxyla, Myrsine guianensis,* and *Weinmannia tomentosa,* representing 51% of all the individuals (Norden, Posada et al., unpublished data).

2.2. In situ Biomass Estimation

Aboveground biomass was estimated in these eight plots using allometric equations for lower and upper montane Andean forests [51,52]. For estimation of AGB in plots located in lower montane Andean forests, we used the following allometric equation [51]:

$$AGB = 0.107314 \cdot DBH^{2.422}$$

where AGB refers to the aboveground biomass and DBH refers to stem diameter at breast height (1.3 m). For AGB estimation in upper montane Andean forests, we used, for a pool of 10 species, the following allometric equation [52]:

$$AGB = 0.190024 \cdot DBH^{2.20295}$$

The reported biomass values correspond to an average of biomass for both equations (Table 1). We measured all stems with a basal diameter (BD) higher than 5 cm, at 5 cm above the soil surface, for a total of 1354 individuals (Table 1). In five plots we also had measures of DBH while in the remaining three plots DBH was estimated with species-specific or genus-specific equations between BD and DBH.

Table 1. Plot ID, forest successional status, number of individuals, number of species, and average above-ground biomass (Mg·ha^{-1}).

Plot ID	Successional Status	Number of Individuals	Number of Species	Above-Ground Biomass (Mg·ha^{-1})
Guatavita 1	Early secondary	168	19	38.2
Guatavita 2	Early secondary	139	15	23.8
Guasca 3	Late secondary	120	16	180.7
Guasca 4	Mid-secondary	185	14	77.2
Guasca 6	Mid-secondary	175	17	54.0
Tabio 7	Mid-secondary	226	23	82.8
Tabio 8	Mid-secondary	203	14	65.9
Torca 13	Late secondary	138	32	111.6

2.3. Remote Sensing-Derived Biomass Estimation

We selected VHR GeoEye-1 and Pleiades-1A imagery in order to cover all the eight permanent plots without cloud presence, and acquired in a period of time close to the field census of the vegetation (Table 2). Both VHR sensors cover the blue, green, red, and near infra-red spectral ranges, while the spatial resolution of multispectral imagery is 1.65 m for GeoEye-1 and 2 m for Pleiades-1A. Every scene covers a 5 km × 5 km area.

Table 2. Remote sensing images and acquisition configurations used in the present study.

Location	Plot ID	Sensor	Date (day/month/year)	Off-Nadir	View Azimuth	Sun Azimuth	Sun Elevation Angle
Guatavita	1,2	Pleiades-1A	14/02/2013	19.1°	179.9°	123.5°	57.7°
Tabio	7,8	Pleiades-1A	14/02/2013	29.7°	180.2°	123.5°	57.7°
Guasca	3,4,6	GeoEye-1	29/12/2013	21.7°	242.8°	145.1°	55.6°
Torca	13	GeoEye-1	29/12/2013	21.4°	239.7°	145.0	55.5°

2.3.1. Data Pre-Processing

Data pre-processing for the VHR imagery involved different steps, and led to the production of five groups of data. First, we performed standard radiometric conversion of digital numbers to radiance using sensor specific calibration coefficients provided in the images' metadata (e.g., [53]). In a second step we applied geometric correction to minimize topographic and sensor geometry image distortions [54]. We used a series of field points taken with a sub-meter resolution GPS (Spectra Precision MobileMapper 120) around the 20 m × 20 m plots, and post-processed using the nearest local geodetic station BOGA. The geometric correction processing was carried out using first order polynomic equations and nearest neighbor resampling [55] in ENVI 4.3. The root mean square error (RMSE) varied from 0.60 to 1.12 m for the four scenes. The first group of VHR data (hereafter Group A) did not follow any further pre-processing.

Four additional data groups were prepared using additional pre-processing steps to account for atmospheric and topographic effects. These steps are very crucial since different VHR scenes captured at different dates and locations were used in the analysis. Therefore, the images were subjected to either: (i) top-of-atmosphere (TOA), atmospheric correction for data Group B; (ii) radiative transfer model atmospheric correction for Group C; ((iii) relative normalization for Group D; and (iv) topographic correction, for Group E.

Top-of-atmosphere (TOA) reflectance (ρ_λ) was computed using the ENVI v5.2 software:

$$\rho_\lambda = \frac{\pi\, L_\lambda d^2}{SE_\lambda \sin\theta}$$

where, L_λ is the Radiance in Watts/(m^2 × Sr); d is the Earth–Sun distance (astronomical distance units); SE_λ is the solar irradiance in Watt/m^2; and θ is the sun elevation angle in degrees.

Atmospheric correction was carried out using the Fast Line-of-sight Atmospheric Analysis of Spectral Hypercubes (FLAASH) method [56]. The FLAASH model corrects wavelengths in the visible-mid infrared spectra and it includes an updated MODTRAN4 radiation transfer module [57]. The FLAASH atmospheric correction used the average height at each of the study sites within each image, and day visibility values derived from the nearest meteorological station (airport of Bogotá El Dorado, about 30 km distance from the plots) as inputs. Historical meteorological data were derived from the Colombian Institute for Hydrology, Meteorology and Environmental Studies—IDEAM (Table 3). According to the climatic data characteristics, we used the sub-Arctic summer (SAS) atmospheric model and the rural aerosol model within FLAASH.

Table 3. Parametrization of the Fast Line-of-sight Atmospheric Analysis of Spectral Hypercubes (FLAASH) atmospheric correction model. SAS is the sub-Arctic summer model.

Locality	Average Visibility (km)	Average Temperature (°C)	Average Height (m)	Aerosol Model	Atmospheric Model
Guasca	13	13.4	3061	Rural	SAS
Torca	11.6	13.8	2716	Rural	SAS
Guatavita	9	13.5	2885	Rural	SAS
Tabio	9	13.6	2614	Rural	SAS

For data group D relative normalization was performed using the dark-object subtraction method followed by the flat bed calibration approach, which does not require pre-knowledge about the atmosphere and/or atmospheric radiative transfer modelling. Consistent dark objects (mostly water bodies) and bright objects (mostly man-made structures and roofs) were used to implement the algorithm [58]:

$$\text{Relative reflectance values} = \frac{\text{Radiance values} - \overline{X}_{\{Dark1\}}}{\overline{X}_{\{Bright1\}}}$$

where $\overline{X}_{\{Dark1\}}$ is the average value of a set of selected dark pixels and $\overline{X}_{\{Bright1\}}$ is the average value of a set of selected bright pixels. First, the radiation scattering was corrected as the dark-object subtraction method suggests [59], using the Dark Subtract module of ENVI 4.3 to choose a set of dark pixels, compute their average value, and subtract it from the radiance scale image. Then, using the flat-field calibration method [60], the corrected radiance values were turned into relative reflectance values by dividing the radiance values by the average value of a set of chosen bright pixels.

Topographic effects on illumination can have a direct effect on the recorded images, especially in the case of large regions with steep slopes [61,62]. We examined the use of two topographic correction methods. The first method is the C-correction algorithm [61] applied on the FLAASH atmospherically corrected data set, and the second method is the empirical algorithm suggested by Meyer and colleagues [63] and implemented on high resolution Quickbird imagery by [64]. We used the most precise digital surface model (DSM) available for the study area, i.e., Intermap Technologies® NEXTMap® World 30™ DSM, with a 30 m ground sampling distance and 10 m LE95 vertical accuracy. The empirical method did not provided meaningful results in two of the study sites; hence only the results of the C-correction method were included in the subsequent analyses.

Finally, at the end of data pre-processing, we had five different data groups as follows:

1. Data group A: data radiometrically calibrated to radiance units.
2. Data group B: TOA reflectance data
3. Data group C: data atmospherically corrected to reflectance units.
4. Data group D: data normalized to relative reflectance values.
5. Data group E: data corrected from topographic effect.

2.3.2. Aboveground Biomass Estimation and Carbon Mapping

Five vegetation indices were selected based on studies of remote sensing-derived biomass estimates, and also according to their applicability with the type of imagery used in this study [30,36,65–68]. Table 4 details the analytic expression of the selected indices and related reference source.

Table 4. Selected vegetation indices and reference source. GREEN, RED and near infrared (NIR) relate to the pixel values for the corresponding bands.

Vegetation Index	Equation	Reference
Normalized Difference Vegetation Index (NDVI)	$NDVI = \frac{(NIR-RED)}{(NIR+RED)}$	[69]
Vegetation index number (VIN)	$VIN = \frac{NIR}{RED}$	[70]
Ratio Vegetation Index (RVI)	$RVI = \frac{RED}{NIR}$	[70]
Normalized Difference Greenness Index (NDGI)	$NDGI = \frac{(GREEN-RED)}{(GREEN+RED)}$	[71]
Transformed Vegetation Index (TVI)	$TVI = \frac{(NDVI+0.5)}{|NDVI+0.5|} \sqrt{|NDVI+0.5|}$	[72]

All the permanent plots were georeferenced with sub-meter precision using the Spectra Precision MobileMapper 120® GPS to derive precise plot boundary polygons. We selected all pixels from inside the plot boundaries (Figure 2), while excluding a few pixels where negative values were observed, a phenomenon commonly observed when radiative-transfer-model-based atmospheric correction is applied. We calculated a single VI value per plot for the five indices by averaging the values of all pixels selected in every plot.

For each plot, the five vegetation indices were then related to the correspondent AGB values estimated from the field observations (Table 1), for all five data groups. To this end, we tested three different models (AGB as a function of VI): linear, semi-log, and log-log. We used both in situ AGB values proportional to the total area of the VHR pixels selected per each plot (Figure 2), and the total AGB of the plot. These two options produced very similar results, hence we only reported the results related to the total plot biomass. The coefficient of determination (R^2) and RMSE were calculated as indications of the models goodness of fit and accuracy, respectively [29].

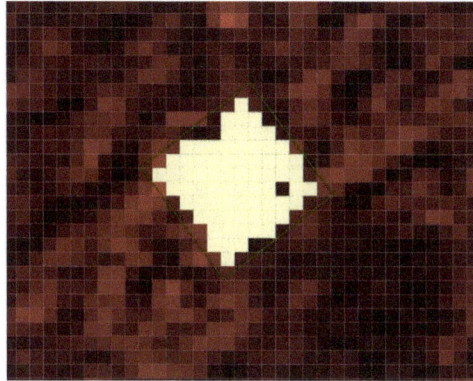

Figure 2. Overlap between plot boundaries (green line) and selected very high resolution pixels, in yellow (GeoEye-1 imagery) for the Guasca, Colombia, study site.

To better illustrate the application of the best fit VI model, we derived a VHR carbon map for secondary forests in one specific test area around an existing monitoring plot. First, we performed a parallelepiped supervised classification [55] of the VHR images of the Tabio area's secondary forests vegetation, using field data and training points delineated on-screen using the GoogleEarth interface. Pre-processing was previously applied to the raw data (radiometric, geometric and FLAASH atmospheric correction). The result was a secondary forests vegetation mask image with 92% user accuracy and 90% producer accuracy using 50 randomly selected and visually validated points analyzed in GoogleEarth. In a second step we applied, to the masked pre-processed image, our best model equation to derive a secondary forest aboveground biomass layer for the region.

Carbon stocks can vary substantially according to vegetation type (e.g., broadleaf, conifers) and biome. Accordingly, we assumed that aboveground carbon = 0.471 × AGB, based on average C content value calculated using 134 tropical angiosperms species from Thomas and Martin's review [73]. Using this relationship, we finally developed a carbon stock map for the Tabio test area.

3. Results

The goodness of fit and accuracy were calculated using R^2 and RMSE, for the best linear, semi-log, and log-log models of the five VIs (Table 5). All p-values were reported in Table 5 so as to indicate the process used for final model selection. For four out of the five vegetation indices, model goodness increased with the implementation of additional pre-processing steps (e.g., applying atmospheric correction (Group B, C, D) on radiance (Group A) images and further apply topographic correction (Group E)). For the normalized difference greenness index (NDGI)—the only index exploiting the Green band—no evident improvement in model goodness was noticeable among the data groups. The importance of the red wavelength as a most critical band in providing vegetation biophysical information is well-supported in the literature (e.g., [44]). Nevertheless, NDGI shows the best R^2 in the radiance and TOA data groups (A, B). Considering both R^2 and RMSE parameters, the best performing models involved the ratio vegetation index (RVI), transformed vegetation index (TVI), and NDVI indices in Group E, while the Vegetation Index Number (VIN) was the worst performing. In absolute terms, the AGB model which performed better was derived from the combination of the topographic corrected data (calculated on FLAASH atmospherically corrected images) and the ratio vegetation index (Figure 3) in a semi-log form ($R^2 = 0.582$, $p = 0.028$):

$$\log \text{AGB} = -3.208 \times \text{RVI} + 2.185$$

Table 5. Goodness of fit and accuracy parameters (R^2, Root Mean Square Error, p-value) for all vegetation indices and data processing groups in the study. Group A: imagery radiometrically calibrated to radiance units; Group B: top-of-atmosphere (TOA) reflectance data; Group C: data atmospherically corrected to reflectance units; Group D: data normalized to relative reflectance values; Group E: data corrected from topographic effect. Root mean square error (RMSE) expressed in Mg ha^{-1}. * Significance level p =0.05.

Index		A R^2	A RMSE	A p*	B R^2	B RMSE	B p*	C R^2	C RMSE	C p*	D R^2	D RMSE	D p*	E R^2	E RMSE	E p*
NDVI	lineal	0.18	41.6	0.292	0.43	42.0	0.317	0.29	34.8	0.079	0.45	38.8	0.171	0.45	34.2	0.069
	semi log	0.25	43.3	0.205	0.53	43.7	0.218	0.53	35.7	0.041	0.56	40.7	0.040	0.56	34.9	0.033
	log-log	0.26	43.0	0.198	0.55	43.5	0.214	0.55	35.6	0.035	0.57	40.2	0.035	0.57	34.8	0.029
RVI	lineal	0.20	41.1	0.262	0.44	41.6	0.291	0.3	34.6	0.075	0.46	38.5	0.161	0.46	33.9	0.066
	semi log	0.27	42.5	0.182	0.55	43.1	0.199	0.55	35.4	0.034	0.58	40.2	0.035	0.58	34.5	0.028
	log-log	0.27	42.8	0.188	0.48	43.3	0.204	0.48	34.4	0.057	0.51	41.9	0.056	0.51	33.5	0.046
VIN	lineal	0.12	43.2	0.405	0.42	43.6	0.445	0.21	35.1	0.083	0.42	40.9	0.252	0.42	35.0	0.082
	semi log	0.18	45.3	0.289	0.3	45.7	0.316	0.39	34.9	0.159	0.3	43.1	0.096	0.3	34.7	0.156
	log-log	0.21	44.7	0.258	0.34	45.2	0.284	0.48	37.4	0.129	0.36	41.5	0.057	0.36	37.0	0.116
NDGI	lineal	0.41	35.3	0.087	0.47	34.9	0.08	0.39	55.2	0.059	0.48	36.0	0.1	0.48	33.1	0.056
	semi log	0.49	36.6	0.054	0.44	36.9	0.045	0.4	33.3	0.072	0.45	36.8	0.091	0.45	33.0	0.069
	log-log	0.51	36.1	0.046	0.52	35.8	0.033	0.48	33.9	0.043	0.52	36.6	0.056	0.52	33.8	0.043
TVI	lineal	0.19	41.4	0.281	0.43	41.9	0.308	0.29	34.7	0.078	0.45	38.7	0.167	0.45	34.1	0.068
	semi log	0.26	43.0	0.196	0.53	43.5	0.211	0.54	35.6	0.039	0.56	40.5	0.038	0.56	34.8	0.031
	log-log	0.26	42.9	0.195	0.54	43.4	0.21	0.54	35.6	0.037	0.57	40.4	0.037	0.57	34.8	0.030

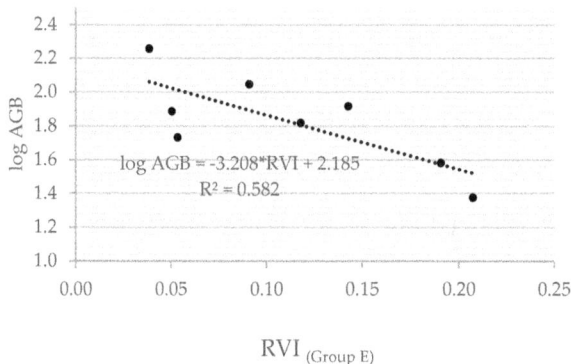

Figure 3. Best performing aboveground biomass (AGB) estimation model based on the ratio vegetation index (RVI) from Group E.

For the purpose of better synthesizing our results, we report AGB model equations only for the semi-log form for all indices and data processing groups (Table 6). In general terms, we found how the synergy betweeen the atmospheric and topographic corrections notably improved model goodness of fit in the AGB estimation models.

Table 6. Aboveground biomass (AGB) semi-log models and model goodness of fit (R^2) and root mean square error (RMSE) for all vegetation indices and five data processing group. * Significance level $p = 0.05$. (Best performing model in bold.)

Data Group	Model	R^2	RMSE (Mg· ha^{-1})	p *
Group A	log AGB = 2.758 × NDGI + 1.098	0.49	36.6	0.054
	log AGB = −2.912 × RVI + 2.875	0.27	42.5	0.182
	log AGB = 2.379 × NDVI + 0.692	0.25	43.3	0.205
	log AGB = 0,2119 × VIN + 1.195	0.18	45.3	0.289
	log AGB = 4.913 × TVI − 3.023	0.26	43.0	0.196
Group B	log AGB = 3.581 × NDGI + 1.173	0.52	36.9	0.045
	log AGB = −4.857 × RVI + 3.009	0.26	43.1	0.199
	log AGB = 3.371 × NDVI − 0.236	0.24	43.7	0.218
	log AGB = 0.1519 × VIN + 1.160	0.17	45.7	0.316
	log AGB = 7.379 × TVI − 5.950	0.25	43.5	0.211
Group C	log AGB = 1.245 × NDGI + 1.522	0.44	33.3	0.072
	log AGB = −3.129 × RVI + 2.191	0.55	35.4	0.034
	log AGB = 1.900 × NDVI + 0.311	0.53	35.7	0.041
	log AGB = 0.004 × VIN + 1.704	0.30	34.9	0.159
	log AGB = 4.363 × TVI − 3.139	0.54	35.6	0.039
Group D	log AGB = 2.837 × NDGI + 1.432	0.40	36.8	0.091
	log AGB = −2.912 × RVI + 2.249	0.55	40.2	0.035
	log AGB = 1.970 × NDVI + 0.343	0.53	40.7	0.040
	log AGB = 0.053 × VIN + 1.362	0.39	43.1	0.096
	log AGB = 4.355 × TVI − 3.044	0.54	40.5	0.038
Group E	log AGB = 1.263 × NDGI + 1.512	0.45	33.0	0.069
	log AGB =−3.208 × RVI + 2.185	**0.58**	**34.5**	**0.028**
	log AGB = 1.949 × NDVI + 0.257	0.56	34.9	0.033
	log AGB = 0.004 × VIN + 1.702	0.30	34.7	0.156
	log AGB = 4.473 × TVI − 3.280	0.57	34.8	0.031

The best performing model for AGB estimation (1) was used to derive maps of aboveground carbon stocks (Mg· ha^{-1}) for the Tabio test area. Aboveground biomass values of secondary forests were calculated for the corresponding VHR scenes (Pleiades-1A and GeoEye-1). Using the relationship of

C stocks = 0.471 × AGB from Thomas and Martin [73], accordingly we mapped the spatial distribution of aboveground C stocks (Mg· ha^{-1}) (Figure 4).

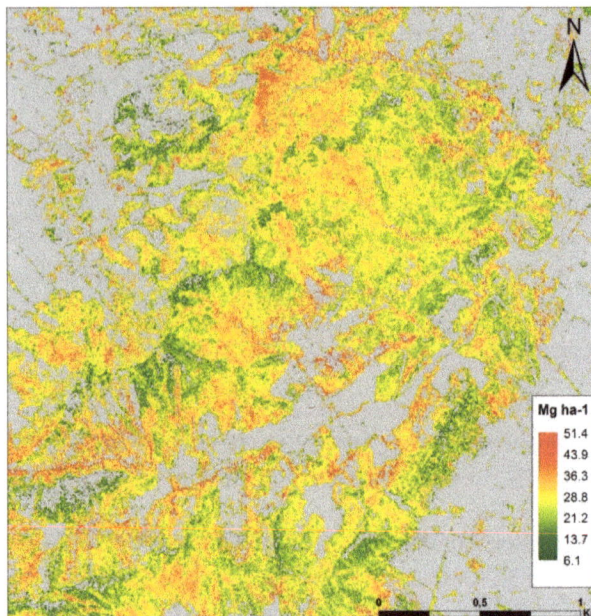

Figure 4. Aboveground carbon stock distribution map for secondary Andean in the Tabio test area near Bogotá, Colombia in Mg· ha^{-1} (based on Pleiades-1A imagery).

4. Discussion

The present study investigated the use of remote sensing to cost-effectively estimate AGB biomass and carbon stocks for Andean successional forests in periurban areas of Bogotá, Colombia. The AGB and carbon stock mapping estimates for a mid-successional stage secondary forest in the test area (ca. 6–51 Mg· ha^{-1}) were within the range reported by other studies of secondary forests in the Neotropics (e.g., [74]). We identified significant analytical relationships between vegetation indices extracted from multiple Geoeye-1 and Pleiades-1A VHR imagery and AGB values estimated using field measurements and allometric equations. A basic underlying assumption of the modelling approach is that larger tree densities (i.e., canopy density) are associated with larger aboveground biomass estimates [65]. Among the different preprocessing workflows used, the study confirmed the importance of the synergy between atmospheric and topographic correction when estimating biomass using satellite-based vegetation indices. At an average height of 2800 m our Andean study area is characterized by the frequent presence of dense aerosols and rugged terrain. As such, these are key factors when deriving the spectral characteristics of Andean Neotropical vegetation, and thus represent significant elements to account for when calculating vegetation indices in these environments. Also, the correction model we applied (FLAASH) has the advantage of being easily adaptable to specific types of environmental conditions, allowing for the proper mitigation of the effects of site-specific atmospheric conditions.

We tested the use of empirical atmospheric correction technique through image-to-image normalization and compared the results with the FLAASH model-based atmospheric correction and at sensor radiance results. We found that image-to-image normalization produced results comparable to the FLAASH module results. The manual selection of less-optimal dark and bright objects in each image could explain the slightly inferior results compared to the FLAASH model output results.

However, we believe image-to-image normalization has the potential of providing better results in the case of the existence of variations in illumination conditions due to sparse clouds [75], which is often the case in the Andeans plains.. More research is needed to test other statistically-based empirical atmospheric correction as well as different topographic correction models [76,77].

In this study we also tested the use of topographic illumination variation correction algorithm [64]. Our results show slight improvement in the AGB model goodness of fit and RMSE when the topographic correction is applied to the images. Although spectrally normalized vegetation indices such as the NDVI are less affected by the topographic-induced illumination variations [78,79], we believe that robust modeling of bidirectional distribution reflection function (BDRF) should provide for a more robust approach to handle the sun-object-sensor geometry variations common in forested land covers [80,81]. However, modeling the BDRF requires multi-view and multi-date images, which can be hard to achieve in our study area due to the persistent cloud cover found throughout the year. We also believe that the presence of a high resolution digital elevation model could have improved the topographic correction process and possibly also the AGB model results.

Overall, the VHR imagery provided satisfactory results for AGB and carbon modelling, considering the limitations posed by the number and size of the plots (20 m × 20 m) available for this analysis. The ratio vegetation index—the simple ratio between the NIR and the RED bands [70]—provided the best results among our AGB estimation models. It is not possible, however, to generalize about the goodness of the model relating RVI to in situ AGB, that is, if this particular index also has an optimal performance in different successional forests and environmental conditions. However, it does represent a significant relationship for high Andean secondary forest vegetation.

The study was limited by the small number of available permanent plots. Further research will involve the establishment of a larger network of regional and urban to rural gradient plots, to better take into account the heterogeneity of diverse species assemblages and their ecological successional stages, together with stochastic small-scale disturbances. Additional sources of error in our approach are also introduced by the selection of specific allometric equations for estimating AGB [82]. Similarly, we did not account for the presence and contribution to total biomass and carbon stocks from nonarboreal vegetation, such as mosses, lianas, ferns, and epiphytes, which can contribute a considerable amount of biomass in tropical secondary forests [83].

5. Conclusions

Secondary forests like those in this study represent a dominant land cover type, to the extent that in several tropical countries their surface area already exceeds that of mature forest cover [84]. In addition, periurban forests that develop after land clearing or reforestation activities or, similarly, after designation as a conservation or protected area, represent key reservoirs for terrestrial biodiversity [85,86], as well as providing a large array of other provision, regulation, and cultural ecosystem services [12,24]. For these reasons, we believe the results of the present study added a relevant piece of information for the characterization of these little known successional forests.

Overall, this study's approach and findings provide key baseline information for landscape level analyses in the Colombian Andes relating to local to regional AGB and carbon stocks estimates for little studied secondary forest types. Also, it provides a comprehensive, rapid, and cost-effective framework for AGB estimation using small field plots coupled with available VHR imagery. This approach can specifically be applied to facilitate periurban and regional mapping assessments of forest ecosystem service provision. It can also be used to better understand the effects of direct and indirect drivers of forest cover change, urbanization rates, and ecosystem service provision, as well as for forest inventory and monitoring of REDD+ (Reducing Emissions from Deforestation and forest Degradation) projects. Methods and mapping framework findings could also be exploited to develop carbon offset protocols to mitigate anthropogenic urban carbon emission from adjacent Bogotá and other cities in the Andes.

Acknowledgments: This research work was funded by Grant DGV188 of the Research Fund of the Universidad del Rosario (FIUR). The Authors acknowledge three anonymous reviewers for contributing to improve the article. Carolina Alvarez is also acknowledged for her help with the biomass calculations.

Author Contributions: N.C., A.A.-E., J.M.P. and F.J.E. conceived and designed the experiments; K.R., J.M.P., A.A.-E. performed the experiments; K.R., N.C., A.A.-E. and J.M.P. analyzed the data; N.C., A.A.-E., J.M.P. and F.J.E. wrote the paper.

Conflicts of Interest: The authors declare no conflict of interest. The founding sponsors had no role in the design of the study; in the collection, analyses, or interpretation of data; in the writing of the manuscript, and in the decision to publish the results.

References

1. Grace, J. Understanding and managing the global carbon cycle. *J. Ecol.* **2004**, *92*, 189–202. [CrossRef]
2. Farquhar, G.D. Carbon Dioxide and Vegetation. *Science* **1997**, *278*, 1411. [CrossRef]
3. UNFCCC. Methodological guidance for activities relating to reducing emissions from deforestation and forest degradation and the role of conservation, sustainable management of forests and enhancement of forest carbon stocks in developing countries. In *Decision 4/CP.15*; UNFCoC Change: Copenhagen, Denmark, 2009.
4. Daily, G.C. *Nature's Services: Societal Dependence on Natural Ecosystems*; Island Press: Washington, DC, USA, 1997.
5. Denman, K.L.; Brasseur, G.; Chidthaisong, A.; Ciais, P.; Cox, P.M.; Dickinson, R.E.; Hauglustaine, D.; Heinze, C.; Holland, E.; Jacob, D.; et al. Couplings Between Changes in the Climate System and Biogeochemistry. In *Climate Change 2007: The Physical Science Basis. Contribution of Working Group I to the Fourth Assessment Report of the Intergovernmental Panel on Climate Change*; Solomon, S., Qin, D.M., Eds.; Cambridge University Press: Cambridge, UK, 2007.
6. Dobbs, C.; Kendal, D.; Nitschke, C.R. Multiple ecosystem services and disservices of the urban forest establishing their connections with landscape structure and sociodemographics. *Ecol. Indic.* **2014**, *43*, 44–55. [CrossRef]
7. Pan, Y.; Birdsey, R.A.; Fang, J.; Houghton, R.; Kauppi, P.E.; Kurz, W.A.; Phillips, O.L.; Shvidenko, A.; Lewis, S.L.; Canadell, J.G.; et al. A large and persistent carbon sink in the world's forests. *Science* **2011**, *333*, 988–993. [CrossRef] [PubMed]
8. Bare, M.C.; Ashton, M.S. Growth of native tree species planted in montane reforestation projects in the Colombian and Ecuadorian Andes differs among site and species. *New For.* **2016**, *47*, 1–23. [CrossRef]
9. Aide, T.M.; Clark, M.L.; Grau, H.R.; López-Carr, D.; Levy, M.A.; Redo, D.; Bonilla-Moheno, M.; Riner, G.; Andrade-Núñez, M.J.; Muñiz, M. Deforestation and reforestation of Latin America and the Caribbean (2001–2010). *Biotropica* **2013**, *45*, 262–271. [CrossRef]
10. Antonio-Fragala, F.; Obregón-Neira, N. Recharge Estimation in Aquifers of the Bogota Savannah. *Ing. Univ.* **2011**, *15*, 145–169.
11. Grau, H.R.; Aide, M. Globalization and land-use transitions in Latin America. *Ecol. Soc.* **2008**, *13*, 16.
12. Escobedo, F.J.; Clerici, N.; Staudhammer, C.L.; Corzo, G.T. Socio-ecological dynamics and inequality in Bogotá, Colombia's public urban forests and their ecosystem services. *Urban For. Urban Green.* **2015**, *14*, 1040–1053. [CrossRef]
13. He, R.; Yang, J.; Song, X. Quantifying the Impact of Different Ways to Delimit Study Areas on the Assessment of Species Diversity of an Urban Forest. *Forests* **2016**, *7*, 42. [CrossRef]
14. Lewis, S.L.; Malhi, Y.; Phillips, O.L. Fingerprinting the impacts of global change on tropical forests. *Philos. Trans. R. Soc. B* **2004**, *359*, 437–462. [CrossRef] [PubMed]
15. Melillo, J.M.; Mcguire, A.D.; Kicklighter, D.W.; Moore, B., III; Vorosmarty, C.J.; Schloss, A.L. Global climate change and terrestrial net primary production. *Nature* **1993**, *363*, 234–240. [CrossRef]
16. Houghton, R.A. Aboveground forest biomass and the global carbon balance. *Glob. Chang. Biol.* **2005**, *11*, 945–958. [CrossRef]
17. Mahli, Y.; Wood, D.; Bakers, T.R.; Wright, J.; Phillips, O.L.; Cochrane, T.; Meir, P.; Chave, J.; Almeida, S.; Arroyo, L.; et al. The regional variation of aboveground life biomass in old-growth Amazonian forests. *Glob. Chang. Biol.* **2006**, *12*, 1107–1138. [CrossRef]
18. Saatchi, S.S.; Houghton, R.A.; Dos Santos Alvalá, R.C.; Soares, J.V.; Yu, Y. Distribution of aboveground live biomass in the Amazon. *Glob. Chang. Biol.* **2007**, *13*, 816–837. [CrossRef]

19. Phillips, J.; Duque, Á.; Scott, C.; Wayson, C.; Galindo, G.; Cabrera, E.; Chave, J.; Peña, M.; Álvarez, E.; Cárdenas, D.; Duivenvoorden, J. Live aboveground carbon stocks in natural forests of Colombia. *For. Ecol. Manag.* **2016**, *374*, 119–128. [CrossRef]

20. Gentry, A.J. Seasonally Dry Tropical Forests. In *Diversity and Floristic Composition of Neotropical Dry Forests*; Bullock, S.H., Mooney, H.A., Medina, E., Eds.; Cambridge University Press: Cambridge, UK, 2009; pp. 146–194.

21. Cuatrecasas, J. Aspectos de la vegetación natural de Colombia. *Perez Arbelaezia* **1989**, *2*, 155–285.

22. SISAC-DANE. *República de Colombia. Encuesta Nacional Agropecuaria. Resultados 1995*; Sistema de Información del Sector Agrario Colombiano SISAC-Departamento Nacional de Estadística DANE: Bogotá, Colombia, 1996.

23. Sánchez-Cuervo, A.M.; Aide, T.M.; Clark, M.L.; Etter, A. Land Cover Change in Colombia: Surprising Forest Recovery Trends between 2001 and 2010. *PLoS ONE* **2012**, *7*, e43943. [CrossRef] [PubMed]

24. Gilroy, J.J.; Woodcock, P.; Edwards, F.A.; Wheeler, C.; Baptiste, B.L.G.; Medina, C.A.; Haugaasen, T.; Edwards, D.P. Cheap carbon and biodiversity co-benefits from forest regeneration in a hotspot of endemism. *Nat. Clim. Chang.* **2014**, *4*, 503–507. [CrossRef]

25. Chave, J.; Aandalo, C.; Brown, S.; Cairns, M.; Chambers, J.C.; Eamus, D.; Fölster, H.; Fromard, F.; Higuchi, N.; Kira, T.; et al. Tree allometry and improved estimation of carbon stocks and balance in tropical forests. *Oecologia* **2005**, *145*, 87–99. [CrossRef] [PubMed]

26. Volkova, L.; Bi, H.; Murphy, S.; Weston, C.J. Empirical Estimates of Aboveground Carbon in Open Eucalyptus Forests of South-Eastern Australia and Its Potential Implication for National Carbon Accounting. *Forests* **2015**, *6*, 3395–3411. [CrossRef]

27. Beets, P.N.; Kimberley, M.O.; Oliver, G.R.; Pearce, S.H.; Graham, J.D.; Brandon, A. Allometric Equations for Estimating Carbon Stocks in Natural Forest in New Zealand. *Forests* **2012**, *3*, 818–839. [CrossRef]

28. Goetz, S.J.; Baccini, A.; Laporte, N.T.; Johns, T.; Walker, W.; Kellndorfer, J.; Houghton, R.A.; Sun, M. Mapping and monitoring carbon stocks with satellite observations: A comparison of methods. *Carbon Balance Manag.* **2009**, *4*. [CrossRef] [PubMed]

29. Lu, D. The potential and challenge of remote sensing-based biomass estimation. *Int. J. Remote Sens.* **2006**, *27*, 1297–1328. [CrossRef]

30. Wang, X.; Shao, G.; Chen, H.; Lewis, B.J.; Qi, G.; Yu, D.; Zhou, L.; Dai, L. An Application of Remote Sensing Data in Mapping Landscape-Level Forest Biomass for Monitoring the Effectiveness of Forest Policies in Northeastern China. *Environ. Manag.* **2013**, *52*, 612–620. [CrossRef] [PubMed]

31. Avitabile, V.; Baccini, A.; Friedl, M.A.; Schmullius, C. Capabilities and limitations of Landsat and land cover data for aboveground woody biomass estimation of Uganda. *Remote Sens. Environ.* **2012**, *117*, 366–380. [CrossRef]

32. Powell, S.L.; Cohen, W.B.; Healey, S.P.; Kennedy, R.E.; Moisen, G.G.; Pierce, K.B.; Ohmann, J.L. Quantification of Live Aboveground Forest Biomass Dynamics with Landsat Time-series and Field Inventory Data: A Comparison of Empirical Modeling Approaches. *Remote Sens. Environ.* **2010**, *114*, 1053–1068. [CrossRef]

33. Labreque, S.; Fournier, R.A.; Luther, J.E.; Piercey, D. A comparison of four methods to map biomass from Landsat-TM and inventory data in western Newfoundland. *For. Ecol. Manag.* **2006**, *226*, 129–144. [CrossRef]

34. Lucas, R.M.; Honzak, M.; do Amaral, I.; Curran, P.; Foody, G.M.; Amaral, S. The contribution of remotely sensed data in the assessment of the floristic composition, total biomass and structure of tropical regenerating forests. In *Regeneracao Florestal: Pesquisas na Amazonia*; Gascon, C., Moutinho, P., Eds.; Inpa Press: Manaus, Brazil, 1998; pp. 61–82.

35. Nelson, R.F.; Kimes, D.S.; Salas, W.A.; Routhier, M. Secondary Forest Age and Tropical Forest Biomass Estimation Using Thematic Mapper Imagery Single-year tropical forest age classes, a surrogate for standing biomass, cannot be reliably identified using single-date TM imagery. *Bioscience* **2000**, *50*, 419–431. [CrossRef]

36. Foody, G.M.; Boyd, D.S.; Cutler, M. Predictive relations of tropical forest biomass from Landsat TM data and their transferability between regions. *Remote Sens. Environ.* **2003**, *85*, 463–474. [CrossRef]

37. Boyd, D.S. The relationship between the biomass of Cameroonian tropical forests and radiation reflected in middle infrared wavelengths (3.0–5.0 mu m). *Int. J. Remote Sens.* **1999**, *20*, 1017–1023. [CrossRef]

38. Lu, D. Aboveground biomass estimation using Landsat TM data in the Brazilian Amazon. *Int. J. Remote Sens.* **2005**, *26*, 2509–2525. [CrossRef]

39. Bannari, A.; Morin, D.; Huette, A.R.; Bonn, F. A review of vegetation indices. *Remote Sens. Rev.* **1995**, *13*, 95–120. [CrossRef]
40. Zhang, C.; Lu, D.; Chen, X.; Zhang, Y.; Maisupova, B.; Tao, Y. The spatiotemporal patterns of vegetation coverage and biomass of the temperate deserts in Central Asia and their relationships with climate controls. *Remote Sens. Environ.* **2016**, *175*, 271–281. [CrossRef]
41. Osuri, A.M.; Madhusudan, M.D.; Kumar, V.S.; Chengappa, S.K.; Kushalappa, C.G.; Sankaran, M. Spatio-temporal variation in forest cover and biomass across sacred groves in a human-modified landscape of India's Western Ghats. *Biol. Conserv.* **2014**, *178*, 193–199. [CrossRef]
42. Viana, H.; Aranha, J.; Lopes, D.; Cohen, W.B. Estimation of crown biomass of Pinus pinaster stands and shrubland above-ground biomass using forest inventory data, remotely sensed imagery and spatial prediction models. *Ecol. Model.* **2012**, *226*, 22–35. [CrossRef]
43. Ji, L.; Peters, A.J. Performance evaluation of spectral vegetation indices using a statistical sensitivity function. *Remote Sens. Environ.* **2007**, *106*, 59–65. [CrossRef]
44. Thenkabail, P.S.; Stucky, N.; Griscom, B.W.; Ashton, M.S.; Diels, D.; van der Meer, B.; Enclona, E. Biomass estimations and carbon stock calculations in the oil palm plantations of African derived savannas using IKONOS data. *Int. J. Remote Sen.* **2004**, *25*, 5447–5472. [CrossRef]
45. Zhou, J.; Zhao, Z.; Zhao, Q.; Zhao, J.; Wang, H. Quantification of aboveground forest biomass using quickbird imagery, topographic variables, and field data. *J. Appl. Remote Sens.* **2013**, *7*, 073484. [CrossRef]
46. Zhu, Y.; Liu, K.; Liu, L.; Wang, S.; Liu, H. Retrieval of Mangrove Aboveground Biomass at the Individual Species Level with WorldView-2 Images. *Remote Sens.* **2015**, *7*, 12192–12214. [CrossRef]
47. Pereira, P.; Zullo, J.; Gonçalves, R.R.; Romani, L.A.S.; Pinto, H.S. Coffee Crop's Biomass and Carbon Stock Estimation With Usage of High Resolution Satellites Images. *IEEE J. Sel. Top. Appl.* **2013**, *6*, 1786–1795.
48. Mendoza, J.E.; Etter, A. Multitemporal analysis (1940–1996) of land cover changes in the southwestern Bogotá Highplain. *Landsc. Urban Plan.* **2002**, *59*, 147–158. [CrossRef]
49. Departamento Administrativo Nacional de Estadística (DANE). Available online: http://www.dane.gov.co/ (accessed on 1 December 2015).
50. Montañez, G.; Arcila, O.; Pacheco, J.C. *Hacia Dónde va la Sabana de Bogotá? Modernización, Conflicto, Ambiente y Sociedad*; Universidad Nacional de Colombia. Centro de Estudios Sociales (SENA): Bogotá, Colombia, 1994.
51. Sierra, C.A.; del Valle, J.I.; Orrego, A.; Moreno, F.H.; Harmon, M.E.; Zapata, M.; Colorado, G.J.; Herrera, M.A.; Lara, W.; Restrepo, D.E.; et al. Total carbon stocks in a tropical forest landscape of the Porce region Colombia. *For. Ecol. Manag.* **2007**, *243*, 299–309. [CrossRef]
52. Pérez, M.C.; Díaz, J.J. Estimación del Carbono Contenido en la Biomasa Forestal Aérea de dos Bosques Andinos en los Departamentos de Santander y Cundinamarca. Master's Thesis, Universidad Distrital Francisco José de Caldas, Bogotá, Colombia, 2010.
53. Chander, G.; Markham, B.L.; Helder, D.L. Summary of current radiometric calibration coefficients for Landsat MSS, TM, ETM+, and EO-1 ALI sensors. *Remote. Sens. Environ.* **2009**, *113*, 893–903. [CrossRef]
54. Lira, J. *Tratamiento Digital de Imágenes Multiespectrales*, 2nd ed.; Universidad Nacional Autónoma de México: México D.F, México, 2010; p. 605.
55. Richards, J.A. *Remote Sensing Digital Image Analysis*; Springer-Verlag: Berlin, Germany, 1999; p. 240.
56. Berk, A.; Bernstein, L.S.; Anderson, G.P.; Acharya, P.K.; Robertson, D.C.; Chetwynd, J.H.; Adler-Golden, S.M. MODTRAN Cloud and Multiple Scattering Upgrades with Application to AVIRIS. *Remote Sens. Environ.* **1998**, *65*, 367–375. [CrossRef]
57. Matthew, M.W.; Adler-Golden, S.M.; Berk, A.; Richtsmeier, S.C.; Levine, R.Y.; Bernstein, L.S.; Acharya, P.K.; Anderson, G.P.; Felde, G.W.; Hoke, M.P.; et al. Status of Atmospheric Correction Using a MODTRAN4-based Algorithm. *Proc. Soc. Photo-Opt. Instrum. Eng.* **2000**, *4049*, 199–207. [CrossRef]
58. Hall, F.G.; Strebel, D.E.; Nickeson, J.E.; Goetz, S.J. Radiometric rectification: Toward a common radiometric response among multidate, multisensor images. *Remote Sens. Environ.* **1991**, *35*, 11–27. [CrossRef]
59. Chavez, P.S. An improved dark-object subtraction technique for atmospheric scattering correction of multiespectral data. *Remote Sens. Environ.* **1988**, *24*, 459–479. [CrossRef]
60. Roberts, D.A.; Yamaguchi, Y.; Lyon, R.J.P. Comparison of various techniques for calibration of AIS data. In Proceedings of the Second AIS Data Analysis Workshop, Pasadena, CA, USA, 6–8 May 1986; Vane, G., Goetz, A.F.H., Eds.; JPL Publ. 86-35; Jet Propulsion Laboratory: Pasadena, CA, USA, 1986; pp. 21–30.

61. Teillet, P.M.; Guindon, B.; Goodenough, D.G. On the slope-aspect correction of multispectral scanner data. *Can. J. Remote Sens.* **1982**, *8*, 84–106. [CrossRef]

62. Soenen, S.A.; Peddle, D.R.; Coburn, C.A. SCS+ C: A Modified Sun-Canopy-Sensor Topographic Correction in Forested Terrain. *IEEE Trans. Geosci. Remote Sens.* **2005**, *43*, 2148–2159. [CrossRef]

63. Meyer, P.; Itten, K.I.; Kellenberger, T.; Sandmeier, S.; Sandmeier, R. Radiometric corrections of topographically induced effects on Landsat TM data in an alpine environment. *ISPRS J. Photogramm.* **1993**, *48*, 17–28. [CrossRef]

64. Wu, J.; Bauer, M.E.; Wang, D.; Manson, S.M. A comparison of illumination geometry-based methods for topographic correction of QuickBird images of an undulant area. *ISPRS J. Photogramm.* **2008**, *63*, 223–236. [CrossRef]

65. Anaya, J.A.; Chuvieco, E.; Palacios-Orueta, A. Aboveground biomass assessment in Colombia: A remote sensing approach. *For. Ecol. Manag.* **2009**, *257*, 1237–1246. [CrossRef]

66. Zheng, D.; Rademacher, J.; Chen, J.; Crow, T.; Bresee, M.; Le Moine, J.; Ryu, S. Estimating aboveground biomass using Landsat 7 ETM+ data across a managed landscape in northern Wisconsin, USA. *Remote Sens. Environ.* **2004**, *93*, 402–411. [CrossRef]

67. Shang, Z.; Zhou, G.; Du, H.; Xu, X.; Shi, Y.; Lü, Y.; Zhou, Y.; Gu, C. Moso bamboo forest extraction and aboveground carbon storage estimation based on multi-source remotely sensed images. *Int. J. Remote Sens.* **2013**, *34*, 5351–5368. [CrossRef]

68. Yan, F.; Wu, B.; Wang, Y. Estimating aboveground biomass in Mu Us Sandy Land using Landsat spectral derived vegetation indices over the past 30 years. *J. Arid Land* **2013**, *5*, 521–530. [CrossRef]

69. Rouse, J.W.; Haas, R.W.; Schell, J.A.; Deering, D.W.; Harlan, J.C. *Monitoring the Vernal Advacement and Retrogradation (Greenwave Effect) of Natural Vegetation*; NASA/GSFCT Type III Final Report; Greenbelt, MD, USA, 1974.

70. Pearson, R.L.; Miller, L.D. Remote mapping of standing crop biomass for estimation of the productivity of the shortgrass prairie, Pawnee National Grasslands, Colorado. In Proceedings of the 8th International Symposium on Remote Sensing of the Environment II, Ann Arbor, MI, USA, 2–6 October 1972; pp. 1355–1379.

71. Chamard, P.; Courel, M.F.; Ducousso, M.; Guénégou, M.C.; Le Rhun, J.; Levasseur, J.E.; Loisel, C.; Togola, M. Utilisation des Bandes Spectrales du vert et du Rouge pour une Meilleure Évaluation des Formations Végétales Actives. In *Télédétection et Cartographie*; AUPELF-UREF: Sherbrooke, QC, Canada, 1991; pp. 203–209.

72. Perry, C.R.; Lautenschlager, L.F. Functional equivalence of spectral vegetation indices. *Remote Sens. Environ.* **1984**, *14*, 169–182. [CrossRef]

73. Thomas, S.C.; Martin, A.R. Carbon content of tree tissues: A synthesis. *Forests* **2012**, *3*, 332–352. [CrossRef]

74. Poorter, L.; Bongers, F.; Aide, T.M.; Almeyda Zambrano, A.M.; Balvanera, P.; Becknell, J.M.; Boukili, V.; Brancalion, P.H.S.; Broadbent, E.N.; Chazdon, R.L.; et al. Biomass resilience of Neotropical secondary forests. *Nature* **2016**, *530*, 211–214. [CrossRef] [PubMed]

75. Klempner, S.L.; Bartlett, B.; Schott, J.R. Ground truth-based variability analysis of atmospheric inversion in the presence of clouds. *Proc. Soc. Photo-Opt. Instrum. Eng.* **2006**, *6301*, 630109. [CrossRef]

76. Kayadibi, Ö. Evaluation of imaging spectroscopy and atmospheric correction of multispectral images (Aster and LandsaT 7 ETM+). In Proceedings of the IEEE 2011 5th International Conference on Recent Advances in Space Technologies (RAST), Istanbul, Turkey, 9–11 June 2011; pp. 154–159.

77. Green, A.A.; Craig, M.D. Analysis of aircraft spectrometer data with logarithmic residuals. In Proceedings of the Third AIS workshop, Pasadena, CA, USA, 8–10 April 1985; pp. 111–129.

78. Galvão, L.S.; Breunig, F.M.; Teles, T.S.; Gaida, W.; Balbinot, R. Investigation of terrain illumination effects on vegetation indices and VI derived phenological metrics in subtropical deciduous forests. *GISci. Remote Sens.* **2016**, *53*, 360–381. [CrossRef]

79. Matsushita, B.; Yang, W.; Chen, J.; Onda, Y.; Qiu, G. Sensitivity of the enhanced vegetation index (EVI) and normalized difference vegetation index (NDVI) to topographic effects: A case study in high-density cypress forest. *Sensors* **2007**, *7*, 2636–2651. [CrossRef]

80. Li, F.; Jupp, D.L.; Thankappan, M.; Lymburner, L.; Mueller, N.; Lewis, A.; Held, A. A physics-based atmospheric and BRDF correction for Landsat data over mountainous terrain. *Remote Sens. Environ.* **2012**, *124*, 756–770. [CrossRef]

81. Hugli, H.; Frei, W. Understanding anisotropic reflectance in mountainous terrain. *Photogramm. Eng. Remote Sens.* **1983**, *49*, 671–683.

82. Viera, S.A.; Alves, F.; Aidar, M.; Spinelli Araújo, L.; Baker, T.; Ferreira Batista, J.L.; Cruz Campos, M.; Barbosa Camargo, P.; Chave, J.; Carvalho Delitti, W.B.; et al. Estimation of biomass and carbon stocks: The case of the Atlantic Forest. *Biota Neotrop.* **2008**, *8*, 21–29. [CrossRef]

83. Stas, S.M. *Above-Ground Biomass and Carbon Stocks in a Secondary Forest in Comparison with Adjacent Primary Forest on Limestone in Seram, the Moluccas, Indonesia*; CIFORHQ 5258; Center for International Forestry Research (CIFOR): Bogor, Indonesia, 2014; p. 19.

84. Food and Agriculture Organization. *The State of the World's Forests*; FAO: Rome, Italy, 2005.

85. Norden, N.; Chazdon, R.L.; Chao, A.; Jiang, Y.H.; Vílchez-Alvarado, B. Resilience of tropical rain forests: Tree community reassembly in secondary forests. *Ecol. Lett.* **2009**, *12*, 385–394. [CrossRef] [PubMed]

86. Chazdon, R.L.; Peres, C.A.; Dent, D.; Sheil, D.; Lugo, A.E.; Lamb, D.; Stork, N.E.; Miller, S.E. The potential for species conservation in tropical secondary forests. *Conserv. Biol.* **2009**, *23*, 1406–1417. [CrossRef] [PubMed]

forests

MDPI

Article

Removal of PM$_{10}$ by Forests as a Nature-Based Solution for Air Quality Improvement in the Metropolitan City of Rome

Federica Marando, Elisabetta Salvatori, Lina Fusaro and Fausto Manes *

Sapienza University of Rome, Department of Environmental Biology, P. le Aldo Moro, 5, Rome 00185, Italy; federica.marando@uniroma1.it (F.M.); elisabetta.salvatori@uniroma1.it (E.S.); lina.fusaro@uniroma1.it (L.F.)
* Correspondence: fausto.manes@uniroma1.it; Tel.: +39-06-4991-2451; Fax: +39-06-4991-2448

Academic Editors: Francisco Escobedo, Stephen John Livesley and Justin Morgenroth
Received: 30 March 2016; Accepted: 11 July 2016; Published: 21 July 2016

Abstract: Nature-based solutions have been identified by the European Union as being critical for the enhancement of environmental qualities in cities, where urban and peri-urban forests play a key role in air quality amelioration through pollutant removal. A remote sensing and geographic information system (GIS) approach was applied to the Metropolitan City (MC) of Rome to assess the seasonal particulate matter (PM$_{10}$) removal capacity of evergreen (broadleaves and conifers) and deciduous species. Moreover, a monetary evaluation of PM$_{10}$ removal was performed on the basis of pollution externalities calculated for Europe. Deciduous broadleaves represent the most abundant tree functional group and also yielded the highest total annual PM$_{10}$ deposition values (1769 Mg). By contrast, PM$_{10}$ removal efficiency (Mg·ha^{-1}) was 15%–22% higher in evergreen than in deciduous species. To assess the different removal capacity of the three functional groups in an area with homogeneous environmental conditions, a study case was performed in a peri-urban forest protected natural reserve (Castelporziano Presidential Estate). This study case highlighted the importance of deciduous species in summer and of evergreen communities as regards the annual PM$_{10}$ removal balance. The monetary evaluation indicated that the overall PM$_{10}$ removal value of the MC of Rome amounted to 161.78 million Euros. Our study lends further support to the crucial role played by nature-based solutions for human well-being in urban areas.

Keywords: urban areas; PM$_{10}$ deposition; urban forests; remote sensing and GIS; tree functional traits

1. Introduction

Improving the air quality in cities is one of the main challenges for the European Union (EU). Air pollution due to particulate matter (PM) is considered to represent one of the main health risks for European citizens [1]. A significant proportion of the population in Europe (73%) currently lives in cities, where pollutant concentrations frequently exceed the limits laid down in air pollution regulations. The number of city dwellers is expected to increase to 82% by the year 2050, i.e., 606 million European citizens will live in cities by then. In this regard, in 2011 around 33% of the urban population lived in areas in which the daily air quality limit value for coarse PM (PM$_{10}$) (50 µg·m^{-3} Directive 2008/50/CE) was exceeded, and if the World Health Organization (WHO) annual air quality guidelines (20 µg·m^{-3}) are considered, the percentage rises to 88% [1].

This scenario of increasing environmental risks in cities calls for new solutions to improve the quality of urban environments. The European Union recently suggested that the properties of natural ecosystems, and the Ecosystem Services (ES) they provide, may become the focus of specific research and innovation policies in order to find new viable solutions to challenges faced by society [2]. These so-called "nature-based solutions" may exert a positive environmental impact, which could form the basis of sustainable urban planning, by reducing energy requirement costs and mitigating climate changes and the causes of stress conditions [3–5]. As defined in the Millennium Ecosystem Assessment [6], ES are divided in supporting, regulating, provisioning and cultural services, and since biodiversity plays a key role in the provision of ES, it also inevitably affects human well-being [7,8].

In a work aimed at illustrating ES provided by different ecosystems in the city of Stockholm, Bolund and Hunnamar [9] identified seven different urban ecosystems: street trees, lawns/parks, urban forests, cultivated lands, wetlands, lakes/sea, and streams/rivers. Indeed, many papers have highlighted the importance of urban parks and gardens, as well as urban and peri-urban forests, which form an interconnected network of green space known as Green Infrastructure (GI) [10], as providers of different types of ES for urban dwellers [11]. These ES include the improvement in urban microclimate [12,13] and psychological benefits [14], as well as the improvement in air quality [7,15–17], thus integrating conventional human technologies [18]. When Nowak et al. [17] recently analyzed the effects of urban forests on air quality and human health in the United States, they found that the improvement in air quality, measured as a percentage of air pollution removal by trees, accounts for less than 1%. However, in highly vegetated areas, trees can improve air quality by as much as 16% [19]. Baumgardner et al. [20] pointed out that around 2% of the ambient PM_{10} in Mexico City is removed from the study area. In a study carried out in the city of Barcelona (Spain), Barò et al. [21] reported that urban forest services reduce PM_{10} air pollution by 2.66%. Moreover, in the Mediterranean city of Tel-Aviv, Cohen et al. [22] observed that an urban park significantly mitigated nitrogen oxides (NO_x) and PM_{10} concentrations, with a greater removal rate being observed in winter, and increased tropospheric ozone levels during summer. The effect of GI on urban air quality thus appears not to be negligible and should be considered in urban planning [23,24]. Indeed, many European cities have a long history not only in the development of the urban fabric, but also in urban green characteristics. In this regard, the major changes that took place in the late 19th century, characterized by rapid urban expansion, largely neglected urban green areas, which means there is now a considerable disparity in the amount of green space available for dwellers in cities across Europe [25]. Within this context, Rome, the capital of Italy, is known to be one of the "greenest" cities in Italy: despite the long-lasting human impact (more than 2750 years) and the marked increase in the urbanized area over the last 60 years, 20% of the overall municipality is still covered by public green areas, which include parks, historical villas, gardens and tree-lined roads, as well as a network of nine natural reserves [26]. Urban forests within the city's boundaries are composed of residual fragments of ancient woodlands that host a wide range of tree species, such as the typical Mediterranean evergreen broadleaves (*Quercus ilex* and *Q. suber*), deciduous *Quercus* woods (*Q. cerris, Q. frainetto*) and conifer plantations (*Pinus pinea*) [7]. There is therefore the need to preserve these existing forests, as well as to improve the urban GI network of Rome, in order to conserve and restore its ES provision, paying particular attention to the effects any initiatives might have on air quality.

The aim of this work was to estimate the seasonal PM_{10} removal capacity of urban and peri-urban forests in the Metropolitan City (MC) of Rome by quantifying the amount of PM_{10} removed by different functional groups of vegetation, i.e., evergreen (broadleaves and conifers) and deciduous forests. We applied a spatially explicit approach, in which the remote sensing of vegetation structure was integrated in the simulation of the PM_{10} deposition fluxes within the different functional groups. Moreover, in order to relate the PM_{10} deposition rates to the varying removal efficiency of the functional groups and to evaluate the PM_{10} removal efficiency of vegetation in a protected area, we present a study case on a peri-urban forest, i.e., the natural reserve of the Castelporziano Presidential Estate. This site is particularly suitable for two reasons: (i) this relatively small area, characterized by the prevalence of forest ecosystems, minimizes the confounding factors deriving from environmental and landscape heterogeneity; and (ii) there is the contemporary presence in this peri-urban forest of all three functional groups of vegetation investigated in this work.

2. Materials and Methods

2.1. Study Areas: The Metropolitan City of Rome and the Castelporziano Presidential Estate

The MC of Rome, Italy (41°54′ N, 12°29′ E), which corresponds to the former Province of Rome, is one of the 14 Italian MCs, administrative units introduced in 2014 (State Law 56/2014). It covers an area of 5352 km^2, which includes extensive and heterogeneous territorial bodies. The MC of Rome, which currently has 4,342,122 inhabitants [27], is characterized by high levels of land use change, accounting for over 50,000 hectares of soil consumption and consisting of a change from non-artificial coverage (non-consumed soil) to artificial coverage (consumed soil), which is defined as the whole sealed and permanently covered surfaces and excludes open natural and semi-natural urban areas [28]. Nevertheless, it also hosts large urban forests and green areas characterized by high levels of natural and historical significance, as well as agricultural areas located within the highly urbanized municipality [29,30]. Beyond the urban inner core of the MC of Rome lie large agricultural surfaces and extensive, heterogeneous forest ecosystems. The MC contains a high degree of biological diversity of tree species found in important natural areas, such as the Regional Park of the Simbruini mountains in the northeast, which is characterized by the widespread presence of deciduous oak species (*Q. cerris*, *Q. frainetto*) and beech woods (*Fagus sylvatica*), the Lepini mountains, which also host typical evergreen broadleaves (*Q. ilex* and *Q. suber*), and the Alban hills in the southeast, where chestnut (*Castanea sativa*) woods prevail, and lastly volcanic mountains (the Tolfa and Sabatini mountains), with mixed broadleaved forests, in its northwestern quadrant. In the southern coastal area of the MC, approximately 20 km from the urban center, lies the Castelporziano Presidential Estate, a natural reserve of around 5900 hectares, characterized by high levels of biodiversity and pristine forests [31,32]. The climate in this estate is strictly Mediterranean and hosts typical Mediterranean ecosystems (Mediterranean maquis, holm oak forests), as well as several deciduous oak communities and pine plantations [33]. Most of the forest cover, which accounts for over 75% of the Castelporziano Presidential Estate, consists of natural or semi-natural forests, many of which are classified as old-growth forest [32].

2.2. Classification of Remotely Sensed Data

In order to assess the urban forest composition of the MC of Rome, the Landsat 8 OLI/TIRS image of 18 July 2015, with a resolution of 30 m^2, was used to produce a map of the main land use categories. After a radiometric calibration and Dark Object Subtraction, using a semi-automatic Land Cover classification implemented in QGIS [34], a supervised classification was performed, using bands 4, 5, 6 and 7, with a maximum likelihood algorithm. The overall accuracy of the classification was then calculated by means of an error matrix. The physiognomic-structural categories of vegetation identified were then grouped into three main functional groups (evergreen broadleaves, deciduous broadleaves and conifers, Table 1), according to a morpho-functional criterion [7].

Table 1. Aggregation scheme of the physiognomic-structural categories of vegetation in the three functional groups.

Physiognomic-Structural Categories of Vegetation	Functional Groups
Conifers prevailing and broadleaved species (*Pinus pinea*, *Quercus* spp.) Reafforestation with Italian stone pine (*Pinus pinea*)	Conifers
Holm oak prevailing (*Quercus ilex*) Mediterranean maquis	Evergreen broadleaves
Deciduous woods prevailing (*Quercus cerris*, *Q. frainetto*, *Q. pubescens*, *Carpinus* spp.) Chestnut woods (*Castanea sativa*) Beech woods (*Fagus sylvatica*)	Deciduous broadleaves

2.3. Temporal Schedule

All the estimations performed for the MC of Rome were calculated according to astronomical seasons, and by accounting for the different phenology of deciduous and evergreen species. While evergreen deposition was calculated throughout the year, for the deciduous functional group we selected a period of 218 days, from 20 March to 24 October, as the vegetative period, which falls between early spring and early autumn.

2.4. Remotely Sensed Leaf Area Index

The Leaf Area Index (LAI) data were retrieved from the Terra Moderate Resolution Imaging Spectroradiometer MODIS MOD15A2H V6 product, with a resolution of 500 m^2 and a temporal resolution ofeight days. Forty-six images for the year 2015 were downloaded from the LPDAAC database and georeferenced into the WGS 84 UTM 33N reference system. Low-quality pixels, identified through MODIS quality control, were removed from the images. The images acquired were then aggregated into seasonal means for the year 2015. The number of missing low-quality pixels, removed previously in the cleaning process, was cut by reducing the temporal resolution. In order to obtain spatial consistency between the MODIS LAI data and the land use classification, spatial interpolation of missing LAI pixels, based on a regularized spline tension algorithm, was applied to avoid an incomplete overlay with the classification. Pixels in the LAI data were missing for the following two reasons: (1) the spatial resolution of the MODIS sensor is lower than that of the Landsat classification; and (2) low-quality pixels were removed (cloud contamination, dead detector).

2.5. Air PM$_{10}$ Concentrations

Hourly concentrations of particulate (PM$_{10}$) for the year 2015 were obtained from 20 monitoring stations data (Regional Environmental Protection Agency, ARPA Lazio) located throughout the MC of Rome. The monitoring stations, which record PM$_{10}$ concentrations in $\mu g \cdot m^{-3}$, are divided in different classes (Legislative Decree 155/2010) on the basis of their location in the MC (Table 2). Annual mean concentrations for the year 2015 are also shown in Table 2. The seasonal mean concentrations were derived from the hourly PM$_{10}$ concentration data. The point concentrations were then spatialized by means of inverse distance weighting (IDW) interpolation in a GIS environment, which yielded the interpolated values on the basis of both the values of and the distance from the nearby concentration data. Deterministic methods such as IDW have provided good results in interpolating sparse observations, and are widely used in pollution models [35–38].

Table 2. Particulate matter (PM$_{10}$) annual mean concentration values and annual range (in $\mu g \cdot m^{-3}$) recorded at the 20 monitoring stations, with the respective class based on the location type within the Metropolitan City (MC) of Rome.

Monitoring Station	Class	Annual Mean Concentration Value ($\mu g \cdot m^{-3}$)	Annual Range ($\mu g \cdot m^{-3}$)
Francia	Urban traffic	32	84
Magna Grecia	Urban traffic	31	72
Ciampino	Urban traffic	32	117
Fermi	Urban traffic	31	66
Tiburtina	Urban traffic	34	93
Civitavecchia Villa Albani	Urban traffic	23	63
Arenula	Urban background	29	85
Cinecittà	Urban background	35	104
Civitavecchia	Urban background	20	44
Villa Ada	Urban background	26	66
Bufalotta	Urban background	29	78
Cipro	Urban background	28	75
Guidonia	Peri-urban traffic	28	72
Cavaliere	Peri-urban background	27	63
Malagrotta	Peri-urban background	24	65
Colleferro-Oberdan	Industrial/peri-urban background	30	101

Table 2. *Cont.*

Monitoring Station	Class	Annual Mean Concentration Value ($\mu g \cdot m^{-3}$)	Annual Range ($\mu g \cdot m^{-3}$)
Colleferro-Europa	Industrial/peri-urban background	34	151
Civ. porto	Industrial	23	56
Allumiere	Rural background	10	31
Castel di Guido	Rural background	22	44

2.6. PM$_{10}$ Deposition

The PM$_{10}$ seasonal concentrations, the LAI MODIS data for the MC of Rome and the surface cover of the three functional groups were used to estimate the amount of PM$_{10}$ dry deposition on vegetation (Figure 1).

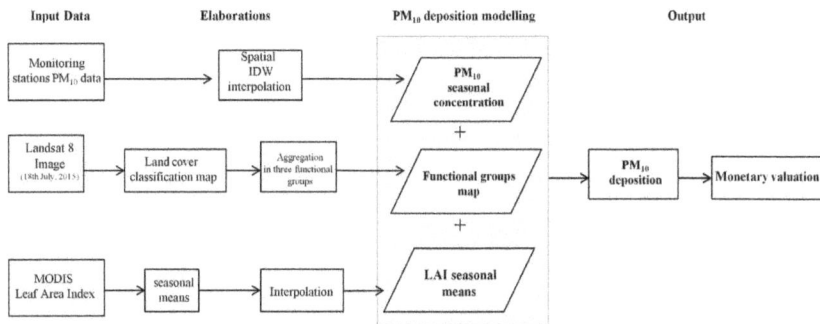

Figure 1. Flowchart of the integrated methodology.

Under the assumption of zero rainfall, the downward deposition rate of PM$_{10}$ was calculated according to Nowak's [39] and Yang's et al. [40] methodology. For PM$_{10}$, the deposition velocity was set at a median value of 0.0064 m·s^{-1}, based on a LAI mean value of 6 [41], and then adjusted to the actual LAI [42,43]. In order to calculate the total amount (Mg) of PM$_{10}$ removed, the fluxes were multiplied for the surface cover of each functional group.

Lastly, in order to calculate PM$_{10}$ removal per hectare (Mg·ha^{-1}), the total amount of PM$_{10}$ was normalized for the surface area of the respective functional group.

2.7. Monetary Evaluation

The monetary value of PM$_{10}$ reduction provided by the functional groups was estimated by using the externality value (cost per Mg) of PM$_{10}$ pollution. Externalities can be described as the estimated social cost of pollution (i.e., human health, environmental impact and material damage) that is not considered in the market price of the goods or services that caused the pollution [19]. By applying the externality value calculated for the European context for PM$_{10}$, which has been previously used in European environmental policies and programs [44,45], we calculated the monetary value for the amount of PM$_{10}$ removed. This value corresponds to 22,990 Euros per Mg, and is calculated on the basis of the value of a life year (VOLY). This value represents the cost to society of the damage caused by pollution to people's health and the environment [46].

3. Results

3.1. Land Cover Map of the Metropolitan City of Rome

Figure 2A shows the land cover map of the Metropolitan City of Rome obtained by classifying a Landsat 8 OLI/TIRS image (18 July 2015).

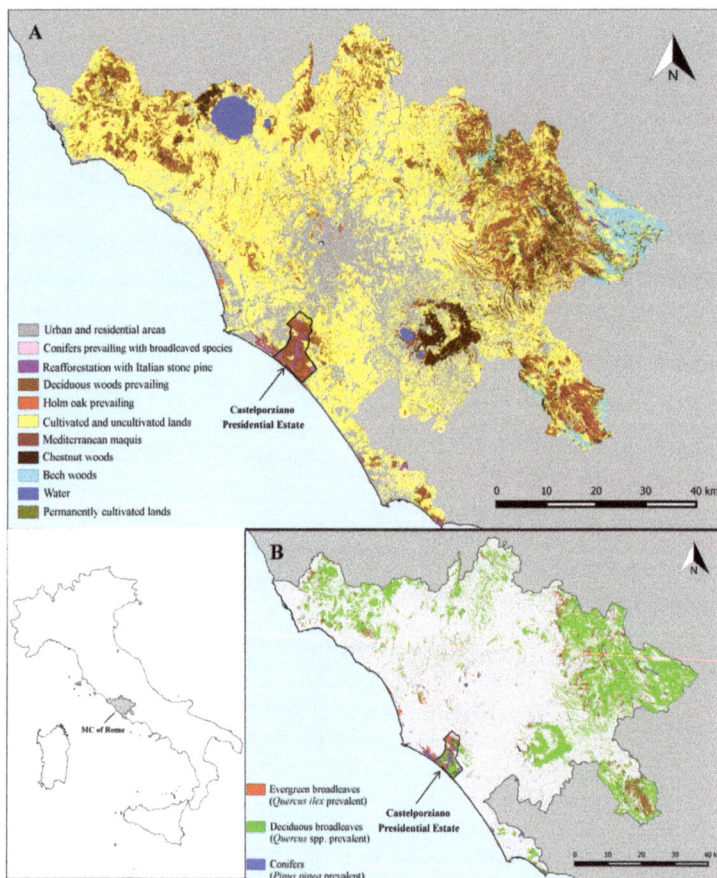

Figure 2. (**A**) Land cover classification of the Metropolitan City (MC) of Rome (18 July 2015 Landsat 8 OLI/TIRS image); (**B**) Map of the three functional groups obtained by means of a morphofunctional aggregation of the physiognomic-structural categories of vegetation of the MC of Rome (see text for further details).

This map, which has an overall classification accuracy of 95.6%, reveals a complex mosaic of 11 different land use classes, seven of which are natural ecosystems with heterogeneous structural and functional traits. The areas covered by cultivated and uncultivated lands and by permanently cultivated lands are also shown, as are urban and residential areas. The main land use types in the MC of Rome are cultivated and uncultivated lands (55%), followed by urbanized and residential areas (22%), while natural ecosystems cover the remaining 22% of the MC.

Figure 2B shows the map of the three vegetation functional groups obtained by means of a morphofunctional aggregation of the woody vegetation of the MC of Rome. Worthy of note is the fact that deciduous broadleaves are the most abundant functional group in the MC (92,927 ha), followed by evergreen broadleaves (21,116 ha) and conifers (2950 ha). The most abundant functional group in the Castelporziano Estate is that of the evergreen broadleaves (2017.53 ha), followed by deciduous broadleaves (1887.84) and, lastly, by conifers (750.06).

3.2. LAI and PM10 Removal Efficiency by Vegetation in the MC of Rome

The LAI values yielded by dense evergreen forests, as well as by large natural and biodiverse areas such as the Castelporziano Presidential Estate in the southern coastal area of the MC, are high throughout the year (Figure 3), peaking in summer (up to ~6.7 $m^2 \cdot m^{-2}$) before gradually dropping to a minimum in winter.

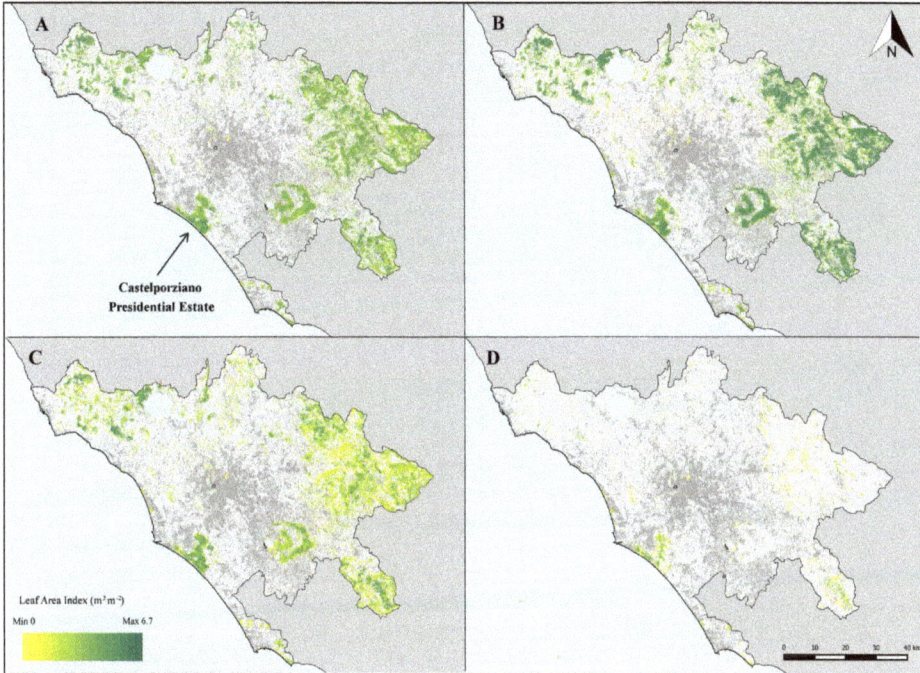

Figure 3. Maps of the seasonal Leaf Area Index (LAI, $m^2 \cdot m^{-2}$) ((**A**) spring; (**B**) summer; (**C**) autumn; (**D**) winter). The autumn LAI values for the deciduous broadleaves shown in panel (**C**) were calculated until 24 October.

The spatial distribution of PM_{10} seasonal deposition rates per surface unit ($g \cdot m^{-2}$) (Figure 4) follows a similar pattern to that of the LAI values. Spring PM_{10} deposition rates in most of the MC (Figure 4A) range approximately from 0.5 to 7.7 $g \cdot m^{-2}$, increasing in the summer months (Figure 4B). The highest values (up to around 9 $g \cdot m^{-2}$) were recorded in the northeastern quadrant of the MC, which is characterized by the widespread presence of deciduous forest stands, whereas lower values are concentrated in the northwestern quadrant of the MC, where PM_{10} concentrations are minimal (data not shown). Autumn deposition values (Figure 4C) are generally lower, presenting, however, peaks of around 12 $g \cdot m^{-2}$ found close to the Castelporziano Estate and in the evergreen broadleaved forests in the southeastern quadrant of the MC. The winter months (Figure 4D) yield lower annual values, though peaks of around 7 $g \cdot m^{-2}$ were recorded in evergreen forest stands in the southeastern quadrant of the MC and in the Castelporziano Presidential Estate. The highest mean LAI values for all three functional groups were observed in summer, with values for deciduous broadleaves (3.84 ± 1.31 $m^2 \cdot m^{-2}$) being followed by evergreen broadleaves (3.04 ± 1.37 $m^2 \cdot m^{-2}$) and conifers (2.65 ± 1.35 $m^2 \cdot m^{-2}$) (Table 3), whereas the lowest were observed in winter (1.26 ± 0.83 $m^2 \cdot m^{-2}$ for evergreen broadleaves and 1.54 ± 1.11 $m^2 \cdot m^{-2}$ for conifers). Deciduous broadleaves yielded higher mean LAI values throughout the vegetative period than the other two functional groups.

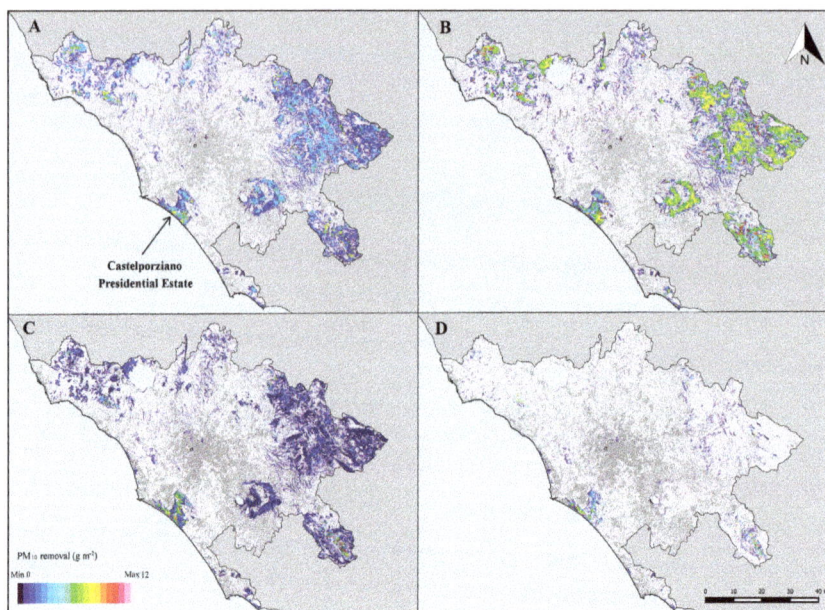

Figure 4. Seasonal particulate matter (PM_{10}) deposition maps ($g \cdot m^{-2}$) estimated for the three functional groups in the MC of Rome ((A) spring; (B) summer; (C) autumn; (D) winter). The autumn deposition values on the deciduous broadleaves shown in panel (C) were calculated until 24 October.

Table 3. Seasonal PM_{10} concentrations ($\mu g \cdot m^{-3}$) and Leaf Area Index (LAI) value ($m^2 \cdot m^{-2}$), calculated for the three functional groups in the MC of Rome. Data are means $\pm$ standard deviation.

	Mean PM_{10} Concentrations ($\mu g \cdot m^{-3}$)			Mean LAI ($m^2 \cdot m^{-2}$)		
	Deciduous	Evergreen	Conifers	Deciduous	Evergreen	Conifers
Spring	23.19 ± 1.04	23.18 ± 1.10	23.47 ± 0.91	3.22 ± 0.76	2.92 ± 0.94	2.58 ± 1.01
Summer	24.49 ± 0.32	24.34 ± 0.67	24.23 ± 0.83	3.84 ± 1.31	3.04 ± 1.37	2.65 ± 1.35
Autumn	20.00 ± 1.71 *	34.24 ± 6.15	33.54 ± 3.95	2.33 ± 1.04 *	1.88 ± 1.04	2.00 ± 1.20
Winter		33.79 ± 7.54	32.71 ± 4.81		1.26 ± 0.83	1.54 ± 1.11

* Autumn mean values for deciduous broadleaves were calculated until 24 October.

The total annual PM_{10} deposition calculated for the three functional groups (in Mg, Table 4) is higher for deciduous broadleaves (5573.86 Mg), and lower for evergreen broadleaves (1293.16 Mg) and conifers (169.88 Mg), which reflects the extent of their surface cover. Table 4 also shows the total and seasonal PM_{10} removal efficiency of each of the three functional groups (in $Mg \cdot ha^{-1}$). Most of the PM_{10} removal by the three functional groups (both as total, in Mg, and in $Mg \cdot ha^{-1}$) occurs in the summer months (3213.52, 489.85, and 54.03 Mg and 0.035, 0.023, and 0.018 $Mg \cdot ha^{-1}$ for deciduous broadleaves, evergreen broadleaves and conifers, respectively), with minimum removal being observed in winter. Total efficiency is comparable for all three functional groups (0.060, 0.061, and 0.058 $Mg \cdot ha^{-1}$ for deciduous broadleaves, evergreen broadleaves and conifers, respectively). The monetary evaluation of the ES of PM_{10} removal yields an overall value of 161.14 million Euros for all the urban and peri-urban forests in the MC of Rome.

Table 4. Seasonal PM_{10} deposition (total, Mg, and per hectare, as $Mg \cdot ha^{-1}$) and related monetary value (in Euros) calculated for the three functional groups in the MC of Rome.

	Deciduous			Evergreen			Conifers		
	Mg	$Mg \cdot ha^{-1}$	Value (€ 10^6)	Mg	$Mg \cdot ha^{-1}$	Value (€ 10^6)	Mg	$Mg \cdot ha^{-1}$	Value (€ 10^6)
Spring	2008.56	0.022	46.18	392.94	0.019	9.03	45.48	0.015	1.05
Summer	3213.52	0.035	73.88	489.85	0.023	11.26	54.03	0.018	1.24
Autumn	351.78 *	0.004 *	8.09	278.26	0.013	6.40	43.57	0.015	1.00
Winter				132.11	0.006	3.04	26.80	0.009	0.62
Total	5573.86	0.060	128.14	1293.16	0.061	29.73	169.88	0.058	3.91

* Autumn mean values for deciduous broadleaves were calculated until 24 October.

3.3. Study Case: Contribution of Castelporziano Presidential Estate Peri-Urban Forest to Air Quality Improvement

Table 5 shows the mean LAI and PM_{10} concentrations for the peri-urban forest of the Castelporziano Presidential Estate. PM_{10} concentrations are very homogeneous between the three functional groups within the Castelporziano Presidential Estate (mean value of approximately 23 and 24 $\mu g \cdot m^{-3}$ in spring and summer, respectively, for all the functional groups), and as an annual mean value (29.29 ± 0.18, 29.29 ± 0.16, 29.22 ± 0.19 for conifers, deciduous broadleaves and evergreen, respectively). The deciduous broadleaves' mean LAI was higher than that of evergreen broadleaves and conifers, particularly in summer (4.25 ± 1.11 $m^2 \cdot m^{-2}$). The LAI values in the Castelporziano Estate for the three functional groups were generally higher throughout the year than those estimated for the whole MC of Rome.

Table 5. Seasonal PM_{10} concentrations ($\mu g \cdot m^{-3}$) and LAI value ($m^2 \cdot m^{-2}$), calculated for the three functional groups in the Castelporziano Estate. Data are means $\pm$ standard deviation.

	Mean PM_{10} Concentrations ($\mu g \cdot m^{-3}$)			Mean LAI ($m^2 \cdot m^{-2}$)		
	Deciduous	Evergreen	Conifers	Deciduous	Evergreen	Conifers
Spring	23.58 ± 0.13	23.52 ± 0.14	23.57 ± 0.14	3.88 ± 0.91	3.49 ± 0.93	3.50 ± 0.77
Summer	24.23 ± 0.12	24.16 ± 0.13	24.22 ± 0.13	4.25 ± 1.11	3.56 ± 1.17	3.63 ± 0.97
Autumn	19.53 ± 0.21 *	32.37 ± 0.50	32.49 ± 0.46	3.99 ± 1.12 *	3.05 ± 1.02	3.46 ± 0.90
Winter		30.95 ± 0.54	31.11 ± 0.51		2.13 ± 0.89	2.63 ± 0.84

* Autumn mean values for deciduous broadleaves were calculated until 24 October.

Table 5 shows the PM_{10} removal values calculated for the three functional groups within the Castelporziano Presidential Estate. Spring and summer removal rates (in $Mg \cdot ha^{-1}$) are higher for deciduous broadleaves (0.032 and 0.040 $Mg \cdot ha^{-1}$); nevertheless, evergreen broadleaves and conifers display high removal rates even in autumn (0.028 and 0.034 $Mg \cdot ha^{-1}$, respectively), and lower removal rates in winter (0.013 and 0.019 $Mg \cdot ha^{-1}$, respectively). As a result, the total PM_{10} removal capacity is higher for evergreen broadleaves and conifers (0.10 and 0.11 $Mg \cdot ha^{-1}$) than for deciduous broadleaves (0.08 $Mg \cdot ha^{-1}$). Table 6 also shows that PM_{10} removal by the peri-urban forest of the Castelporziano Presidential Estate yields an overall value of 10.44 million Euros.

Table 6. Annual PM_{10} deposition (total, Mg, and per hectare, as $Mg \cdot ha^{-1}$) and related monetary value, calculated for the three functional groups in the Castelporziano Estate.

	Deciduous			Evergreen			Conifers		
	Mg	$Mg \cdot ha^{-1}$	Value (€ 10^6)	Mg	$Mg \cdot ha^{-1}$	Value (€ 10^6)	Mg	$Mg \cdot ha^{-1}$	Value (€ 10^6)
Spring	60.74	0.032	1.40	53.18	0.026	1.22	19.48	0.026	0.45
Summer	75.90	0.040	1.75	58.87	0.029	1.35	22.02	0.029	0.51
Autumn	18.68 *	0.010 *	0.43 *	56.06	0.028	1.29	25.80	0.034	0.59
Winter				27.18	0.013	0.62	14.55	0.019	0.33
Total	155.32	0.08	3.57	195.28	0.10	4.49	81.85	0.11	1.88

* Autumn mean values for deciduous broadleaves were calculated until 24 October.

4. Discussion

The PM_{10} deposition values obtained for the year 2015 are comparable to those reported in previous studies performed in the Metropolitan City of Rome [15,16,47]. Manes et al. [16] also previously investigated the role played by GI in PM_{10} abatement in 10 MCs of Italy for the year 2003. The removal values reported previously are slightly lower than those that emerge from this study, though it should be borne in mind that 2015 was characterized by intense episodes of PM_{10} pollution, with almost 35 days in which air pollutant concentrations exceeded the threshold of 50 $\mu g \cdot m^{-3}$ in the city of Rome, which is the limit imposed by the Italian government (Legislative Decree 155/2010) for the protection of human health [48], whereas PM_{10} air concentrations modeled for the year 2003 were generally lower. What emerges from this study is the elevated efficiency of deciduous species in PM_{10} removal during the spring and summer months resulting from a higher LAI: the evaluation of the seasonal PM_{10} removal trend showed that deciduous broadleaves are the species that most effectively removes PM_{10} from the atmosphere during the vegetative period, which is in keeping with their phenology. Indeed, as expected, the data yielded both by the MC of Rome and the Castelporziano Presidential Estate showed that the summer PM_{10} removal rates are higher. This finding is in agreement with those of Silli et al. [47], who reported a PM_{10} removal peak during the summer season and differences in the PM_{10} removal capacity between the three functional groups in an urban park in the center of Rome. Nevertheless, the Castelporziano study case allowed us to more accurately define the varying removal capacity of the three functional groups in a territory characterized by relatively homogeneous environmental conditions. If we consider the total removal values of the three functional groups in the Castelporziano study case, what emerges is the importance for PM_{10} abatement of evergreen species, as also showed in an experimental study performed in an evergreen broadleaved urban forest [49]. Indeed, we observed that the total annual PM_{10} removal efficiency of evergreen species (evergreen broadleaves and conifers) is 20% to 27% higher than that of deciduous broadleaves on an annual scale. This suggests that evergreen communities have a greater impact on air quality amelioration on a year-long basis. Furthermore, since PM_{10} pollution levels usually rise in winter [49–51], we presume that increasing evergreen species cover in highly polluted areas would, given the ability of such species to abate pollutant levels throughout the year, help to prevent or mitigate pollution peaks. It is noteworthy that although the estimated mean PM_{10} air concentrations in the MC of Rome are not markedly different from those in the Castelporziano Presidential Estate, the latter yielded higher removal rates for all the functional groups considered. Indeed, the mean LAI values in the natural reserve, which reflect tree ecophysiological conditions [52], were higher for all three functional groups. This discrepancy between the overall MC of Rome, which is classified as one of the 'greenest' cities in Europe [25], and the Castelporziano Presidential Estate is likely due to the harsh conditions to which trees in the urban environment are exposed, including urban heat island that reduce photosynthesis and transpiration, limited nutrient availability in the soil, overbuilding and other biotic and abiotic stress factors [53–56]. Bearing this in mind, foresight management of GI aimed at preserving a suitable environment for vegetation would improve its functional status, and consequently enhance its removal capacity. Further studies are also needed to shed more light on PM deposition processes related to PM with other aerodynamic diameters, such as the particularly harmful fine PM ($PM_{2.5}$) and ultrafine PM_1. Nowak et al. [17] reported that urban trees remove substantially less $PM_{2.5}$ than PM_{10}. Moreover, in their study on an urban park, Silli et al. [47] reported that vegetation contributed to a greater extent to the abatement of PM_{10} (12.84%) than to that of a finer PM fraction ($PM_{2.5}$, 2.56%), confirming reports by Yin et al. [57] for urban parks in China. Modeling ecological processes do, however, have certain limitations. In particular, the simulation of PM_{10} deposition on vegetation entails approximating PM_{10} deposition velocity to the plant canopy (V_d), which depends on other, more complex parameters besides LAI, such as wind speed, relative humidity and air temperature [58]. Although a more accurate modeling of this parameter is beyond the scope of our work, a finer local-scale analysis, as previously also highlighted by Escobedo and Nowak [42], is warranted to better quantify air pollution removal, and consequently to assess its monetary value, in order to provide management alternatives to policy-makers. It should also be borne in mind that the use of a moderate resolution sensor such as MODIS, particularly in a Mediterranean environment, which is characterized by a high degree of landscape heterogeneity, may affect LAI values regarding the contribution made by different vegetation types to the signal received by the sensor, or may underestimate LAI values in small patches of vegetation [59]. The monetary assessment thus depends on the accuracy of the biophysical modeling, but has certain intrinsic limitations. Indeed, the monetary evaluation does not, but should, take into account the cost of the management and maintenance of the GI, particularly in Mediterranean areas, where particularly stressful summer conditions require additional management practices to improve ES provisions by urban green, such as irrigation, phytosanitary treatments and pruning [54].

5. Conclusions

The removal of PM_{10} by urban and peri-urban forests, which is performed above all by deciduous species in the summer months, but also on a more constant basis by the evergreen community, would contribute considerably to the improvement of air quality in the MC of Rome, an area characterized by a high population density, relatively high air pollution levels and marked land use changes related to agricultural and industrial practices. Indeed, around 22% of the 535,200 hectares that make up the MC of Rome are covered by a forest ecosystem whose PM_{10} removal corresponded to an overall monetary value of 161.78 million Euros for the year 2015. Since citizens can benefit from multiple ES provided by natural ecosystems, urban development strategies should increasingly be aimed at enhancing the natural and artificial GI network so as to comply with ES provision and human health recommendations. "The European Green City Index", a research project sponsored by Siemens (2009) [60], showed that out of the 30 leading European cities in 30 different countries, the City of Rome was placed 17th for air quality and 23rd for environmental governance, which refers to the strategies adopted to improve and monitor environmental performance. Social policies must develop and plan funding aimed at promoting infrastructures with a low environmental impact (with low emission levels of greenhouse gases and other pollutants) and nature-based solutions by taking into account ES values.

Acknowledgments: This research was conducted with funding from: MIUR, ROME, Project PRIN 2010-2011 "TreeCity"; Sapienza University of Rome Ateneo Research Project, year 2015—prot. C26A15PWLH; Ministero della Salute, Centro Nazionale per la Prevenzione ed il Controllo delle Malattie—CCM Project: "Metodi per la valutazione integrata dell'impatto ambientale e sanitario (VIIAS) dell'inquinamento atmosferico".

Author Contributions: Fausto Manes designed the experiment. Federica Marando analyzed the data. Federica Marando, Fausto Manes, Elisabetta Salvatori and Lina Fusaro wrote the article and critically revised the intellectual content. All authors have read and approved the final manuscript after minor modifications.

Conflicts of Interest: The authors declare no conflict of interest.

References

1. European Environment Agency (EEA). *Air Quality in Europe—2015 Report*; EEA Report No 5/2015; European Environment Agency: Copenhagen, Denmark, 2015.
2. European Commission, Directorate-General for Research and Innovation. *Towards an EU Research and Innovation Policy Agenda for Nature-Based Solutions & Re-Naturing Cities. Final Report of the Horizon 2020 Expert Group on 'Nature-Based Solutions and Re-Naturing Cities'*; Available online: http://bookshop.europa.eu/en/towards-an-eu-research-and-innovation-policy-agenda-for-nature-based-solutions-re-naturing-cities-pbKI0215162/ (assessed on 23 March 2016).
3. Bowler, D.E.; Buyung-Ali, L.; Knight, T.M.; Pullin, A.S. Urban greening to cool towns and cities: A systematic review of the empirical evidence. *Landsc. Urban Plan.* **2010**, *97*, 147–155. [CrossRef]
4. Gill, S.E.; Handley, A.R.; Ennos, A.R.; Pauleit, S. Adapting cities for climate change: The role of the green infrastructure. *Built Environ.* **2007**, *33*, 115–133. [CrossRef]
5. Maes, J.; Jacobs, S. Nature-Based Solutions for Europe's Sustainable Development. *Conserv. Lett.* **2015**. [CrossRef]

6. MA, Millennium Ecosystem Assessment. *Ecosystems and Human Well-Being: Current State and Trends*; Island Press: Washington, DC, USA, 2005.
7. Manes, F.; Incerti, G.; Salvatori, E.; Vitale, M.; Ricotta, C.; Costanza, R. Urban ecosystem services: Tree diversity and stability of tropospheric ozone removal. *Ecol. Appl.* **2012**, *22*, 349–360. [CrossRef] [PubMed]
8. Van den Berg, M.; van Poppel, M.; van Kamp, I.; Andrusaityte, S.; Balseviciene, B.; Cirach, M.; Danileviciute, A.; Ellis, N.; Hurst, G.; Masterson, D.; et al. Visiting green space is associated with mental health and vitality: A cross-sectional study in four European cities. *Health Place* **2016**, *38*, 8–15. [CrossRef] [PubMed]
9. Bolund, P.; Hunhammar, S. Ecosystem services in urban areas. *Ecol. Econ.* **1999**, *29*, 293–301. [CrossRef]
10. Tzoulas, K.; Korpela, K.; Venn, S.; Yli-Pelkonen, V.; Kazmierczak, A.; Niemela, J.; James, P. Promoting ecosystem and human health in urban areas using Green Infrastructure: A literature review. *Landsc. Urban Plan.* **2007**, *81*, 167–178. [CrossRef]
11. Roy, S.; Byrne, J.; Pickering, C. A systematic quantitative review of urban tree benefits, costs, and assessment methods across cities in different climatic zones. *Urban For. Urban Green.* **2012**, *11*, 351–363. [CrossRef]
12. Sung, C.Y. Mitigating surface urban heat island by a tree protection policy: A case study of The Woodland, Texas, USA. *Urban For. Urban Green.* **2013**, *12*, 474–480. [CrossRef]
13. Coronel, A.S.; Feldman, S.R.; Jozami, E.; Facundo, K.; Piacentini, R.D.; Dubbeling, M.; Escobedo, F.J. Effects of urban green areas on air temperature in a medium-sized Argentinian city. *AIMS Environ. Sci.* **2015**, *2*, 803–826.
14. Lee, A.C.K.; Maheswaran, R. The health benefits of urban green spaces: A review of the evidence. *J. Public Health* **2011**, *33*, 212–222. [CrossRef] [PubMed]
15. Manes, F.; Silli, V.; Salvatori, E.; Incerti, G.; Galante, G.; Fusaro, L.; Perrino, C. Urban ecosystem services: Tree diversity and stability of PM_{10} removal in the metropolitan area of Rome. *Ann. Bot.* **2014**, *4*, 19–26.
16. Manes, F.; Marando, F.; Capotorti, G.; Blasi, C.; Salvatori, E.; Fusaro, L.; Ciancarella, L.; Mircea, M.; Marchetti, M.; Chirici, G.; et al. Regulating Ecosystem Services of forests in ten Italian metropolitan Cities: Air quality improvement by PM_{10} and O_3 removal. *Ecol. Indic.* **2016**, *67*, 425–440. [CrossRef]
17. Nowak, D.J.; Hirabayashi, S.; Bodine, A.; Greenfield, E. Tree and forest effects on air quality and human health in the United States. *Environ. Pollut.* **2014**, *193*, 119–129. [CrossRef] [PubMed]
18. Kroeger, T.; Escobedo, F.J.; Hernandez, J.L.; Varela, S.; Delphin, S.; Delphin, S.; Fisher, J.R.B.; Waldron, J. Reforestation as a novel abatement and compliance measure for ground-level ozone. *Proc. Natl. Acad. Sci. USA* **2014**, *111*, E4204–E4213. [CrossRef] [PubMed]
19. Nowak, D.J.; Crane, D.E.; Stevens, J.C. Air pollution removal by urban trees and shrubs in the United States. *Urban For. Urban Green.* **2006**, *4*, 115–123. [CrossRef]
20. Baumgardner, D.; Varela, S.; Escobedo, F.J.; Chacalo, A.; Ochoa, C. The role of a peri-urban forest on air quality improvement in the Mexico City megalopolis. *Environ. Pollut.* **2012**, *163*, 174–183. [CrossRef] [PubMed]
21. Baró, F.; Chaparro, L.; Gomez-Baggethun, E.; Langemeyer, J.; David, J.; Terradas, J. Contribution of Ecosystem Services to Air Quality and Climate Change Mitigation Policies: The Case of Urban Forests in Barcelona, Spain. *Ambio* **2014**, *43*, 466–479. [CrossRef] [PubMed]
22. Cohen, P.; Potchter, O.; Schnell, I. The impact of an urban park on air pollution and noise levels in the Mediterranean city of Tel-Aviv, Israel. *Environ. Pollut.* **2014**, *195*, 73–83. [CrossRef] [PubMed]
23. Niemelä, J.; Saarela, S.R.; Söderman, T.; Kopperoinen, L.; Yli-Pelkonen, V.; Väre, S.; Kotze, D.J. Using the ecosystem services approach for better planning and conservation of urban green spaces: A Finland case study. *Biodivers. Conserv.* **2010**, *19*, 3225–3243. [CrossRef]
24. Lafortezza, R.; Davies, C.; Sanesi, G.; Konijnendijk, C.C. Green Infrastructure as a tool to support spatial planning in European urban regions. *iForest Biogeosci. For.* **2013**, *6*, 102. [CrossRef]
25. Fuller, R.A.; Gaston, K.J. The scaling of green space coverage in European cities. *Biol. Lett.* **2009**, *5*, 352–355. [CrossRef] [PubMed]
26. Attorre, F.; Francesconi, F.; Pepponi, L.; Provantini, R.; Bruno, F. Spatio-temporal analyses of parks and gardens of Rome. *Stud. Hist. Gard. Des. Landsc.* **2003**, *23*, 293–306. [CrossRef]
27. Demo Istat. Available online: http://demo.istat.it/bilmens2016gen/index.html (accessed on 19 July 2016).

28. ISPRA. Available online: http://www.isprambiente.gov.it/files/pubblicazioni/rapporti/Rapporto_218_15. pdf (accessed on 30 March 2016).
29. Capotorti, G.; Del Vico, E.; Lattanzi, E.; Tilia, A.; Celesti-Grapow, L. Exploring biodiversity in a metropolitan area in the Mediterranean region: The urban and suburban flora of Rome (Italy). *Plant Biosyst.* **2013**, *147*, 174–185. [CrossRef]
30. Blasi, C.; Capotorti, G.; Marchese, M.; Marta, M.; Bologna, M.A.; Bombi, P.; Bonaiutoc, M.; Bonnesc, M.; Carrusc, G.; Cifelli, F.; et al. Interdisciplinary research for the proposal of the Urban Biosphere Reserve of Rome Municipality. *Plant Biosyst.* **2008**, *142*, 305–312. [CrossRef]
31. Salvati, L.; Tombolini, I. Cropland vs. forests: Landscape composition and land-use changes in Peri-urban Rome (1949–2008). *WSEAS Trans. Environ. Dev.* **2013**, *9*, 278–289.
32. Pignatti, S.; Capanna, E.; Porceddu, E. Castelporziano, Research and Conservation in a Mediterranean Forest Ecosystem: Presentation of the Volume. *Rend. Lincei* **2015**, *26*, 265–266. [CrossRef]
33. Manes, F.; Grignetti, A.; Tinelli, A.; Lenz, R.; Ciccioli, P. General features of the Castelporziano test site. *Atmos. Environ.* **1997**, *31*, 19–25. [CrossRef]
34. Congedo, L.; Macchi, S. Investigating the relationship between land cover and vulnerability to climate change in the Dar es Salaam. 2013. Available online: http://www.planning4adaptation.eu/Docs/events/ WorkShopII/WorkingPaper_Activity2_1_complete.pdf (accessed on 29 March 2016).
35. Declercq, F.A.N. Interpolation methods for scattered sample data: Accuracy, spatial patterns, processing time. *Cartogr. Geogr. Inform.* **1996**, *23*, 128–144. [CrossRef]
36. Moore, K.; Neugebauer, R.; Lurmann, F.; Hall, J.; Brajer, V.; Alcorn, S.; Tager, I. Ambient ozone concentrations cause increased hospitalizations for asthma in children: An 18-year study in Southern California. *Environ. Health Perspect.* **2008**, *116*, 1063. [CrossRef] [PubMed]
37. Babak, O.; Deutsch, C.V. Statistical approach to inverse distance interpolation. *Stoch. Env. Res. Risk Assess* **2009**, *23*, 543–553. [CrossRef]
38. Xu, X.; Sharma, R.K.; Talbott, E.O.; Zborowski, J.V.; Rager, J.; Arena, V.C.; Volz, C.D. PM$_{10}$ air pollution exposure during pregnancy and term low birth weight in Allegheny County, PA, 1994–2000. *Int. Arch. Occup. Environ. Health* **2011**, *84*, 251–257. [CrossRef] [PubMed]
39. Nowak, D.J. Air pollution removal by Chicago's urban forest. In *Chicago's Urban Forest Ecosystem: Results of the Chicago Urban Forest Climate Project*; McPherson, E.G., Nowak, D.J., Rowntree, R.A., Eds.; General Technical Report NE-186; USDA Forest Service: Radnor, PA, USA, 1994; pp. 63–81.
40. Yang, J.; McBride, J.; Zhoub, J.; Sun, Z. The urban forest in Beijing and its role in air pollution reduction. *Urban For. Urban Green.* **2005**, *3*, 65–78. [CrossRef]
41. Lovett, G.M. Atmospheric deposition of nutrients and pollutants in North America: An ecological perspective. *Ecol. Appl.* **1994**, *4*, 629–650. [CrossRef]
42. Escobedo, F.J.; Nowak, D.J. Spatial heterogeneity and air pollution removal by an urban forest. *Landsc. Urban Plan.* **2009**, *90*, 102–110. [CrossRef]
43. Hirabayashi, S.; Kroll, C.N.; Nowak, D.J. *I-Tree Eco Dry Deposition Model Descriptions*; United States Forest Service: Syracuse, NY, USA, 2015.
44. European Environment Agency (EEA). *Costs of Air Pollution from European Industrial Facilities 2008–2012—An Updated Assessment*; EEA Technical report No 20/2014; European Environment Agency: Copenhagen, Denmark, 2014.
45. Bickel, P.; Friedrich, R. *ExternE: Externalities of Energy: Methodology 2005 Update*; European Commission: Luxembourg, Luxembourg, 2005.
46. Currie, B.A.; Bass, B. Estimates of air pollution mitigation with green plants and green roofs using the UFORE model. *Urban Ecosyst.* **2008**, *11*, 409–422. [CrossRef]
47. Silli, V.; Salvatori, E.; Manes, F. Removal of airborne particulate matter by vegetation in an urban park in the city of Rome (Italy): An ecosystem services perspective. *Ann. Bot.* **2015**, *5*, 53–62.
48. ISPRA. Available online: http://www.isprambiente.gov.it/public_files/XI-Rapporto-sulla-qualit%C3%A0-ambiente-urbano-2-mar.pdf (accessed on 30 March 2016).
49. Cavanagh, J.A.E.; Zawar-Reza, P.; Wilson, J.G. Spatial attenuation of ambient particulate matter air pollution within an urbanised native forest patch. *Urban For. Urban Green.* **2009**, *8*, 21–30. [CrossRef]
50. Yang, K.L. Spatial and seasonal variation of PM$_{10}$ mass concentrations in Taiwan. *Atmos. Environ.* **2002**, *36*, 3403–3411. [CrossRef]

51. Cattani, G.; di Bucchianico, A.D.M.; Dina, D.; Inglessis, M.; Notaro, C.; Settimo, G.; Viviano, G.; Marconi, A. Evaluation of the temporal variation of air quality in Rome, Italy from 1999 to 2008. *Ann. Ist. Super. Sanità* **2010**, *46*, 242–253. [PubMed]

52. Fusaro, L.; Salvatori, E.; Mereu, S.; Silli, V.; Bernardini, A.; Tinelli, A.; Manes, F. Researches in Castelporziano test site: Ecophysiological studies on Mediterranean vegetation in a changing environment. *Rend. Lincei* **2015**, *26*, 473–481. [CrossRef]

53. Attorre, F.; Bruno, M.; Francesconi, F.; Valenti, R.; Bruno, F. Landscape changes of Rome through tree-lined roads. *Landsc. Urban Plan.* **2003**, *49*, 115–128. [CrossRef]

54. Fusaro, L.; Salvatori, E.; Mereu, S.; Marando, F.; Scassellati, E.; Abbate, G.; Manes, F. Urban and peri-urban forests in the metropolitan area of Rome: Ecophysiological response of *Quercus ilex* L. in two Green Infrastructures in an Ecosystem Services perspective. *Urban For. Urban Green.* **2015**, *14*, 1147–1156. [CrossRef]

55. Manes, F.; De Santis, F.; Giannini, M.A.; Vazzana, C.; Capogna, F.; Allegrini, I. Integrated ambient ozone evaluation by passive samplers and clover biomonitoring mini-stations. *Sci. Total Environ.* **2003**, *308*, 133–141. [CrossRef]

56. Manes, F.; Astorino, G.; Vitale, M.; Loreto, F. Morpho-functional characteristics of Quercus ilex L. leaves of different age and their ecophysiological behaviour during different seasons. *Plant Biosyst.* **1997**, *131*, 149–158. [CrossRef]

57. Yin, S.; Shen, Z.; Zhou, P.; Zou, X.; Che, S.; Wang, W. Quantifying air pollution attenuation within urban parks: An experimental approach in Shanghai, China. *Environ. Pollut.* **2011**, *159*, 2155–2163. [CrossRef] [PubMed]

58. Mohan, S.M. An overview of particulate dry deposition: Measuring methods, deposition velocity and controlling factors. *Int. J. Environ. Sci. Technol.* **2016**, *13*, 387–402. [CrossRef]

59. Sprintsin, M.; Karnieli, A.; Berliner, P.; Rotenberg, E.; Yakir, D.; Cohen, S. Evaluating the performance of the MODIS Leaf Area Index (LAI) product over a Mediterranean dryland planted forest. *Int. J. Remote Sens.* **2009**, *30*, 5061–5069. [CrossRef]

60. Siemens Annual Report 2009. Available online: http://www.siemens.com/annual-report_2009 (accessed on 30 March 2016).

![forests logo] *forests*

MDPI

Article

Quantifying Tree and Soil Carbon Stocks in a Temperate Urban Forest in Northeast China

Hailiang Lv [1,2], Wenjie Wang [1,*], Xingyuan He [1], Lu Xiao [3], Wei Zhou [1,3] and Bo Zhang [3]

[1] Northeast Institute of Geography and Agricultural Ecology, Chinese Academy of Sciences, Changchun 130102, China; lvhailang@iga.ac.cn (H.L.); hexingyuan@iga.ac.cn (X.H.); bsstzw@163.com (W.Z.)
[2] University of Chinese Academy of Sciences, No. 19A Yuquan Road, Beijing 100049, China
[3] Key Laboratory of Forest Plant Ecology, Northeast Forestry University, Harbin 150040, China; xiaolu1212@foxmail.com (L.X.); w9426426@163.com (B.Z.)
* Correspondence: wjwang225@hotmail.com; Tel./Fax: +86-431-8554-2336

Academic Editors: Francisco Escobedo, Stephen John Livesley and Justin Morgenroth
Received: 30 June 2016; Accepted: 6 September 2016; Published: 10 September 2016

Abstract: Society has placed greater focus on the ecological service of urban forests; however, more information is required on the variation of carbon (C) in trees and soils in different functional forest types, administrative districts, and urban-rural gradients. To address this issue, we measured various tree and soil parameters by sampling 219 plots in the urban forest of the Harbin city region. Averaged tree and soil C stock density (C stocks per unit tree cover) for Harbin city were 7.71 ($\pm$7.69) kg C·m^{-2} and 5.48 ($\pm$2.86) kg C·m^{-2}, respectively. They were higher than those of other Chinese cities (Shenyang and Changchun), but were much lower than local natural forests. The tree C stock densities varied 2.3- to 3.2-fold among forest types, administrative districts, and ring road-based urban-rural gradients. In comparison, soil organic C (SOC) densities varied by much less (1.4–1.5-fold). We found these to be urbanization-dependent processes, which were closely related to the urban-rural gradient data based on ring-roads and settlement history patterns. We estimated that SOC accumulation during the 100-year urbanization of Harbin was very large (5 to 14 thousand tons), accounting for over one quarter of the stored C in trees. Our results provide new insights into the dynamics of above- and below-ground C (especially in soil) during the urbanization process, and that a city's ability to provide C-related ecosystem services increases as it ages. Our findings highlight that urbanization effects should be incorporated into calculations of soil C budgets in regions subject to rapid urban expansion, such as China.

Keywords: carbon storage; SOC density; urban-rural gradients; soil carbon; Harbin

1. Introduction

Urban forests are one of the most important types of green infrastructures in cities [1]; they provide many important ecosystem services. Carbon stocks of urban forests influence local climate, carbon cycles, energy use, and climate change [2]. Consequently, the C stock capacity of urban forests is of great importance and is receiving increasing focus [2,3]. Researchers worldwide have studied and evaluated carbon stocks and sequestration by urban trees [2–8]. Cities in China like Beijing [4,5], Xiamen [6], Hangzhou [7] and cities worldwide in the US [2,8] and Europe [9] have available information regarding carbon stocks and sequestration in urban forests. These studies mostly quantified the current amount of carbon stored in the urban forest and its ecological services while focusing less on the C variations and long-term dynamics [6,9]. Ren et al. [6] indicated that urban sprawl negatively affected the surrounding forests, and human disturbance played the dominant role in influencing the carbon stocks and density of forest patches close to human activities [6]. Trees and forests in cities have long been influenced by humans. Thus, a better understanding of the long-term dynamics of urban vegetation is essential in determining its ecosystem services and improving its management [9].

C stock density is the C stocks in trees and/or soils per unit area. The distribution of C stock density is spatially uneven in urban forests, and is affected by human activities, such as road construction and real estate development [10]. A 110-year study has shown that changes in tree stocks were not constant across the urban area but varied with the current intensity of urbanization [9]. Urbanization gradients (e.g., history of settlements, which could to some extent substitute long-term dynamics and land use patterns) may possibly impact the spatial variation of C stored in tree biomass [11,12] and soils [13,14]. By dividing urban forests according to different administrative districts, functional forest types, ring roads (periphery transportation routes to ease traffic flows), and history of settlements (based on the time at which urban areas were settled), it may be possible to statistically detect these variations in C [14].

Compared with tree biomass C dynamics, fewer studies have focused on the C dynamics of soils in urban forests [15–17], even though soil C stocks are three times greater than the vegetation biomass C stocks [16]. Soil characteristics are influenced not only by plant-soil interaction and afforestation [18], but human activities, such as real estate development and road construction activities. These activities may have stronger influences on surface soils than tree biomass [19]. Studies have shown that the age of parks, in general, may be highly influential on soil variables, such as C sequestration [20]. Further, a study of 67 yards with home age ranging from 3 years to 87 years found that the relationship between soil C and home age was positive at 0–15 cm ($p = 0.0003$, $R^2 = 0.19$) [21]. We therefore hypothesized that at a large scale, the urbanization level or age of a city may have a decisive influence on soil C in urban forests. Knowledge of how urbanization levels (ring roads and history of settlements), urban forest types, and administrative districts affect variation of C in soils is important to better estimate and maximize the carbon sequestration function of urban forests [15,17,22]. The urbanization level of a city can be characterized by urban-rural gradient belts [23], ring-road regions (urban sprawl with ring-road development) [14] and differences in the years since urban settlement [24]. Accordingly, these approaches are used in this study. To date, few studies have quantified urbanization influences on soil C [25–27]. Relations between variations of C and urbanization levels must be clearly clarified [25].

Harbin is the most populated city in northeastern China. It is an ideal area for identifying variations of C in urban forests. There are sufficient data available to evaluate how urbanization gradients affect tree and soil C dynamics. Long-term records on urbanization process are available dating back to the 1900s. Few studies have investigated the C stocks or variations in tree biomass and soils of urban forests in Harbin [28–30]. Furthermore, these studies did not consider the heterogeneous distribution of C or its relations with urban-rural gradients.

Our study aimed to determine: (1) the C stock density of urban forest trees and soils in comparison with other nearby cities and natural forests; (2) spatial variations in trees: C stock density, DBH, and basal area (3) spatial variations in SOC density, SOC content and soil bulk density; and (4) the implications of scientific estimates of urban forest C stocks and how results could be maximized along urban–rural gradients.

2. Materials and Methods

2.1. Study Area

The study area is located in the urban area of Harbin City (45°45′ N; 126°38′ E). Harbin is the capital of Heilongjiang Province in north-eastern China, and is an important city in this region. The average elevation of Harbin city is 151 m above sea level. The municipal district covers an area of 10,198 km^2. The built-up area within the fourth ring road covers an area of 345.31 km^2. The urban area contains 3.95 million people. The mean temperature in January is only -18.4 °C (-1.1 °F), and mean temperature in July is 23.0 °C (73.4 °F) based on climate data from 1971 to 2000. Annual precipitation is 524 millimeters (20.6 in). The frost-free period lasts 140 days, while the ice period lasts 190 days [31]. The most prevalent soils across Harbin are black soil (Luvic Phaeozem, FAO). There are also chernozem (Haplic Chernozem, FAO), and meadow soils (Eutric Vertisol, FAO) [32,33].

Harbin was founded in 1898, with the arrival of the Chinese Eastern Railway. Consequently, the city has a long history and clear regional boundaries [34]. The initial urban area of Harbin City was constructed near to the Middle East railway in 1896. After a commercial port was opened in 1907, Harbin gradually became a single city with multiple towns, and urban areas gradually expanded. In 1932, Harbin fell under the control of Japanese invaders. During this period, the Japanese built military constructions in Harbin, until liberation in 1945. After the founding of China in 1949, Harbin progressively proceeded with urban planning. After the 1980s reform and opening-up policy, Harbin entered a period of rapid development with a much faster rate of urbanization. Urban area settlement time is used to represent the level of urbanization. In China and around the underdeveloped regions in the world, cities like Harbin tend to expand radially from the old city territory [35].

2.2. Field Survey

The urban forest in Harbin has been previously classified into 4 types of forest: (1) roadside forest (RF), which is distributed on either side of the road or railway; (2) ecological public welfare forest (EF), which mainly refers to shelterbelt forests for farmland and nursery forests that provide greening infrastructure development; (3) landscape and relaxation forest (LF), which is primarily forest parks and public parks; and (4) affiliated forest (AF), which refers to forests that are affiliated with different universities, public institutes and large community districts [36]. A stratified random sampling method was adopted to locate sampling plots in different regions of the city. These plots were selected based on their forest coverage and size of surveyed regions (administrative districts, ring-road region, and urban settlement time). The sampling plots are presented in Figure 1.

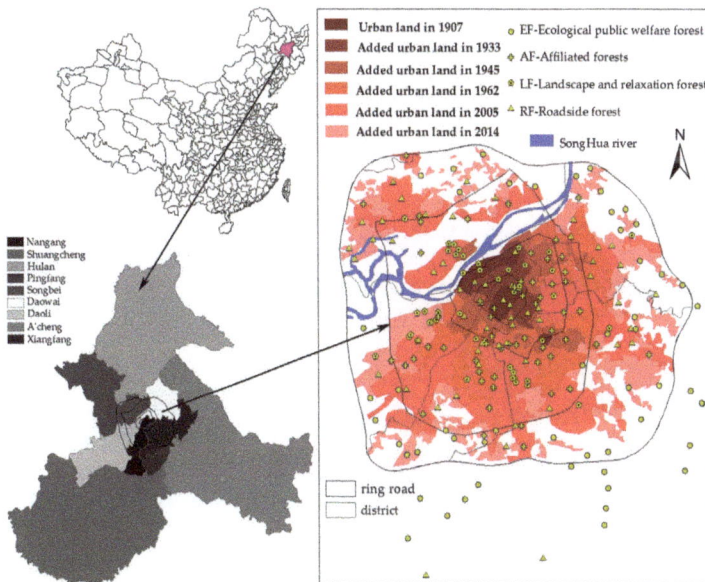

Figure 1. Location of the study area showing Harbin City in northeastern China, and the distribution of sampling plots in the Harbin City region.

Field surveys were conducted from August to September 2014. For each plot, the longitude, latitude, and altitude at the center of each location were recorded. The diameter at breast height

(1.3 m, DBH) and basal area (1.3 m, at breast height) of trees with a diameter greater than 1 cm were recorded. The species name, genus name, and family name of each tree were also recorded. Tree height was measured using a Nikon forestry PRO550 (Jackson, MS, USA). Soil samples were collected at the same time using a 100 cm^3 cutting ring (4 cutting rings per plot, M&Y Instrument Technology Co. Ltd., Shanghai, China). The shapes of the plots used in this study were dependent on forest type. RF and EF were surveyed as rectangular plots, with the width being fixed (8 to 20 m), while the length was adjusted to obtain survey areas of at least 400 m^2. LF was usually quite large in area; thus, plots were fixed at 20 m × 20 m. AF often had an irregular shape, and the plots were fixed as rectangular or square shapes to obtain an area of about 400 m^2. This type of survey method guaranteed a similar survey area for the forest types. Further, each plot was under tree coverage.

A total of 219 plots were surveyed, which includes 58 plots for AF, 42 plots for LF, 62 plots for RF, and 56 plots for EF. The study area includes 5 administrative districts, and out of the 219 surveyed plots, Songbei district (SB) contained 34 plots, Nangang district (NG) contained 52, Xiangfang district (XF) contained 59, Daoli district (DL) contained 33, and Daowai district (DW) contained 31. In the 219 plots, 16 plots fell within the first ring road (built in 1990 and abbreviated as "First"), 32 plots were distributed between the 1st and 2nd roads (built in 2001, Second), 77 plots were distributed between the 2nd and 3rd ring roads (built in 2016, Third), 74 plots were distributed between the 3rd and 4th ring roads (built in 2009, Fourth), and the remaining 20 plots were distributed outside the of 4th ring (in the planning, Outside or Fifth in regression analysis). The furthest plot was 10 kilometers from the 4th ring road, which is located within the planning 5th ring road. Among the 219 plots, 7 plots had a 100-year history, 11 plots had an 80-year history, 27 plots had a 70-year history, 26 plots had a 50-year history, 43 plots had a 10-year history, and 52 plots were located in the newly built-up region. The remaining 53 plots were located in rural regions.

2.3. Calculation of Tree Carbon Stock Densities

The tree C stock density (kg C·m^{-2}) was measured as C stocks in trees per unit tree cover in each plot. Aboveground dry weight biomass of trees was estimated using tree biomass allometric growth equations obtained from published literature studies (Table **??**) that were geographically near our study area [37]. We adjusted the equations based on the following rules: If no species-specific allometric equation was found, an equation for the same genus or family of the species was used [38]. If no equations for a genus or a family were found, a generalized equation was used [39]. If no tree root biomass equation was found, a root-shoot ratio of 0.26 was used [40]. Urban trees tend to have less aboveground tree biomass than trees in a natural forest because of pruning and maintenance; thus, the tree biomass estimate was multiplied by a factor of 0.8 for trees with a diameter greater than 30 cm [36]. Individual tree biomass was converted to C by multiplying by a factor of 0.5. The mean C stock density for each plot was calculated using the following equation:

$$CDj = \frac{\sum\limits_{i=1}^{n} Di}{Aj} \tag{1}$$

where CD_j is the average C stock density for the *j*th plot; D_i is the C stocks for the *i*th tree-shrub species; *n* is the tree-shrub species number; and Aj is the plot area for the *j*th plot.

2.4. Soil Organic C, Bulk Density Measurement, and SOC Density Calculation

Soils were sampled within 1 m of the main tree species for each plot (the main tree species are listed in Table A1). Only the surface soil was sampled (0–20 cm soil depth). Soil bulk density was measured using the ring-cup (100 mL) method. The ring-cup was inserted into the soil with a plastic hammer. Then, the intact soil with 100 mL fixed volume was taken out. Fixed volume intact soil (400 cm^3, 4 cutting rings per plot) was kept in a cloth soil bag, and air-dried in a dry ventilated room to constant weight for laboratory analysis [41].

Soil dry mass is the fully air-dried soil mass in a cloth bag after at least 2 months. It is less than 1 percent different from oven dried (105 °C) soil, and can be used as the experimental samples for SOC content analysis.

Soil organic C was determined by the heated dichromate/titration method [42], with this method being described previously [41].

SOC density was the product of SOC content, soil bulk density, and sampling soil depth (20 cm in this study).

2.5. Analysis of the Urbanization Gradients, Mainly Represented as Ring Road and Urban Settlement Time

According to the city ring road development, 5 regions (1st ring road, 2nd ring road, 3rd ring road, 4th ring road, and outside 5th ring road) were used to represent urban-rural gradients from the central urban (first ring road region-downtown older city) to rural area (outside ring road region-rural younger city, Table 1). Previous studies [14,43] have used ring roads as a proxy for the degree of urbanization [43] or urban-rural gradients [14].

Table 1. Urban forest classification based on urban settlement time and ring roads in Harbin City.

Urban Forest Classification	Classified Regions	Description	Urban-Rural Gradients
Ring road region (fast road or express way)	First	Within 1st ring road	Urban
	Second	Distribute between 1st ring and 2nd ring road	↓
	Third	Distribute between 2nd ring and 3rd ring road	
	Forth	Distribute between 3rd ring and 4th ring road	
	Outside	Outside of forth ring road	Rural
Urban settlement time (history of settlements)	100-year	Urban area constructed before 1906	Urban
	80-year	urban area constructed between 1933 and 1907	↓
	70-year	urban area constructed between 1945 and 1934	
	50-year	urban area constructed between 1962 and 1946	
	10-year	urban area constructed between 2005 and 1963	
	0-year	urban area constructed during 2006 and 2014	Rural

Note: the black arrows represent the urbanization degree from high to low.

According to settlement time, 6 urban regions have been classified with settlement times of 0, 10, 50, 70, 80, and 100 years, which represent human disturbance history. The specific classification method is displayed in Table 1. We determined settlement time breakpoints according to the urban development process of Harbin [34]. Previous studies [15,44] have performed similar research and used the ages of residential areas or parks as a factor to assess the influence on soil C and nitrogen [15] or heavy metals [44].

DaoLi district, DaoWai district, and XiangFang district are all old districts, and may represent a high level of urbanization. NanGang district is younger than those three districts, and represents a medium level of urbanization. SongBei district is a new district, and represents a low level of urbanization (Figure 1).

2.6. Statistical Analysis

One-way ANOVA and Duncan's new multiple range method were used to examine differences in C stock density in tree biomass and soil, SOC contents, soil bulk density, diameter at breast height (DBH), and basal area at breast height (simplified as basal area) among different urban forest types, urbanization gradients of ring-road regions and urban areas with different settlement history, and different administrative districts. These were performed using SPSS (version 19.0, 2010, IBM, Armonk, NY, USA) and MS Excel (14.0.4760.1000, 2010, Microsoft, Redmond, WA, USA). The significance level was 0.05.

Spatial variation in tree C stock density, SOC density, SOC content, soil bulk density, diameter at breast height (DBH), and basal area were quantified and mapped using ArcGIS 10.0 software (Esri, Beijing, China).

3. Results

3.1. Variation in Forest Types with Respect to Tree Biomass and Soil C-Related Parameters

Tree C stock density, SOC density, SOC content, basal area, and DBH differed significantly among different forest types (Figure 2, Table A2). The highest tree C stock density (15.50 kg·m^{-2}), basal area (47.76 m^2·ha^{-1}), and DBH (25.31 cm) in EF were 1.7 to 3 times those of other forest types. The tree C stock density, basal area, and DBH in AF, RF, and LF were not significantly different ($p > 0.05$, Figure 2 left).

The peak SOC density (6.85 kg·m^{-2}) and SOC content (25.62 g·kg^{-1}) in LF were 1.4 to 1.5 times higher than those of the other 3 forest types, whereas the lowest soil bulk density (1.35 g·cm^{-3}) also occurred in LF. the SOC density, soil bulk density, and SOC content of AF, RF, and EF did not differ significantly ($p > 0.05$, Figure 2 right).

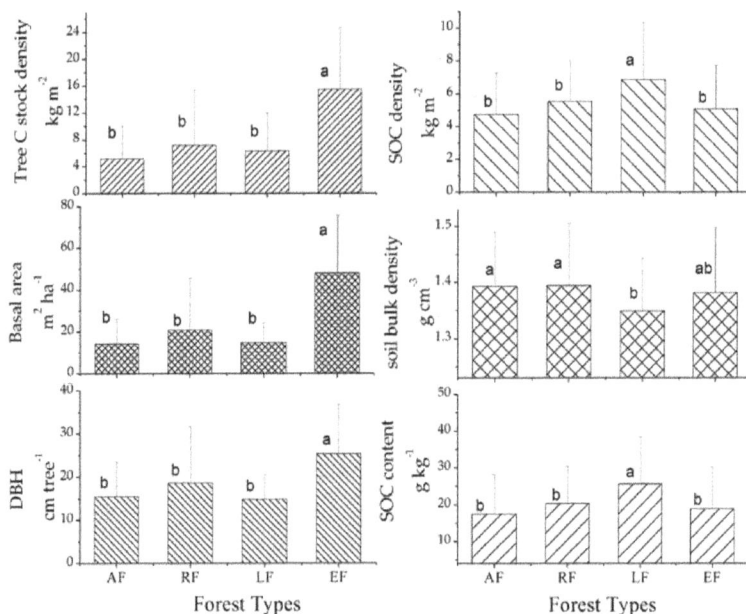

Figure 2. Differences in tree C stock density, basal area, and diameter at breast height (DBH) (**left**), plus differences in soil organic carbon (SOC) density, SOC content, and soil bulk density (**right**) among different forest types in Harbin City. Note: AF, affiliated forest; RF, roadside forest; LF, landscape and relaxation forest; EF, ecological public welfare forest. Error bars showed standard deviation, and lowercase letters (a and b) indicate significant differences ($p < 0.05$) based on the one-way ANOVA test statistics. The same letters (a and a, ab) indicate no significant differences, and different letters (a and b) indicate significant differences.

3.2. Variation in Ring Roads with Respect to Tree Biomass and Soil C-Related Parameters

Tree C stock density, DBH, and basal area differed significantly among different ring-roads (Figure 3 left, Table A2). They were greater in the outer ring roads (forth and fifth ring road) than inner ring roads (first, second, and third ring road) ($p < 0.05$). In comparison, SOC density, soil bulk density, and SOC content were greater in the inner ring roads (first and second ring road) than outer ring roads (third, forth, and fifth ring road).

The highest tree C stock density, basal area, and DBH in the fifth (outside of forth) ring road (17.82 kg·m^{-2}, 56.00 m^2·ha^{-1}, and 29.30 cm respectively) were 2 to 3 times those of the other 4 ring roads. No correlations were detected for tree C stock density, basal area or DBH along this urban-rural gradients (Figure 3 left).

There were no significant differences in SOC density, soil bulk density, and SOC content among different ring roads (Table A2). However, SOC density logarithmically declined with increasing ring road number ($R^2 = 0.87$, $p < 0.05$, Figure 3 right), declining from 6.91 kg·m^{-2} to 5.01 kg·m^{-2}. Soil bulk density linearly declined with increasing ring road number ($R^2 = 0.9$, $p < 0.05$), declining from 1.40 g·cm^{-3} to 1.37 g·cm^{-3}. The SOC content exponentially declined with increasing ring road number ($R^2 = 0.92$, $p < 0.05$), declining from 24.88 g·kg^{-1} to 18.59 g·kg^{-1}.

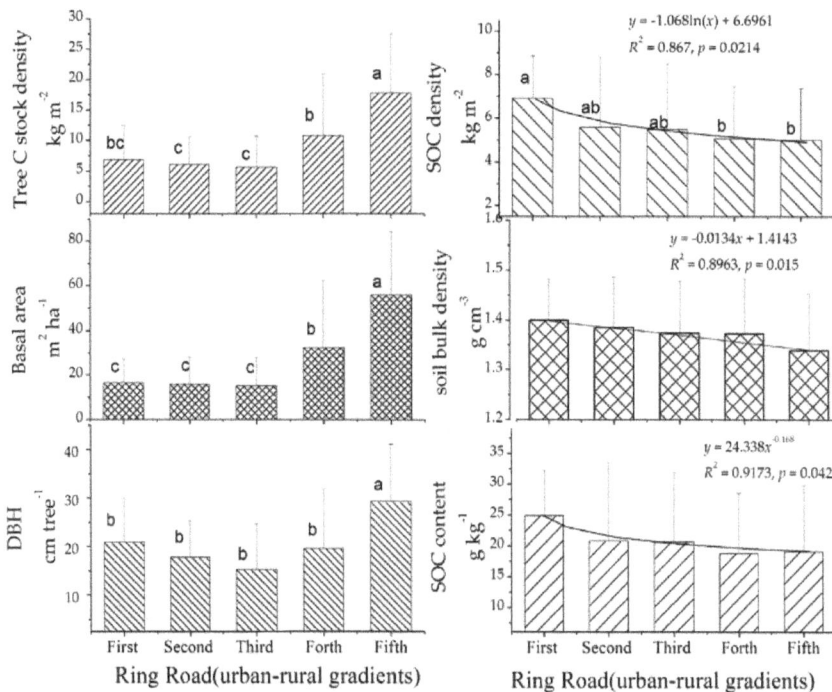

Figure 3. Differences in tree C stock density, basal area, and diameter at breast height (DBH) (**left**), plus differences in soil organic carbon (SOC) density, SOC content, and soil bulk density (**right**) across ring-road-based urban-rural gradients in Harbin City. The same letters (e.g., a and a, ab) indicate no significant differences, and different letters (e.g., a and b, or b and c) indicate significant differences ($p < 0.05$) based on the one-way ANOVA test statistics.

3.3. Variation in History of Settlement Region with Respect to Tree Biomass and Soil C-Related Parameters

Average DBH differed significantly among urban areas with different settlement times ($p < 0.05$; Figure 4, Table A2). No significant differences were found in other parameters. Even though there were no significant differences in SOC density and SOC content between different urban settlement times, the regression analysis showed that they both linearly increased with urban settlement time

($p < 0.05$). SOC density increased from 5.06 kg·m^{-2} to 6.94 kg·m^{-2} with an increasing rate of 15.4 g·m^{-2}·year^{-1}, and SOC content increased from 18.62 g·kg^{-1} to 25.38 g·kg^{-1}, with an increasing rate of 0.055 g·kg^{-1}·year^{-1} (Figure 4, right).

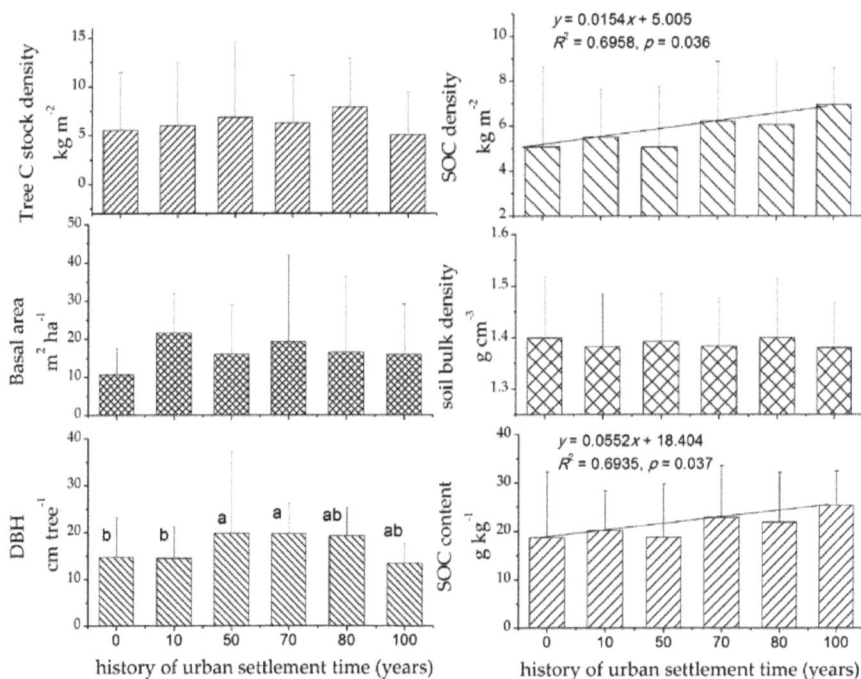

Figure 4. Differences in tree C stock density, basal area, and diameter at breast height (DBH) (**left**), plus differences in soil organic carbon (SOC) density, SOC content, and soil bulk density (**right**) across the history of urban settlement time-based urban-rural gradients in Harbin City. The same letters (e.g., a and a, ab) indicate no significant differences, and different letters (e.g., a and b, or b and c) indicate significant differences ($p < 0.05$) based on the one-way ANOVA test statistics.

3.4. Variation in Administrative Districts with Respect to Tree Biomass and Soil C-Related Parameters

Tree C stock density, SOC density, basal area, and DBH differed significantly among different administrative districts ($p < 0.05$; Figure 5, Table A2), and no significant differences were found in other parameters. Compared with SOC density (1.47-fold), much greater variation was found in tree C stock density (1.47-fold versus 2.25-fold, respectively). The highest tree C stock density in the DW district (10.40 kg·m^{-2}) was 2-fold higher than that in the DL district (4.62 kg·m^{-2}). DBH and basal area were highest in the XF district (20.95 cm, 32.06 m^2·ha^{-1}) and lowest in the DL district (13.43 cm, 12.14 m^2·ha^{-1}). The highest SOC density and SOC content in the old district (high level of urbanization) of DL (6.52 kg·m^{-2}, 23.94 g·kg^{-1}) were 1.47 times those of the young district (low level of urbanization) of SB (4.42 kg·m^{-2}, 16.11 g·kg^{-1}).

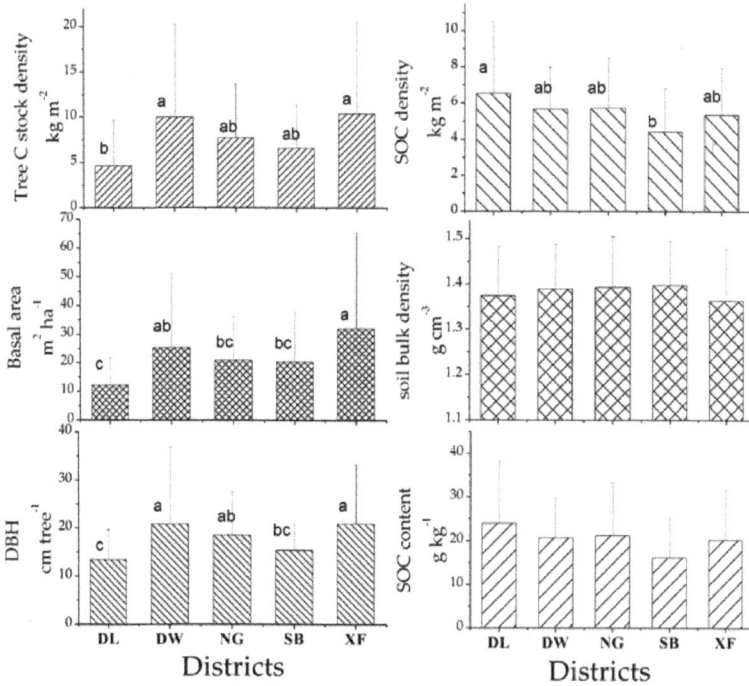

Figure 5. Tree biomass differences in tree C stock density, basal area, and diameter at breast height (DBH) (**left**), plus differences in soil organic carbon (SOC) density, SOC content, and soil bulk density (**right**) among different administrative districts in Harbin City. Note: DL, Daoli District; DW, Daowai District; NG, Nangang District; SB, Songbei District; XF, Xiangfang District. The same letters (e.g., a and a, a and ab) indicate no significant differences, and different letters (e.g., a and b, or b and c) indicate significant differences ($p < 0.05$) based on the one-way ANOVA test statistics.

3.5. Spatial Distribution Map: Visual Confirmation of Variation

The variations of C in tree biomass and soils had different spatial distributions. Specifically, spatial variations of tree C stock density, DBH and basal area were similar in that these parameters of EF in the outer ring roads were greater than those in the inner ring roads for all forest types (Figure 6). However, the spatial variations of SOC density and SOC content of LF and RF in the inner ring road regions were greater than the outer ring roads for all forest types. Compared with soil C variations, the spatial distribution of tree C was more uneven (Figure 6).

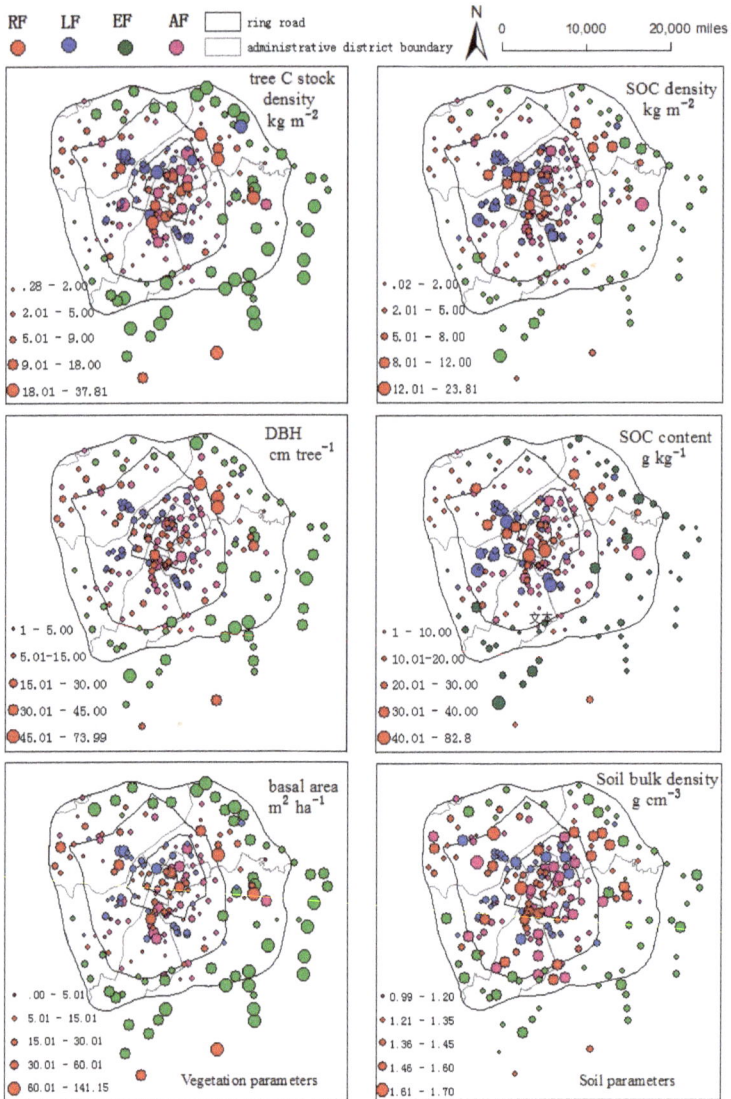

Figure 6. Variation in the spatial distribution of urban forest in Harbin City with respect to tree C stock density, basal area, and diameter at breast height (DBH) (**left**), plus variation in the spatial distribution of soil organic carbon (SOC) density, SOC content, and soil bulk density (**right**). Note: AF, affiliated forest; RF, roadside forest; LF, landscape and relaxation forest; EF, ecological public welfare forest. Label colors represent urban forest types, while label size represents the value range of each parameter.

4. Discussion

4.1. Tree and Soil C Stock Density of Harbin City Versus Other Cities and Natural Forests

The tree C stock density of Harbin's urban forests ranged from 0.36 to 37.81 kg C·m^{-2}, and averaged 7.71 kg C·m^{-2}. This average value was much higher than Los Angeles and Sacramento,

USA, reported by McPherson et al. [8], and cities in China like Beijing, Xiamen, and Hangzhou (Table 2). However, this average value was within the range of 3.14 kg C·m^{-2} for South Dakota and 14.14 kg C·m^{-2} in Omaha, NE, as reported by Nowak et al. [2]. The total C sequestration offsets by street trees and urban forests varied from 0.2% in Beijing [4] to 18.57% in Hangzhou [7], relative to their annual CO_2-equivalent emissions from total energy consumption.

The average C stock density value in Harbin was higher than that in other nearby urban areas, such as 3.32 kg C·m^{-2} in Shenyang [27] and 4.41 kg C·m^{-2} in Changchun [8] (Table 2). These cities have the same climate and same soil type as Harbin City. However, this C stock density value was much lower than that documented for nearby natural forests located at the same latitude (45° N) [45]. The larger tree C stock density in Harbin City compared with other nearby cities may be because of the larger trees in Harbin. Tree DBH >30 cm accounted for 13.89% of total trees in Harbin, of which, 4.85% had a DBH >45 cm, and only 18.26% of the total trees in Harbin had a DBH of <7.6 cm (Table 3). In comparison, 50% of trees in Shenyang City had a DBH of <7.6 cm [37]. Furthermore, almost 70% of trees in Changchun had a DBH of <15.2 cm [14].

Table 2. Tree C stock density, soil organic carbon (SOC) density, SOC content, and soil bulk density in Harbin City compared with other cities and local natural forests.

City	Tree C Stocks kg·m^{-2}	SOC Density kg·m^{-2}	SOC Content g·kg^{-1}	Soil Bulk Density g·cm^{-3}	Soil Depth cm	Citations
Beijing	3.19	–	–	–	–	Tang et al., 2005 [4]
Xiamen	2.08	–	–	–	–	Ren et al., 2012 [6]
Hangzhou	3.03	–	–	–	–	Zhao et al., 2010 [7]
Los Angeles, Sacramento	0.82; 1.54	–	–	–	–	McPherson et al., 2013 [8]
Other cities in US	3.14–14.14	1.5–16.3	–	–	–	Nowak et al., 2013 [2]; Pouyat et al., 2006 [16]
Shenyang	3.32	–	19.29	–	20	Liu and Li, 2012 [37]; Wang et al., 2009 [46]
Changchun	4.41	–	22	–	20	Zhang et al., 2015 [14]; Wang et al., 2011 [47]
Local natural forest	10.54	7.52	33.42	1.08	>20	Zhang, 2010 [45]; Wang et al., 2014 [48], Wang et al., 2013 [49]
Harbin	7.71	5.48	19.98	1.38	20	This research

– means no data.

Table 3. Urban forest tree diameter distribution in major cities in northeastern China.

City	<7.6	7.6–15.2	15.3–30.5	>30.6	Citations
Shenyang	50%	32%	15%	3%	Liu and Li, 2012 [37]
Changchun	18.14%	48.94%	26.07%	6.85%	Zhang et al., 2015 [14]
Harbin	18.26%	36.99%	30.86%	13.89%	This research

The surface soil (0–20 cm) organic C stock density of urban forests in Harbin ranged from 0.31 to 23.8 kg·m^{-2} (average: 5.48 kg·m^{-2}). This value was within the range of 1.5 kg·m^{-2} for Washington, DC, and 16.3 kg·m^{-2} for Chicago, IL, but lower than that recorded for nearby natural forests (Table 2). The SOC content of Harbin was 19.98 g·kg^{-1} on average, which was between Shenyang and Changchun (Table 2). Soil bulk density data are not available for these two nearby cities; thus, we could only assume that their SOC density was similar to that of Harbin City because of the similar SOC content in these two cities. Due to the limits of urban management, we only collected surface soils (surface to 20 cm depth) from Harbin, which could lead to an underestimate of the SOC stocks in its urban forest.

4.2. Tree C Stock Variation along the Urban-Rural Gradients

Urban forests have decreased with increasing urbanization [50]. Meanwhile, many afforestation movements have been initiated in cities [51]. Since the 1980s, large areas of urban green spaces in China have been occupied due to accelerating urbanization and real estate development. However, China simultaneously implemented the concept of Forest Cities (designated by the State Forestry Administration of the People's Republic of China) to increase urban forest coverage. By 2014, 96 cities across the country were awarded the title of "national forest city" [52]. Parallel deforestation and afforestation in the same urban region showed no noticeable or inconsistent spatial patterns in Harbin. Consequently, variation of tree C in urban space exhibited a casual-like process at different urbanization levels. For instance, tree C stock density showed no relations with urban-rural gradients. There was as much as 2.3- to 3.2-fold tree C variations in different forest types, ring roads, and administrative districts (Figures 2, 3 and 5, $p < 0.05$) in Harbin city. Regression analysis showed no significant relations between tree C and urbanization levels in this city (Figures 3 and 4 left).

4.3. SOC Density Changes along Urban-Rural Gradients

In contrast to tree C, the spatial distribution of soil C was more like an urbanization-dependent process, which increased with rural-urban gradients (represented by decreasing ring road or increasing history of settlements) in the research region (Figures 3 and 4). These results support those obtained by previous studies [15,17,25]. Pouyat et al. [17] showed that the highest organic matter content is found in the surface soil of urban areas (97 ± 3.3 g·kg^{-1}), followed by suburban areas (83 ± 2.6 g·kg^{-1}), and rural areas (73 ± 4.3 g·kg^{-1}). In addition, the authors showed that forest floor mass decreased with increasing distance to central urban areas ($y = -0.0295x + 8.2385$, $R^2 = 0.54$) [17]. Koerner and Klopatek [25] showed that SOC content and SOC stocks were higher in the surface soils (0–10 cm and 10–20 cm) of downtown regions compared with suburban and rural regions, regardless of whether the canopy was *Larrea tridentata* or interplant space. Our results showed that urban forest soil appears to accumulate C over settlement time. For instance, there was a 37.92% increase in SOC density between the area of outside ring-road region (rural) and the first ring road region (central urban, Figure 3), representing an accumulation rate of 15.4 g·C·m^{-2}·year^{-1} (55.2 g·SOC·kg^{-1}·year^{-1}) over the 100 years of urbanization in Harbin City (Figure 4). By surveying soil to 100 cm depth, a previous study reported much larger SOC accumulation in Baltimore City, USA, that used housing age as a predictor of soil C with the accumulation rate of 82 g·C·m^{-2}·year^{-1} [15]. Our study supports these preceding studies, showing that the urbanization has an incremental consequence on SOC density.

The mechanisms that dominate higher SOC content in downtown regions are unclear. Many factors influence SOC accumulation in downtown regions, with various studies investigating how to clarify the underlying mechanism. Researchers have shown that pools of labile C are lower, whereas pools of passive C are higher, in urban trees stands relative to rural forest stands [53]. In addition, lower microbial tree biomass [53] and lower litter quality in urban forests (decays 25% more slowly than litter in rural forest) [54] may lead to reduced C decomposition rates and consequent SOC accumulation in urban region. These phenomena may lead to higher SOC in urban forests compared with rural forests [17]. However, urban conditions (including higher temperature) may on the other hand accelerate litter decomposition [17,55], which is not conducive to SOC accumulation. In fact, the urbanization of arid and semiarid ecosystems leads to enhanced C cycling rates that may alter regional C budgets [27]. To sum up, certain conclusions may be drawn about SOC accumulation over time, but we cannot be certain of the mechanisms that drive this pattern due to the complex, sometimes opposing factors controlling C stocks in urbanized regions [15].

4.4. Implications for Urbanization-Induced C Stocks in China

Since 1978, China has experienced a rapid and unprecedented process of urbanization, with urban areas almost exponentially expanding outwards in many cities in parallel with ring road construction

(Figure A1). For instance, the urban area in Harbin expanded by 65% over a 30-year period, from 200 km^2 in the 1980s to 330 km^2 in the 2010s. Further, in the 1980s, Harbin had only two ring roads, which increased to four ring-roads in the 2010s (Figure A1). With urban sprawl, the city area became a great source of carbon owing to human-related C emissions, and urban forests could serve as an important C stocks in cities. A basic assumption for previous estimations is the independence of SOC density to urbanization. Our results indicate that soil C is higher in the older central urban area than rural urban area. In future estimations of C stocks in urban forests, this urbanization-induced SOC accrual should be fully considered for a precise estimation. The amount of SOC accumulation in urban forests during urban sprawl was estimated from the SOC density accumulation rate (Figure 4 right for settlement time; Figure 3 right for ring-road expansion), along with the built-up area, urban settlement time (assuming that variation in SOC was due to urbanization), and average forest coverage (10%). The urbanization gradients (ring road expansion and/or history of urban settlement) in Harbin have led to about 5 to 14 thousand tons of SOC accumulating in the 0–20 cm soil layer. Previous studies showed that the total aboveground tree C stocks in the urban forests of Harbin are about 19 thousand tons [30]. Our data demonstrated that SOC accumulation in the soil during urbanization (about 100 years) in Harbin is about 26%–74% of this total tree C stocks [30]. This result shows the importance of underground SOC accumulation for estimating the total C stocks in urban areas, especially locations and regions subject to rapid urbanization [17]. The fact that SOC density increases during urbanization should be considered when making robust estimates of how urbanization affects the C sequestration function.

5. Conclusions

Urban forests act as a sink for carbon by storing carbon in tree biomass and accumulating carbon in soils. The C stocks of urban forests are of great importance because they influence local climate, carbon cycles, energy use and climate change. Urban forest tree biomass and SOC density in Harbin were higher than other local cities, but much lower than local natural forests. There was a 2- to 3-fold spatial variation in tree C stock density among forest types, administrative districts and ring roads, which randomly varied with urbanization gradients (e.g., ring roads and history of settlements). In comparison, regular urbanization-dependent changes were observed in soil C, as demonstrated by the close relationship between SOC density and urbanization gradients (ring roads and urban settlement history). A city's ability to provide C-related ecosystem services, especially soil C, increases as it ages. Urbanization favors C sequestration in soil with an accumulation rate of 15.4 g·C·m^{-2}·year^{-1}. We estimated that urbanization-induced SOC accumulation is about 5 to 14 thousand tons in the whole region, which is of importance when comparing it with total tree C. Future studies on urban forest C dynamics should focus on urbanization effects especially for belowground C dynamics.

Acknowledgments: This study was supported financially by the One Hundred Talents Program in Chinese Academy of Sciences (Grand No. Y3H1051001), Outstanding Youth Fund from Heilongjiang Province (JC201401), basic research fund for national universities from the Ministry of Education of China (2572014EA01), key project from Chinese Academy of Sciences (Grand No. KFZD-SW-302), and NSFC project (Grand No. 31670699). Thanks are also due to Wei Chenhui, Ren Manli, Pei Zhongxue, and Zhang Dan for their help during the field survey and data processing. We would like to thank Editage (www.editage.com) for English language editing.

Author Contributions: Hailiang Lv, Wenjie Wang, and Xingyuan He conceived and designed the experiments; Hailiang Lv, Lu Xiao, Wei Zhou, and Bo Zhang performed the experiments; Hailiang Lv and Wenjie Wang analyzed the data; Lu Xiao, Wei Zhou, and Bo Zhang contributed reagents/materials/analysis tools; Hailiang Lv wrote the paper.

Conflicts of Interest: The authors declare no conflict of interest.

Appendix

Table A1. Allometric relations used in this paper.

Latin Name	Tree Biomass Equations	Citations
Pinus tabuliformis	$B_{ag} = B_{stem} + B_{branch} + B_{leaf}$; $B_{stem} = 0.11 \times D^{2.34}$; $B_{branch} = 0.01 \times D^{2.58}$; $B_{leaf} = 0.0049 \times D^{2.48}$; $B_r = 0.64 \times D^{2.1}$	Ma [56]; Liu and Li [37]
Ulmus	$B_{ag} = B_{stem} + B_{branch} + B_{leaf}$; $B_{stem} = 0.043 \times D^{2.87}$; $B_{branch} = 0.0074 \times D^{2.67}$; $B_{leaf} = 0.0028 \times D^{2.50}$	Chen and Guo [57]; Liu and Li [37]
Picea	$B_{ag} = B_{stem} + B_{branch} + B_{leaf}$; $B_{stem} = 0.057 \times D^{2.48}$; $B_{branch} = 0.012 \times D^{2.41}$; $B_{leaf} = 0.083 \times D^{2.37}$; $B_r = 0.0088 \times D^{2.54}$	Chen and Guo [57]; Liu and Li [37]
Betula platyphylla	$B_{ag} = 102.159 \times D^{2.367}/1000$; $B_r = 101.358 \times D^{2.518}/1000$	Wang [39]
Populus	$B_{ag} = 101.826 \times D^{2.558}/1000$; $B_r = 101.025 \times D^{2.56}/1000$	Wang [39]
Pinus koraiensis	$B_{ag} = 102.236 \times D^{2.144}/1000$; $B_r = 101.296 \times D^{2.376}/1000$	Wang [39]
Larix gmelinii	$B_{ag} = 101.977 \times D^{2.451}/1000$; $B_r = 101.085 \times D^{2.57}/1000$	Wang [39]
Acer	$B_{ag} = 101.930 \times D^{2.535}/1000$; $B_r = 102.112 \times D^{1.981}/1000$	Wang [39]
Fraxinus	$B_{ag} = 102.136 \times D^{2.408}/1000$; $B_r = 101.396 \times D^{2.467}/1000$	Wang [39]
Juglans mandshurica	$B_{ag} = 102.235 \times D^{2.287}/1000$; $B_r = 101.226 \times D^{2.397}/1000$	Wang [39]
Phellodendronamurense	$B_{ag} = 101.942 \times D^{2.332}/1000$; $B_r = 101.024 \times D^{2.617}/1000$	Wang [39]
Tilia	$B_{ag} = 101.606 \times D^{2.668}/1000$; $B_r = 101.273 \times D^{2.452}/1000$	Wang [39]
Quercus mongolica	$B_{ag} = 102.002 \times D^{2.456}/1000$; $B_r = 101.482 \times D^{2.356}/1000$	Wang [39]
Pinus sylvestris L. var. sylvestriformis	$B_{ag} = B_{stem} + B_{br} + B_{leaf}$; $B_r = 200.0322 \times D^{1.495}/1000$; $B_{stem} = 0.0159368 \times D^{2.949} + 0.6300862 \times D^{0.759}$; $B_{branch} = 0.0557699 \times D^{2.483}$; $B_{leaf} = 0.1090 \times D^{4.293}/1000$	Zou et al. [58]
Pinus sylvestris var. mongolica Litv.	$B_{ag} = B_{stem} + B_{branch} + B_{leaf}$; $B_{stem} = 0.0439 \times (D^2 H)^{0.8852}$; $B_{branch} = 0.02388 \times D^{4.1912} H^{-2.3076}$; $B_{leaf} = 0.1082 \times D^{2.7169} H^{-1.3955}$	Jia et al. [59]
Platycladus	$B_{ag} = B_{stem} + B_{branch} + B_{leaf}$; $B_{stem} = 0.013 \times (D^2 H)^{0.5969} + 0.0036 \times (D^2 H)^{0.6758}$; $B_{branch} = 0.00274 \times (D^2 H)^{0.5973} + 0.004965 \times (D^2 H)^{0.5975} + 0.00055 \times (D^2 H)^{0.5879}$; $B_{leaf} = 0.003787 \times (D^2 H)^{0.5976}$	Chang et al. [60]
Padusracemosa	$B_{ag} = 0.00009 \times D^{2.696}$; $B_r = 0.035 \times D^{2.641}/1000$	Li [61]
Rosaceae	$B_{ag} = 10 - 0.6657 \times D^{1.7041}$	Wu [62]
Tree generalized equation	$B_{ag} = 101.945 \times D^{2.467}/1000$; $B_{total} = 102.033 \times D^{2.469}/1000$; $B_r = B_{total} - B_{ag}$	Wang [40]
Acer ginnala	$B_{ag} = 0.527 \times D^{2.217}/1000$; $B_r = 0.149 \times D^{2.261}/1000$	Li [61]
Syringa reticulata	$B_{ag} = 0.395 \times D^{2.3}/1000$; $B_r = 0.129 \times D^{2.302}/1000$	Li [61]
Euonymus alatus	$B_{ag} = 0.095 \times D^{2.655}/1000$; $B_r = 0.089 \times D^{2.291}/1000$	Li [61]
Rhamnusschneideri	$B_{ag} = 0.169 \times D^{2.555}/1000$; $B_r = 0.092 \times D^{2.314}/1000$	Li [61]
Viburnum sargenti	$B_{ag} = 0.141 \times D^{2.649}/1000$; $B_r = 0.245 \times D^{1.994}/1000$	Li [61]
Tree Shrub generalized equation	$B_{ag} = 0.182 \times D^{2.487}/1000$; $B_r = 0.089 \times D^{2.37}/1000$	Li [61]

D means DBH (cm), H means height (m), B_{ag} means aboveground biomass (kg), B_r means root biomass.

Table A2. Tree and soil carbon stocks differences in different forest types, ring roads, administrative districts, urban area settlement time of urban forests in Harbin City: One-way ANOVA results.

Source	Dependent Variable	df	F	Sig.
Forest types	SOC density kg·m^{-2}	3	5.181	<0.01 **
	Tree C stocks density kg·m^{-2}	3	23.026	<0.001 ***
	SOC content g·kg^{-1}	3	4.735	<0.01 **
	Soil bulk density g·cm^{-3}	3	1.813	0.146
	DBH cm	3	7.403	<0.001 ***
	Basal area m^2·ha^{-1}	3	31.598	<0.001 ***

Table A2. *Cont.*

Source	Dependent Variable	df	F	Sig.
Ring roads (urban-rural gradients)	SOC density kg·m^{-2}	4	1.535	0.193
	Tree C stocks density kg·m^{-2}	4	12.670	<0.001 ***
	SOC content g·kg^{-1}	4	1.095	0.360
	Soil bulk density g·cm^{-3}	4	0.284	0.207
	DBH cm	4	7.403	<0.001 ***
	Basal area m^2·ha^{-1}	4	18.103	<0.001 ***
History of settlements	SOC density kg·m^{-2}	5	1.116	0.354
	Tree C stocks density kg·m^{-2}	5	0.383	0.860
	SOC content g·kg^{-1}	5	0.972	0.437
	Soil bulk density g·cm^{-3}	5	0.167	0.975
	DBH cm	5	2.325	<0.05 *
	Basal area m^2·ha^{-1}	5	0.529	0.754
Districts	SOC density kg·m^{-2}	5	2.373	<0.05 *
	Tree C stocks density kg·m^{-2}	5	6.059	<0.001 ***
	SOC content g·kg^{-1}	5	2.047	0.073
	Soil bulk density g·cm^{-3}	5	0.695	0.533
	DBH cm	5	11.683	<0.01 **
	Basal area m^2·ha^{-1}	5	6.723	<0.01 **

Note: * represent 0.05 significant level, ** represent 0.01 significant level, and *** represent 0.001 significant level.

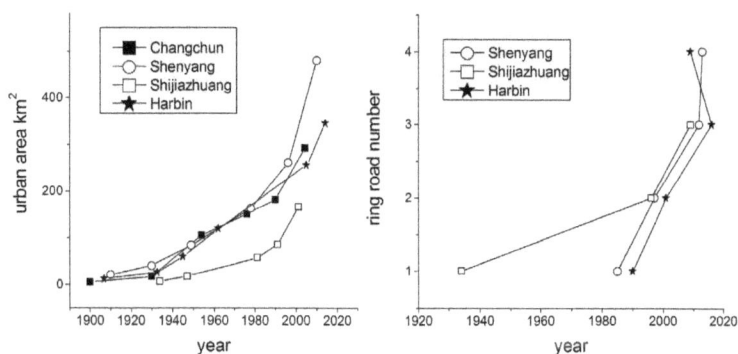

Figure A1. Rapid urbanization of sample cities in China, data acquired and anlyzed based on papers [35,63,64].

References

1. Jim, C.Y.; Chen, W.Y. Ecosystem services and valuation of urban forests in China. *Cities* **2009**, *26*, 187–194. [CrossRef]
2. Nowak, D.J.; Greenfield, E.J.; Hoehn, R.E.; Lapoint, E. Carbon storage and sequestration by trees in urban and community areas of the United States. *Environ. Pollut.* **2013**, *178*, 229–236. [CrossRef] [PubMed]
3. Escobedo, F.; Varela, S.; Zhao, M.; Wagner, J.E.; Zipperer, W. Analyzing the efficacy of subtropical urban forests in offsetting C emissions from cities. *Environ. Sci. Policy* **2010**, *13*, 362–372. [CrossRef]
4. Yang, J.; McBride, J.; Zhou, J.X.; Sun, Z.Y. The urban forest in Beijing and its role in air pollution reduction. *Urban Forest. Urban Green.* **2005**, *3*, 65–78. [CrossRef]
5. Tang, Y.J.; Chen, A.P.; Zhao, S.Q. Carbon storage and sequestration of urban street trees in Beijing, China. *Front. Ecol. Evol.* **2016**, *4*, 53.
6. Ren, Y.; Yan, J.; Wei, X.H.; Wang, Y.J.; Yang, Y.S.; Hua, L.Z.; Xiong, Y.Z.; Niu, X.; Song, X.D. Effects of rapid urban sprawl on urban forest carbon stocks: Integrating remotely sensed, GIS and forest inventory data. *J. Environ. Manag.* **2012**, *113*, 447–455. [CrossRef] [PubMed]
7. Zhao, M.; Kong, Z.H.; Escobedo, F.J.; Gao, J. Impacts of urban forests on offsetting carbon emissions from industrial energy use in Hangzhou, China. *J. Environ. Manag.* **2010**, *91*, 807–813. [CrossRef] [PubMed]

8.   McPherson, E.G.; Xiao, Q.; Aguaron, E. A new approach to quantify and map carbon stored, sequestered and emissions avoided by urban forests. *Landsc. Urban Plan.* **2013**, *120*, 70–84. [CrossRef]

9.   Díaz-Porras, D.F.; Gaston, K.J.; Evans, K.L. 110 years of change in urban tree stocks and associated carbon storage. *Ecol. Evol.* **2014**, *4*, 1413–1422. [CrossRef] [PubMed]

10. Shakeel, T.; Conway, T.M. Individual households and their trees: Fine-scale characteristics shaping urban forests. *Urban For. Urban Green.* **2014**, *13*, 136–144. [CrossRef]

11. Yoon, T.K.; Park, C.W.; Lee, S.J.; Ko, S.; Kim, K.N.; Son, Y.; Son, Y. Allometric equations for estimating the aboveground volume of five common urban street tree species in Daegu, Korea. *Urban For. Urban Green.* **2013**, *12*, 344–349. [CrossRef]

12. Timilsina, N.; Escobedo, F.J.; Staudhammer, C.L.; Brandeis, T. Analyzing the causal factors of C stores in a subtropical urban forest. *Ecol. Complex.* **2014**, *20*, 23–32. [CrossRef]

13. Edmondson, J.L.; O'Sullivan, O.S.; Inger, R.; Potter, J.; McHugh, N.; Gaston, K.J.; Leake, J.R. Urban Tree Effects on Soil Organic C. *PLoS ONE* **2014**, *9*. [CrossRef] [PubMed]

14. Zhang, D.; Zheng, H.; Ren, Z.; Zhai, C.; Shen, G.; Mao, Z.; He, X. Effects of forest type and urbanization on C storage of urban forests in Changchun, Northeast China. *Chin. Geogr. Sci.* **2015**, *25*, 147–158. [CrossRef]

15. Raciti, S.M.; Groffman, P.M.; Jenkins, J.C.; Pouyat, R.V.; Fahey, T.J.; Pickett, S.T.; Cadenasso, M.L. Accumulation of C and Nitrogen in Residential Soils with Different Land-Use Histories. *Ecosystems* **2011**, *14*, 287–297. [CrossRef]

16. Pouyat, R.V.; Yesilonis, I.D.; Nowak, D.J. Carbon storage by urban soils in the United States. *J. Environ. Qual.* **2006**, *35*, 1566–1575. [CrossRef] [PubMed]

17. Pouyat, R.; Groffman, P.; Yesilonis, I.; Hernandez, L. Soil C pools and fluxes in urban ecosystems. *Environ. Pollut.* **2002**, *116*, S107–S118. [CrossRef]

18. Oldfield, E.E.; Felson, A.J.; Wood, S.A.; Hallett, R.A.; Strickland, M.S.; Bradford, M.A. Positive effects of afforestation efforts on the health of urban soils. *For. Ecol. Manag.* **2014**, *313*, 266–273. [CrossRef]

19. Zhao, Y.G.; Zhang, G.L.; Zepp, H.; Yang, J.L. Establishing a spatial grouping base for surface soil properties along urban–rural gradient—A case study in Nanjing, China. *Catena* **2007**, *69*, 74–81. [CrossRef]

20. Setälä, H.M.; Francini, G.; Allen, J.A.; Hui, N.; Jumpponen, A.; Kotze, D.J. Vegetation Type and Age Drive Changes in Soil Properties, Nitrogen, and Carbon Sequestration in Urban Parks under Cold Climate. *Front. Ecol. Evol.* **2016**, *4*. [CrossRef]

21. Huyler, A.; Chappelka, A.H.; Prior, S.A.; Somers, G.L. Influence of aboveground tree biomass, home age, and yard maintenance on soil carbon levels in residential yards. *Urban Ecosyst.* **2014**, *17*, 787–805. [CrossRef]

22. Zhang, H.X.; Zhuang, S.Y.; Qian, H.Y.; Wang, F.; Ji, H.B. Spatial Variability of the Topsoil Organic C in the Moso Bamboo Forests of Southern China in Association with Soil Properties. *PLoS ONE* **2015**, *10*. [CrossRef] [PubMed]

23. Chen, H.; Zhang, W.; Gilliam, F.S.; Liu, L.; Huang, J.; Zhang, T.; Mo, J. Changes in soil C sequestration in *Pinus massoniana* forests along an urban-to-rural gradient of southern China. *Biogeosciences* **2013**, *10*, 6609–6616. [CrossRef]

24. Drayton, B.; Primack, R.B. Plant Species Lost in an Isolated Conservation Area in Metropolitan Boston from 1894 to 1993. *Conserv. Biol.* **1996**, *10*, 30–39. [CrossRef]

25. Koerner, B.A.; Klopatek, J.M. C fluxes and nitrogen availability along an urban-rural gradient in a desert landscape. *Urban Ecosyst.* **2010**, *13*, 1–21. [CrossRef]

26. Rao, P.; Hutyra, L.R.; Raciti, S.M.; Finzi, A.C. Field and remotely sensed measures of soil and biomass C and nitrogen across an urbanization gradient in the Boston metropolitan area. *Urban Ecosyst.* **2013**, *16*, 593–616. [CrossRef]

27. Kaye, J.P.; McCulley, R.L.; Burke, I.C. Carbon fluxes, nitrogen cycling, and soil microbial communities in adjacent urban, native and agricultural ecosystems. *Glob. Chang. Biol.* **2005**, *11*, 575–587. [CrossRef]

28. Gao, Y. Study on Soil of Harbin Urban Biomass. Master's Thesis, Northeast Forestry University, Harbin, China, July 2002.

29. Hu, H.; Chen, X.; Xu, S. Soil nutrients and plant cultivation Countermeasures of Urban green sapce in Harbin. *Res. Environ. Sci.* **2013**, *20*, 110–114.

30. Ying, T.; Li, M.; Fan, W.Y. Estimation of C storage of Urban Forests in Harbin city. *J. Northeast For. Univ.* **2009**, *9*, 33–35.

31. Zhang, X.M.; Chen, L.; Ji, J.Z.; Wang, J.; Wang, Y.B.; Guo, W.; Lan, B.W. Climate change and its effect in Harbin from 1881 to 2010. *J. Meteorol. Environ.* **2011**, *27*, 13–20.

32. Overview of Harbin, People's Government Portal Website of Harbin City. Available online: http://www.hlj. xinhuanet.com/wq/2011--06/09/c_13919750.htm (assessed on 20 August 2016).

33. Chang, J. The Study on Investigation of Soil in Urban Green Space and Effects for the Improving in Harbin. Master's Thesis, Northeast Forestry University, Harbin, China, 2015.

34. He, Y.; Yu, B. *Yesterday-Today-Tomorrow: An Overview of Harbin Landscape*; Heilongjiang People's Publishing House: Harbin, China, 2011.

35. Xiao, J.; Shen, Y.; Ge, J.; Tateishi, R.; Tang, C.; Liang, Y.; Huang, Z. Evaluating urban expansion and land use change in Shijiazhuang, China, by using GIS and remote sensing. *Landsc. Urban Plan.* **2006**, *75*, 69–80. [CrossRef]

36. He, X.; Liu, C.; Chen, W.; Guan, Z.; Zhao, G. Discussion on urban forest classification. *Chin. J. Ecol.* **2004**, *23*, 175–178.

37. Liu, C.; Li, X. Carbon storage and sequestration by urban forests in Shenyang, China. *Urban For. Urban Green.* **2012**, *11*, 121–128. [CrossRef]

38. Davies, Z.G.; Edmondson, J.L.; Heinemeyer, A.; Leake, J.R.; Gaston, K.J. Mapping an urban ecosystem service: Quantifying above-ground C storage at a city-wide scale. *J. Appl. Ecol.* **2011**, *48*, 1125–1134. [CrossRef]

39. Wang, C.K. Biomass allometric equations for 10 co-occurring tree species in Chinese temperate forests. *For. Ecol. Manag.* **2006**, *222*, 9–16. [CrossRef]

40. Nowak, D.J.; Crane, D.E. C storage and sequestration by urban trees in the USA. *Environ. Pollut.* **2002**, *116*, 381–389. [CrossRef]

41. Wang, W.W.; Qiu, L.; Zu, Y.G.; Su, D.X.; An, J.; Wang, H.Y.; Zheng, G.Y.; Sun, W.; Chen, X.Q. Changes in soil organic C, nitrogen, pH and bulk density with the development of larch (*Larix gmelinii*) plantations in China. *Glob. Chang. Biol.* **2011**, *17*, 2657–2676.

42. Bao, S. *Soil Agro-Chemistrical Analysis*; China Agriculture Press: Beijing, China, 2000.

43. Huang, D.C.; Su, Z.M.; Zhang, R.Z.; Koh, L.P. Degree of urbanization influences the persistence of *Dorytomus* weevils (Coleoptera: Curculionoidae) in Beijing, China. *Landsc. Urban Plan.* **2010**, *96*, 163–171. [CrossRef]

44. Chen, T.B.; Zheng, Y.M.; Lei, M.; Huang, Z.C.; Wu, H.T.; Chen, H.; Fan, K.K.; Yu, K.; Wu, X.; Tian, Q.Z. Assessment of heavy metal pollution in surface soils of urban parks in Beijing, China. *Chemosphere* **2005**, *60*, 542–551. [CrossRef] [PubMed]

45. Zhang, Q.Z. Carbon Density and Carbon Sequestration Capacity of Six Temperate Forests in Northeast China. Master' Thesis, Northeast Forestry University, Harbin, China, 2010.

46. Wang, Q.B.; Duan, Y.Q.; Wei, Z.Y. Spatial variability of urban soil organic C in Shenyang. *Chin. J. Soil Sci.* **2009**, *40*, 252–257.

47. Wang, Y.; Li, C.Y.; Li, C.L.; Han, Y.N.; Li, Y.; Zhang, J.J. Preliminary Study on the Quantitative Characteristics of Organic C in Soils under Different Land Use Patterns in Changchun City. *J. Jilin Agric. Univ.* **2011**, *33*, 51–56.

48. Wang, W.J.; Wang, H.M.; Zu, Y.G. Temporal changes in SOM, N, P, K, and their stoichiometric ratios during reforestation in China and interactions with soil depths: Importance of deep-layer soil and management implications. *For. Ecol. Manag.* **2014**, *325*, 8–17. [CrossRef]

49. Wang, W.J.; Su, D.X.; Qiu, L.; Wang, H.M.; An, J.; Zheng, G.; Zu, Y. Concurrent changes in soil inorganic and organic carbon during the development of larch, *Larix gmelinii*, plantations and their effects on soil physicochemical properties. *Environ. Earth Sci.* **2013**, *69*, 1559–1570. [CrossRef]

50. DeFries, R.S.; Rudel, T.; Uriarte, M.; Hansen, M. Deforestation driven by urban population growth and agricultural trade in the twenty-first century. *Nat. Geosci.* **2010**, *3*, 178–181. [CrossRef]

51. Nielsen, A.B.; Jensen, R.B. Some visual aspects of planting design and silviculture across contemporary forest management paradigms–perspectives for urban afforestation. *Urban For. Urban Green.* **2007**, *6*, 143–158. [CrossRef]

52. Let City Embrace Forest, Working Hard to Build a Forest City in Our Country. Available online: http://www.gov.cn/xinwen/2015--11/24/content_2971646.htm (accessed on 24 Novermber 2015).

53. Groffman, P.M.; Pouyat, R.V.; McDonnell, M.J.; Pickett, S.T.; Zipperer, W.C. Carbon pools and trace gas fluxes in urban forest soils. In *Advances in Soil Science: Soil Management and Greenhouse Effect*; CRC Press, Inc.: Boca Raton, FL, USA, 1995; pp. 147–158.

54. Carreiro, M.M.; Howe, K.; Parkhurst, D.F.; Pouyat, R.V. Variation in quality and decomposability of red oak leaf litter along an urban-rural gradient. *Biol. Fertil. Soils* **1999**, *30*, 258–268. [CrossRef]
55. Davidson, E.A.; Janssens, I.A. Temperature sensitivity of soil C decomposition and feedbacks to climate change. *Nature* **2006**, *440*, 165–173. [CrossRef] [PubMed]
56. Ma, Y.A. Study on the bimass of Chinese Pine forests. *J. Beijing For. Univ.* **1989**, *11*, 1–10.
57. Chen, C.; Guo, X. Research on biomass of broad-leaved Korean pine forest. *For. Investig. Des.* **1984**, 10–19.
58. Zou, C.; Pu, J.; Xu, W. Biomass and productivity of *Pinus sylvestriformis* plantation. *Chin. J. Appl. Ecol.* **1995**, *6*, 123–127.
59. Jia, W.; Jiang, S.; Li, F.R. Biomass of Single Tree of *Pinus sylvestris* var. *mongolica* in Eastern Heilongjiang. *J. Liaoning For. Sci. Technol.* **2008**. [CrossRef]
60. Chang, X.X.; Che, K.J.; Song, C.F.; Li, B.X. Biomass and Nutrient Element Accumulation of *Sabina prez ewalskii* Foret Community. *J. Northwest For. Coll.* **1997**, *12*, 24–29.
61. Li, X. Biomass Allometry and Allocation of Common Understory in a Natural Secondary Forest in Maoershan, Northeast China. Master's Thesis, Notheast Forestry University, Harbin, China, 2010.
62. Wu, F. Appraisal of Carbon Storage in Urban Forest Patches and Its Distribution Pattern in Maanshan City, Master's Thesis, Anhui Agricultural University, Hefei, China, 2012.
63. Kuang, W.H.; Zhang, S.W.; Zhang, Y.Z.; Sheng, Y. Analysis of Urban Land Utilization Spatial Expansion Mechanism in Changchun City since 1900. *Acta Geogr. Sin.* **2005**, *60*, 841–850.
64. Sun, Y.; Liu, Z.; Wang, Q.B.; Liu, H.B. Spatial Structure Evolution of Urban Land Use in Shenyang during 1910–2010. *Prog. Geogr.* **2012**, *31*, 1204–1211.

Section 4:
Management and Planning

![forests logo] *forests*

MDPI

Article

Tree Mortality Undercuts Ability of Tree-Planting Programs to Provide Benefits: Results of a Three-City Study

Sarah Widney [1,2,*], Burnell C. Fischer [1,2,3] and Jess Vogt [2,4]

1   School of Public and Environmental Affairs, Indiana University, Bloomington, IN 47405, USA; bufische@indiana.edu
2   Bloomington Urban Forestry Research Group, Indiana University, Bloomington, IN 47408, USA
3   The Vincent and Elinor Ostrom Workshop in Political Theory and Policy Analysis, Indiana University, Bloomington, IN 47408, USA
4   Department of Environmental Science and Studies, College of Science and Health, DePaul University, Chicago, IL 60614, USA; jessica.m.vogt@gmail.com
*   Correspondence: sewidney@indiana.edu; Tel.: +1-252-531-6727

Academic Editors: Francisco Escobedo, Stephen John Livesley and Justin Morgenroth
Received: 16 December 2015; Accepted: 9 March 2016; Published: 11 March 2016

Abstract: Trees provide numerous benefits for urban residents, including reduced energy usage, improved air quality, stormwater management, carbon sequestration, and increased property values. Quantifying these benefits can help justify the costs of planting trees. In this paper, we use i-Tree Streets to quantify the benefits of street trees planted by nonprofits in three U.S. cities (Detroit, Michigan; Indianapolis, Indiana, and Philadelphia, Pennsylvania) from 2009 to 2011. We also use both measured and modeled survival and growth rates to "grow" the tree populations 5 and 10 years into the future to project the future benefits of the trees under different survival and growth scenarios. The 4059 re-inventoried trees (2864 of which are living) currently provide almost $40,000 (USD) in estimated annual benefits ($9–$20/tree depending on the city), the majority (75%) of which are increased property values. The trees can be expected to provide increasing annual benefits during the 10 years after planting if the annual survival rate is *higher* than the 93% annual survival measured during the establishment period. However, our projections show that with continued 93% *or lower* annual survival, the increase in annual benefits from tree growth will not be able to make up for the loss of benefits as trees die. This means that estimated total annual benefits from a cohort of planted trees will decrease between the 5-year projection and the 10-year projection. The results of this study indicate that without early intervention to ensure survival of planted street trees, tree mortality may be significantly undercutting the ability of tree-planting programs to provide benefits to neighborhood residents.

Keywords: planted trees; i-Tree Streets; tree survival; tree growth; tree benefits; ecosystem services

1. Introduction

1.1. Benefits of Urban Trees

The urban forest provides many benefits (*i.e.*, ecosystem services) for urban residents, from stormwater mitigation and air pollutant removal to reduced crime rates and better psychological well-being [1–7]. Nonprofit organizations and municipalities plant substantial numbers of young trees, sometimes in large tree-planting or canopy campaigns, to increase the provisioning of these benefits for urban residents (e.g., Philly Plant One Million Campaign (in Philadelphia, Pennsylvania), MillionTreesNYC (in New York City, New York), Mile High Million (in Denver, Colorado), *etc.*). These entities incur significant costs to plant and maintain trees [8,9], yet there is often little post-planting monitoring to assess whether the trees survive after planting. Additionally, the benefits of the trees are often presumed rather than measured or quantified.

1.2. Value of Quantifying Benefits

Quantification of the benefits provided by the urban forest can be used to help justify the costs of planting and maintaining trees and encourage investment in green infrastructure, and is also useful for evaluating goals associated with tree planting. For example, tree-planting campaigns are often undertaken to reduce combined sewer overflows or sequester carbon and reduce energy usage to mitigate the effects of climate change (e.g., Plant One Million 2015, City Plants 2015). Estimation of the amount (or monetary value) of stormwater intercepted or carbon sequestered by the planted trees could be used to evaluate whether the tree-planting programs are meeting their goals in a cost-effective manner. Assigning a monetary value to tree benefits also allows people to better understand the value of trees.

However, trees do not provide as many benefits when they are young and small as they do when they are large, mature, and healthy (Figure 1). Therefore, an estimation of benefits at the time the trees are planted will significantly undervalue the benefits the trees will provide in the future. An accurate prediction of the benefits trees will provide when they are mature would be more useful, for example, when conducting a cost-benefit analysis of a tree-planting program. Yet there are few to no programs that allow the user to employ locally relevant, empirically generated growth and survival rates to estimate the future benefits of planted trees.

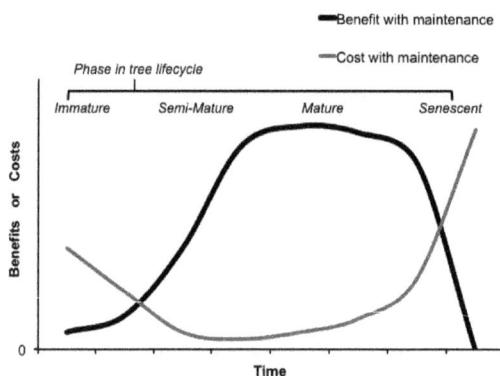

Figure 1. Hypothetical benefits and costs over a tree's lifetime [10]. Used with the permission of the International Society of Arboriculture.

1.3. Costs Associated with Trees

Street trees provide benefits to urban residents, but there are also costs associated with these trees. In general, the costs of trees are not as well quantified or monetized as the benefits of trees [10]. A recent review of the cost of maintaining the urban forest categorizes the costs of trees as direct costs of planting and maintaining trees, costs of repairs to infrastructure damaged by street trees, costs of negative externalities associated with trees (*i.e.*, ecosystem disservices), and opportunity costs [10], but others include infrastructure damage as an ecosystem disservice (e.g., [11,12]). Direct costs include all of the costs associated with planting and maintaining trees, including the purchase of the tree, cost of planting the tree, and subsequent pruning, watering, leaf removal, program administration, and eventual tree and stump removal/disposal [8–10]. Typical infrastructure interference costs are the costs of pavement and sewer repair and power outages caused by falling limbs or trees [10,13,14]. Some important negative externalities, or ecosystem disservices, associated with trees include emissions of biogenic VOCs, release of carbon dioxide (CO_2) during maintenance activities and from leaf and wood decomposition, and allergies caused by pollen release [10,12]. Finally, opportunity costs include the lost space that could have been dedicated to parking, bike lanes, sidewalk cafes, or other uses of the public right-of-way, and the lost money that could have been spent on another program [10].

In a true cost-benefit analysis (CBA), where the object is to obtain the net benefits of a project or program (e.g., a project to plant a certain number of trees), all of the above costs must be monetized and weighed against all of the benefits that trees provide. We acknowledge the utility of conducting CBAs: (1) to determine whether the net benefits derived, in strictly numeric, monetary terms, justify the investment; or (2) to compare and select the socially optimal investment strategy among a set of options (*i.e.*, select the program with the highest net benefits, or select the highest benefit-to-cost ratio) [15]. However, the aim of this study is not to conduct a CBA, but to analyze how the total (not net) benefits (as calculated by the i-Tree Streets software) change as the trees grow during the decade or so after transplanting and estimate how the benefits are affected by different growth and survival scenarios.

1.4. i-Tree Streets

To facilitate quantification and monetization of urban tree benefits, the U.S. Forest Service, Davey Resource Group, and other organizations partnered to produce i-Tree, a suite of programs that can be used to estimate the benefits provided by the urban forest [16,17]. This paper will focus on the use of i-Tree Streets (Version 5.1.5), which is designed to estimate the monetary value of the benefits of public and private street trees based on inventory data. We use i-Tree Streets to estimate the current benefits provided by recently planted trees in three U.S. cities. In addition, we use growth and survival rates measured for each study city to project the samples of trees 5 and 10 years into the future and estimate the annual benefits of the resulting populations in i-Tree Streets. We also model different survival and growth rates to elucidate the effects of increased survival and faster growth on the estimated benefits of the subsequent population. We acknowledge that the benefit estimates we report are used for comparison, as a translation of tree population characteristics (including growth and survival rates) into monetary values (in US Dollars) as outputs from i-Tree, and may not reflect actual benefits as valued and experienced by community residents. Based on this study, we conclude that survival is more important than growth in determining future benefits derived from a population of street trees.

2. Materials and Methods

2.1. Study Sites

The Bloomington Urban Forestry Research Group (BUFRG) at Indiana University (Bloomington, Indiana, USA [18]) was funded by the U.S. Forest Service's National Urban and Community Forestry Advisory Council (NUCFAC) to conduct a study of tree planting projects supported by nonprofit organizations in urban settings in five U.S. cities [19]. The overall goal of the project was to evaluate the outcomes of neighborhood tree-planting programs. We partnered with the Alliance for Community Trees (ACTrees) and five of its nonprofit member organizations, Trees Atlanta, The Greening of Detroit, Keep Indianapolis Beautiful, Inc., the Pennsylvania Horticultural Society, and Forest ReLeaf of Missouri, to conduct this research. The study included a re-inventory (conducted during the summer of 2014) of trees planted in projects funded by these nonprofits from 2009 to 2011. This paper will focus on the cities that had large numbers (>1000) of trees re-inventoried during 2014: Detroit, Michigan; Indianapolis, Indiana; and Philadelphia, Pennsylvania.

2.2. Planted Tree Re-Inventory

Teams of citizen scientists (volunteers and/or high school and college students) trained by BUFRG researchers conducted a planted tree re-inventory [20,21] in summer 2014. The trees in the sample were planted through neighborhood tree-planting projects funded by our nonprofit partners in 2009, 2010, or 2011. All re-inventoried trees were street trees planted in the public right-of-way or very near the public right-of-way (*i.e.*, an adjacent front yard). Trees were re-inventoried in 25 randomly selected neighborhoods per city. For the purposes of this research, a "neighborhood" was defined as a census block group. We randomly selected 25 block groups from the list of all block groups in the city that contained one or more trees planted as part of a planting project of greater than 20 trees.

The re-inventory protocol includes a suite of variables and methods to collect information on the tree, the surrounding growing environment, and evidence of care and maintenance (see [20,21] for complete variable list). The relevant variables for the present study were tree species, caliper (diameter at 6 in (15 cm) above the base of the tree), diameter at breast height (DBH; taken 4.5 ft (1.37 m) up the trunk of the tree), and overall condition rating. The overall condition ratings assigned to trees were good, fair, poor, dead, sprouts, stump, shrub, or absent (see Table 1 for overall condition rating definitions). Condition ratings were not used directly to calculate tree benefits in i-Tree, but were used to calculate survival rates and identify trees that were more likely to die and remove these trees from the population first in benefits projections (see Sections 2.8 and 2.10 below). We had locations and species of every tree planted between 2009 and 2011 by the partner nonprofit in each of the cities, which facilitated the re-inventory and enabled data collectors to assign the condition rating of "absent" if the tree was missing or had been replaced with a different species. Within each randomly selected neighborhood, we assessed survival status for all trees and collected the suite of re-inventory variables for at least every other tree in the neighborhood as time allowed.

Table 1. Explanation of overall condition ratings used in the planted tree re-inventory protocol [21].

Rating	Explanation
Good	Full canopy, minimal to no mechanical damage to trunk, no branch dieback over 5 cm (2″) in diameter, no suckering (root or water sprouts), form is characteristic of species.
Fair	Thinning canopy, new growth in medium to low amounts, tree may be stunted, significant mechanical damage to trunk (new or old), insect/disease is visibly affecting the tree, form not representative of species, premature fall coloring on foliage, needs training pruning.
Poor	Tree is declining, visible dead branches over 5 cm (2″) in diameter in canopy, significant dieback of other branches in inner and outer canopy, severe mechanical damage to trunk usually including decay from damage, new foliage is small, stunted or minimum amount of new growth, needs priority pruning of dead wood.
Dead	Standing dead tree, no signs of life with new foliage, bark may be beginning to peel.
Sprouts	Only a stump of a tree is present, with one or more water sprouts of 45 cm (18″) or greater in height growing from the remaining stump and root system.
Stump	Only a stump remains, no water sprouts greater than 45 cm (18″) high present.
Shrub	Existing vegetation is a shrub growth habit rather than tree growth habit, either because a shrub-form was planted or because the species has been pruned into the shape of a shrub (e.g., many crape myrtle, *Lagerstroemia* sp.).
Absent	No tree present, not even a stump remains visible in the location where the tree should have been; this category should also be used for trees that have obviously been replaced (are the incorrect species, much smaller than they should be given the planting date, *etc.*) and there is no evidence of the original tree.

In addition to locations and species of trees, our nonprofit partners also provided us with an approximation of the caliper-at-planting and exact date (e.g., 30 October 2009) or season (e.g., Fall 2009) the tree was planted. We merged these data with the re-inventory data in Stata [22] to obtain a complete data set with planting and re-inventory variables for each re-inventoried tree.

2.3. How i-Tree Works

i-Tree Streets (Version 5.1.5) estimates the monetary value of benefits in five categories: energy, carbon, air quality, stormwater, and property value benefits. The benefit estimates are influenced by: (1) the size (diameter at breast height, or DBH) and species of the trees; (2) the annual rainfall and number of heating and cooling days for the climate zone; and (3) the prices assigned to benefits, including the cost of stormwater management, the cost of energy for heating and cooling, and the average home resale value. Crown size, leaf area, and growth of common street tree species in each climate zone are based on data collected in the reference city for each climate zone, published in the i-Tree regional community tree guides (Table 2). The climate zones relevant to this study were the Lower Midwest, which includes Indianapolis, and the Northeast, which includes Detroit and Philadelphia [8,9]. The Northeast climate zone is characterized as having more rainfall and more air pollution than the Lower Midwest, meaning that air quality and stormwater benefits are expected to be relatively more important in the Northeast than in the Lower Midwest. Growth of trees is also modeled differently in the two climate zones, as i-Tree models growth using logistic equations in the Lower Midwest and uses linear and logistic equations in the Northeast [8,9]. The difference in growth models affects the annual benefit estimates particularly for large (>30 cm DBH) trees (see Discussion Section 4.2 for more details).

Table 2. Reference cities and community tree guide citations for the i-Tree climate zone used for each study city.

Climate Zone	Study Cities	Reference City	Community Tree Guide Citation
Lower Midwest	Indianapolis, IN	Indianapolis, IN	Peper, *et al.* 2009 [8]
Northeast	Detroit, MI and Philadelphia, PA	Queens, NY	McPherson, *et al.* 2007 [9]

i-Tree models tree species that were not measured in the reference city as a similar species and uses the midpoint of each size class interval to represent all the trees in that size class. Most benefit types are calculated based on estimated annual leaf area increase [8]. Annual leaf area increases and other aspects of tree structure are estimated from the measured DBHs based on predictive equations [23,24]. The monetary value of benefits per leaf area for each benefit type is determined by empirical data collected by U.S. Forest Service researchers in various US cities [1,25,26] and adjusted based on local prices (set by the user).

2.4. Preparation for i-Tree

Some transformations were required to prepare re-inventory data for i-Tree. First, all dead, sprout, stump, shrub, or absent trees, trees with no DBH recorded, and trees with no species recorded were removed from the data set. A new data set was created with only the species name, DBH, and overall condition rating for each remaining entry. Tree species names were replaced with the corresponding i-Tree species codes (originally developed by the U.S. Department of Agriculture for all plants; http://plants.usda.gov). If i-Tree did not have a species code for a certain tree species, that tree was assigned the species code of a closely related species (*i.e.*, a species in the same genus) or assigned a general designation such as other broadleaf deciduous medium (BDM OTHER). For some genera, such as *Malus* and *Lagerstroemia*, i-Tree does not have species designations, and every tree in that genus is listed as, for example, *Malus* sp. This simplification avoids the complication of genera that have many cultivars that are genetic hybrids or crosses of multiple species.

2.5. Generating Benefit Estimates in i-Tree Streets

Because of the nature of our data (street tree inventory data), we chose to use i-Tree Streets, which uses species and diameter at breast height (DBH) of trees to estimate the benefits they provide. For each city, we set the climate zone, which determines the temperature and precipitation models used (see Table 2 for climate zone used for each study city). i-Tree allows the user to indicate whether the inventory type is "complete" or a sample, and we used the "complete" option to get an estimate of benefits only for the sampled trees. We did not modify the default prices for costs associated with trees, but we did customize the benefit prices for each city. The home prices were obtained from Trulia.com (Detroit and Philadelphia) and Zillow.com (Indianapolis). Electricity and natural gas prices were obtained from DTE Energy [27] (Detroit), Citizens Energy Group [28] (Indianapolis), and the Pennsylvania Public Utility Commission [29] (Philadelphia).

For each city, we defined size classes as 0–5 cm, 5–10 cm, 10–15 cm, 15–20 cm, 20–25 cm, 25–30 cm, and 30–1000 cm (i-Tree does not allow the user to change the upper limit of the largest size class). Trees with a DBH of exactly the breakpoint between size classes are included in the larger size class (*i.e.*, a 5 cm tree will be included in the 5–10 cm size class).

Once all the parameters were set for the city, we imported the inventory data. For each city, we generated "All Benefits-Costs Reports" and a "Population Summary" report. These reports were exported from i-Tree and saved with metadata to keep track of the parameters used. We considered the main outcomes to be the total annual benefit estimates and the amount of annual benefit estimates of each type, both reported in monetary terms as US Dollar values.

2.6. Estimating Costs

Although we do not conduct an explicit cost-benefit analysis in this study (see Section 1.3 above), we do compare i-Tree-calculated benefits to two different estimates of the costs associated with the trees in this study. Average annual per-tree costs are obtained from the i-Tree community tree guides [8,9]. These cost estimates include planting, pruning, removal/disposal, infrastructure damage, irrigation, cleanup, liability/legal, and administrative costs. The total expected costs for a tree that lives 40 years are averaged over that 40-year period, so the estimated annual costs are assumed to be constant [8,9]. (Note that this assumption is very different from the actual likely distribution of costs over the lifetime of a tree as schematized in Figure 1 above.) We also present estimates of the initial planting and watering costs (*i.e.*, establishment costs—the major monetary costs associated with young, small trees) obtained from one of our nonprofit partners to provide a more relevant cost to benefit comparison for the recently planted trees in Indianapolis.

Ecosystem disservices are not explicitly accounted for in either of our cost estimates; however, i-Tree subtracts biogenic VOC emissions from the total air quality benefits and subtracts CO_2 released through maintenance activities and decomposition from the total carbon benefits [8,9], so those costs are factored into the benefit estimates we present.

2.7. Calculating Cumulative Benefits

Since we assumed linear growth rates (a constant increase in diameter per year) for 5- and 10-year projections, we can use the midpoint of each 5-year time period multiplied by 5 years to calculate cumulative benefits accrued over the time period (see Table 3). The cumulative benefits accrued during the first time period between planting and re-inventory (2014) were estimated as half the annual benefits at the time of re-inventory multiplied by 4 years, since the trees were between 3 and 5 years old at re-inventory. Cumulative benefits for the 5-year projection (2019; 8–10 years after planting) are calculated as the cumulative benefits at re-inventory, plus the annual benefits at year 2017 multiplied by 5 years. Cumulative benefits for the 10-year projection (2024; 13–15 years after planting) are calculated as the cumulative benefits for the 5-year projection (2019), plus the annual benefits at year 2022 multiplied by 5 years.

Table 3. Time, calendar year, and age of trees at each timepoint used in benefit projections.

Time Point	Year	Age of Trees (Time in Ground)
Planting	2009–2011	0
Re-inventory	2014	3–5 years
*Midpoint * 1*	*2017*	*6–8 years*
5-year projection	2019	8–10 years
*Midpoint * 2*	*2022*	*11–13 years*
10-year projection	2024	13–15 years

* Midpoints are only used in calculating cumulative benefits for 5- and 10-year projections and are not shown in tables and figures in the results.

2.8. Calculating Annual Survival Rates

We calculated growth and survival rates of the recently planted trees separately for each city. The survival rate was based on the condition rating assigned to the tree: good, fair, or poor trees were rated as alive and absent, dead, shrub, stump, or sprout trees were rated as dead. To account for the fact that not all trees were the same age, we calculated an annual survival rate, I_{annual}, defined as

$$I_{annual} = \sqrt[t]{I_t} \tag{1}$$

where t is the number of years since planting and I_t is the cumulative survival rate for all trees planted in a given year [30]. We used half years to designate the difference between fall and spring plantings. For simplicity we are assuming that trees were planted at the beginning of their indicated planting season and were re-inventoried more than halfway through the summer (Though we had exact planting dates for two of the three study cities, we used a seasonal grouping for all cities in this project for consistency across cities.), so t = 3.5 years since planting for trees planted in Spring 2011, and t = 3 years since planting for trees planted in Fall 2011. We calculated the annual survival rate separately for trees planted in Spring 2009, Fall 2009, Spring 2010, Fall 2010, Spring 2011, and Fall 2011, and used an average weighted by the number of trees planted each season to represent the overall annual survival rate for the city.

2.9. Calculating Growth Rates

We calculated growth of trees as relative growth rate, defined as

$$relative\ growth\ rate = \frac{\ln C_2 - \ln C_1}{Number\ of\ growing\ seasons\ since\ planting} \tag{2}$$

where C_1 and C_2 are measurements of tree caliper at the time of planting and time of re-inventory, respectively (after [31] as adapted by [32]). The 2014 growing season was counted towards growth, so the oldest trees (those planted in Spring 2009) had 6 growing seasons since planting and the youngest trees (those planted in Fall 2011) had 3.

2.10. Projecting Future Mortality

In each city, we modeled survival of the re-inventoried trees 5 and 10 years into the future with three different scenarios: one in which the annual survival rate was the same as that found for the first 3–6 seasons of growth (establishment-phase survival rate), one in which the annual survival rate was 96.4%, corresponding to annual survival found in the literature review by Roman and Scatena [30], and one in which no trees died after 2014 (the no additional mortality scenario). The "no additional mortality" scenario is not realistic, but it represents how the population will look if all of the currently living trees survive for the next 5 or 10 year period. It is likely that the survival rate will increase after the trees are established because annual mortality is highest during the establishment phase [33–35]; hence we modeled scenarios in which survival rates were higher than in the establishment phase. We also predicted that trees that were in poor condition at the time of re-inventory were more likely to die, so those trees were eliminated from the dataset first in modeled mortality. Within condition rating, trees were eliminated randomly (*i.e.*, not by size or species).

To randomize mortality, we used Stata [22] to assign a random number between 0 and 1 to each tree in the data set. Then, all trees with a random number higher than the future survival proportion were eliminated from the data set. For example, if 58% of all current trees were expected to be alive in 5 years based on average annual survival, all trees assigned a random number greater than 0.58 were eliminated from the data set.

2.11. Projecting Future Growth

After simulating future mortality, we applied the average caliper growth rate for all the trees in each city to the DBH of each "surviving" tree to predict the tree's new DBH at 5 and 10 years after re-inventory. We did not modify the caliper growth rate before applying it to the trees' DBHs because we expect caliper growth rate to be very similar to DBH growth rate for trees of this size over this time period (*i.e.*, 5–10 years). Because trees can be expected to grow faster after they have become established [35], we also modeled growth rates 40% higher than the establishment phase growth rate. We modeled only one faster growth rate because in preliminary model testing, we determined the 40% faster growth rate to be the only "fast" growth rate that made a noticeable difference in the average DBH of the trees while still being realistic.

These models of survival and growth were applied to the re-inventoried trees to "grow" the tree sample 5 and 10 years into the future, resulting in data sets that have fewer trees than they did in 2014 because of future mortality (or the same number of trees as in 2014 in the no additional mortality scenario) and the trees are larger than they were in 2014. We then estimated the benefits in i-Tree Streets (using the methods described in Section 2.5 above) to predict the future annual benefits the trees will provide under different growth and survival scenarios (Table 4).

Table 4. Combinations of survival and growth scenarios used in benefit projections.

Survival	Growth
Establishment-phase	Average
Establishment-phase	40% faster than average
96.4% annual	Average
96.4% annual	40% faster than average
No additional mortality	Average
No additional mortality	40% faster than average

3. Results

3.1. Re-Inventory Results

A total of 4059 trees were re-inventoried in summer 2014. The re-inventory sampled at least 10% of the trees planted from 2009 to 2011 for each city (Table 5).

Table 5. Number of trees surveyed compared to the number of trees planted from 2009 to 2011 for the three study cities.

City	Total Number of Trees Re-Inventoried	Number of Trees Planted 2009–2011	Percent of Planted Trees Re-Inventoried
Detroit	1241	6777	18%
Indianapolis	1076	11,294	10%
Philadelphia	1742	6894	32%

100% of the trees in randomly selected block groups were inventoried in Detroit and Philadelphia, while 52% of trees were re-inventoried in Indianapolis. Data collection teams were instructed to inventory every other tree in a block group unless they estimated they would have sufficient time to inventory all planted trees.

3.1.1. Current Benefits

Based on estimates from i-Tree, these 4059 re-inventoried trees provide a total of $46,377 in annual benefits. The value ranged from $8.91 per tree in Detroit to $20.25 per tree in Indianapolis (Table 6). At the time of re-inventory, the cumulative benefits provided by the sample trees (*i.e.*, the sum of the annual benefits provided each year since planting) ranged from $17,096 in Detroit to $31,268 in Indianapolis (Table 6).

Table 6. Summary of estimated total annual benefits, average annual benefits per tree, and cumulative benefits for the three study cities in 2014.

City	Total Annual Benefits	Average Annual Benefits Per Tree	Cumulative Benefits
Detroit	$8,548	$8.91	$17,096
Indianapolis	$15,635	$20.25	$31,268
Philadelphia	$15,556	$15.70	$31,112

Property value benefits were the predominant benefit type in Indianapolis and Philadelphia, while energy benefits predominated in Detroit (Figure 2). Ninety-five percent of estimated annual benefits in Indianapolis were property value benefits because property value increase is the predominant benefit type in the Lower Midwest climate zone i-Tree models.

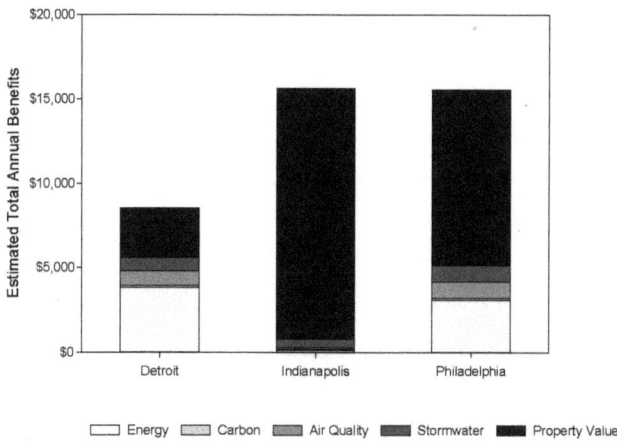

Figure 2. Estimated total annual benefits, by type, for each city based on 2014 re-inventory data.

3.1.2. Growth and Survival Rates for Each City

The annual and cumulative survival rates were similar for Detroit and Indianapolis and lower for Philadelphia (Table 7). Relative growth rate ranged from 1.18 cm/year in Indianapolis to 1.48 cm/year in Detroit.

Table 7. Survival and growth of re-inventoried trees in the three study cities.

City	Cumulative Survival Rate	Annual Survival Rate	Relative Growth Rate
Detroit	79%	93%	1.48 cm/year
Indianapolis	80%	93%	1.18 cm/year
Philadelphia	59%	87%	1.19 cm/year

3.2. Projected Populations

The three different annual survival scenarios result in different proportions of trees remaining in the sample population 5 and 10 years in the future (Figure 3). With continued establishment-phase annual survival of 93% in Detroit and Indianapolis, only 40% of the planted trees in those cities will be alive in 10 years, compared to 80% of planted trees alive in 10 years with no additional mortality. With continued establishment-phase annual survival of 87% in Philadelphia, only 29% of planted trees will be alive in 5 years and only 15% will be alive in 10 years, compared to 59% of planted trees alive in 10 years with no additional mortality.

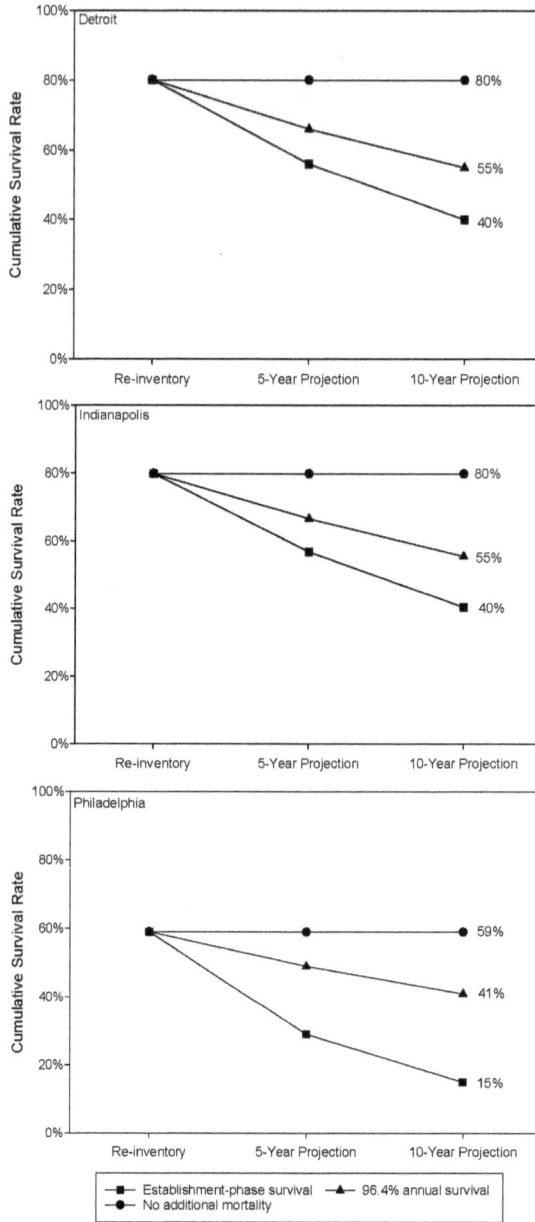

Figure 3. Cumulative survival rates of re-inventoried trees in 5 and 10 years with the three different survival scenarios in the three study cities.

The two different growth scenarios (average *vs.* 40% faster growth) resulted in different average DBHs 5 and 10 years in the future (Figure 4). For each study city, the difference between the two growth rates is an approximately 2.5 cm difference in average DBH after 5 years and an approximately 5 cm difference in average DBH after 10 years.

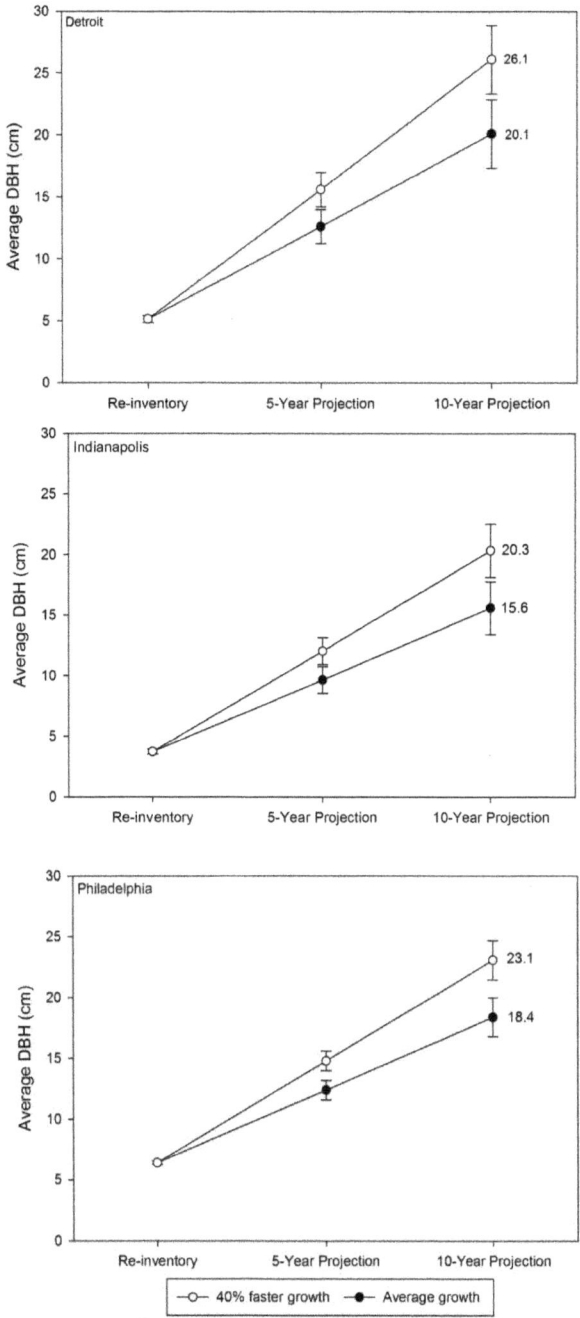

Figure 4. Average diameter at breast height (DBH) at re-inventory and projected 5 and 10 years into the future with average growth and 40% faster growth for each study city. Error bars represent 95% confidence intervals based on the standard error of the growth rate in each city.

3.3. Projected Annual Benefits

3.3.1. Effect of Survival Scenario

The survival scenarios had a larger effect on projected total annual benefits than the growth scenarios. For both Detroit and Indianapolis, the no additional mortality scenario resulted in double the amount of projected total annual benefits of the establishment-phase survival scenario in the 10-year projection (Figure 5, Table 8). The difference was even more apparent for Philadelphia, where the no additional mortality scenario resulted in 3.5 times the amount of projected total annual benefits of the establishment-phase survival scenario in the 10-year projection (Table 8).

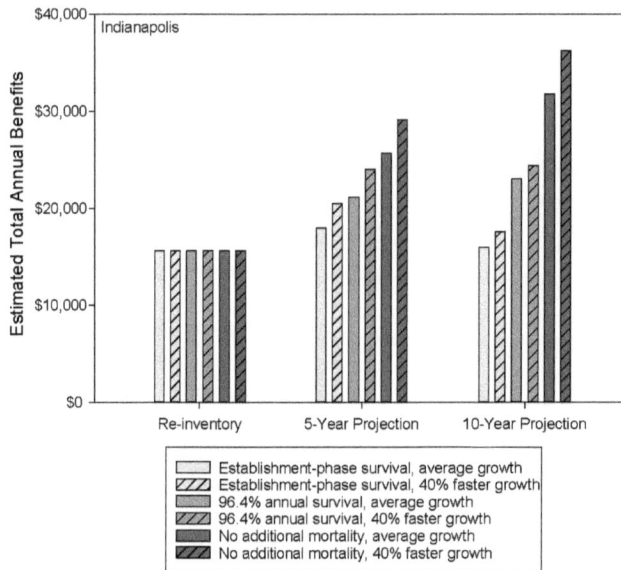

Figure 5. Estimated total annual benefits of re-inventoried trees in Indianapolis at the time of re-inventory and projected 5 and 10 years into the future.

Table 8. Projected total annual benefits of re-inventoried trees in Detroit, Indianapolis, and Philadelphia 10 years in the future with all combinations of survival and growth scenarios.

City	Establishment-Phase Survival		96.4% Annual Survival		No Additional Mortality	
	Average growth	40% faster growth	Average growth	40% faster growth	Average growth	40% faster growth
Detroit	$16,622	$21,999	$22,564	$29,899	$32,476	$42,972
Indianapolis	$16,004	$17,616	$23,021	$24,424	$31,798	$36,209
Philadelphia	$9,667	$11,947	$24,484	$30,161	$34,558	$42,831

If the tree populations maintain the establishment-phase annual survival rates and the average growth rates continue, high tree mortality results in fewer trees conferring benefits; thus, the projected populations in Indianapolis and Philadelphia provide fewer benefits in the future (Figure 5). However, if the tree populations maintain an annual survival rate of 96.4%, the total annual benefits provided by the trees can be expected to increase over the next 10 years. If no more trees die, the total annual benefits provided by the tree populations can be expected to double in the next 10 years (Figure 5).

3.3.2. Effect of Growth Scenario

The growth rate scenarios had an effect on projected total annual benefits, but the effect was less pronounced than the effect of the survival scenario. For Indianapolis, average growth and no additional mortality resulted in $31,798 in total annual benefits in 10 years, while 40% faster growth and no additional mortality resulted in $36,209 in total annual benefits in 10 years (Figure 5, Table 8). The growth scenario had a bigger effect in Detroit, where average growth and no additional mortality resulted in $32,476 in total annual benefits in 10 years, while 40% faster growth and no additional mortality resulted in $42,972 in total annual benefits in 10 years (Table 8).

3.4. Projected Cumulative Benefits

Due to their additive nature, cumulative benefits increase substantially when the tree populations are projected into the future (Figure 6). Cumulative benefits increase by a factor of roughly 4 from the time of re-inventory to the 5-year projection, and then double again from the 5- to 10-year projection (Figure 6). Projected cumulative benefits in 10 years range from $154,000 to $271,000 in Detroit, from $200,000 to $307,000 in Indianapolis, and from $158,000 to $320,000 in Philadelphia (Table 9).

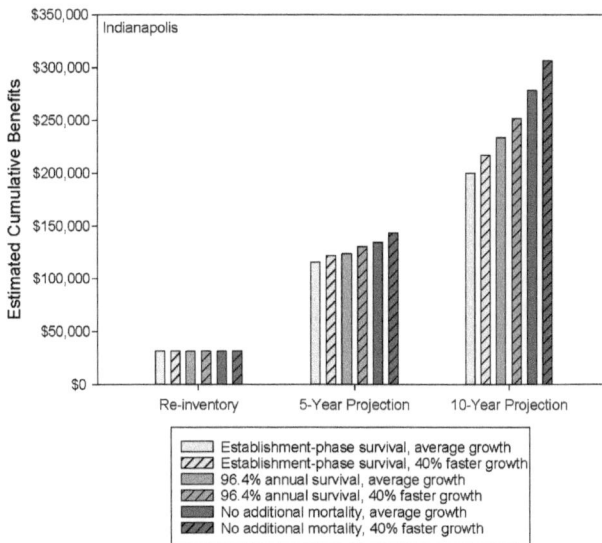

Figure 6. Estimated total cumulative benefits of re-inventoried trees in Indianapolis at the time of re-inventory and projected 5 and 10 years into the future.

Table 9. Projected total cumulative benefits of re-inventoried trees in Detroit, Indianapolis, and Philadelphia 10 years in the future with all combinations of survival and growth scenarios (rounded to the nearest thousand).

City	Establishment-Phase Survival		96.4% Annual Survival		No Additional Mortality	
	Average growth	40% faster growth	Average growth	40% faster growth	Average growth	40% faster growth
Detroit	$154,000	$184,000	$181,000	$219,000	$221,000	$271,000
Indianapolis	$200,000	$217,000	$234,000	$252,000	$278,000	$307,000
Philadelphia	$158,000	$173,000	$238,000	$268,000	$280,000	$320,000

3.5. Projected Per-Tree Benefits

The average benefits per tree reflect the effect of tree size in determining future benefits. For all cities studied, faster growth results in higher average benefits per tree (Figure 7, Table 10). Under average growth conditions, the average value per tree 10 years in the future ranges from $34.87 (Philadelphia) to $40.30 (Indianapolis); the average value under 40% faster growth conditions ranges from $43.22 (Philadelphia) to $46.26 (Detroit).

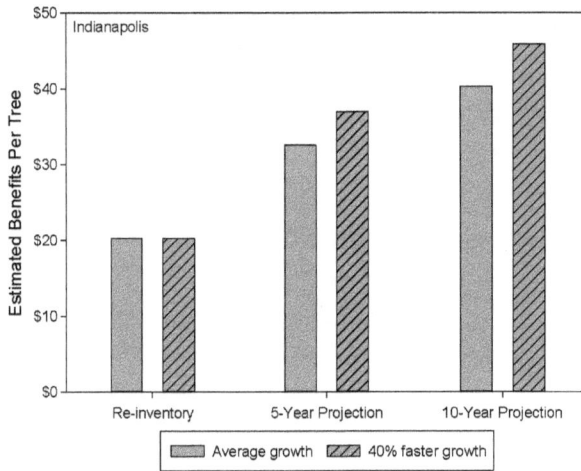

Figure 7. Estimated annual benefits per tree in Indianapolis at the time of re-inventory and projected 5 and 10 years into the future with average growth and 40% faster growth.

Table 10. Estimated total annual benefits per tree 5 and 10 years in the future under average growth and 40% faster growth conditions in all study cities.

		Average Growth		40% Faster Growth	
City	Re-Inventory	5-Year Projection	10-Year Projection	5-Year Projection	10-Year Projection
Detroit	$8.91	$21.88	$34.96	$26.97	$46.26
Indianapolis	$20.25	$32.56	$40.30	$36.94	$45.89
Philadelphia	$15.70	$25.01	$34.87	$28.76	$43.22

3.6. Estimated Costs and Net Benefits

The i-Tree regional community tree guides estimate the average costs per tree, assuming a 40-year lifetime, to be $34/year for a large public tree in the Northeast [9] and $24/year for a large public tree in the Lower Midwest [8]. Using these estimated annual costs, the average annual benefits (Table 10) exceed average annual costs (*i.e.*, net annual benefits are positive) after 13–15 years of growth (the 10-year projection) in Detroit and Philadelphia and after 8–10 years of growth (the 5-year projection) in Indianapolis.

Keep Indianapolis Beautiful, Inc. (Indianapolis, IN, USA) estimates their average per-tree costs are $100 per tree for planting plus $100 per year of watering for a total of $300 per tree for a tree watered for a 2-year establishment period (*personal communication*, N. Faris (Keep Indianapolis Beautiful, Inc.), 21 September 2015). Using this estimated cumulative per-tree cost, the net benefits per planted tree in Indianapolis are still negative after 13–15 years of growth, even with no additional mortality and 40% faster than average growth (Table 11).

Table 11. Cumulative benefits and net benefits per planted tree in Indianapolis, IN, 10 years in the future with all combinations of survival and growth scenarios.

Indianapolis (Per 1076 Planted Trees)	Establishment-Phase Survival		96.4% Annual Survival		No Additional Mortality	
	Average Growth	40% Faster Growth	Average Growth	40% Faster Growth	Average Growth	40% Faster Growth
Cumulative benefits for all trees	$200,000	$217,000	$234,000	$252,000	$278,000	$307,000
Average cumulative benefits per planted tree	$186	$202	$217	$234	$258	$285
Net benefits per planted tree after 10 years	−$114	−$98	−$83	−$66	−$42	−$15

4. Discussion

4.1. Future Benefit Increases

The total benefits provided by the sample of planted trees may increase or decrease depending on the survival rate in the future. These street trees can be expected to provide increasing annual benefits during the next 5–10 years if the annual survival rate is higher than the 93% annual survival measured during the establishment period. However, our projections show that with continued 93% or lower annual survival, the increase in annual benefits from tree growth will not be able to make up for the loss of benefits as trees die. This means that estimated total annual benefits from the populations of trees will decrease between the 5-year projection and the 10-year projection for Indianapolis and Philadelphia (Figure 5). In contrast, with higher survival rates, the benefits provided by the populations of trees can be expected to increase over the next 10 years in all cities (Figure 5). Somewhere between 93% and 96.4% annual survival represents the "tipping point" where the benefits added due to growth exceed the loss of benefits due to mortality.

4.2. Comparison of Results among Cities

The general patterns in how benefits change over time are not universally applicable across all cities studied. Notably, the estimated total annual benefits provided by the street tree population in 2014 in Detroit were substantially lower than the benefit estimates for the other cities. This difference is the result of a much lower average home resale value in Detroit ($37,000 compared to $128,000 and $155,000 for Indianapolis and Philadelphia, respectively). The unimportance of property value benefits in Detroit relative to other cities is likely responsible for the different trend in benefits over time in Detroit (namely, no decrease in estimated benefits from 5-year projection to 10-year projection, even in the low survival scenario). The trees in Detroit also had a higher growth rate than the trees in Indianapolis and Philadelphia (1.48 cm/year compared to 1.18 and 1.19 cm/year, respectively; Table 7), so growth was more able to make up for tree mortality.

In addition, the cities are located in different i-Tree climate zones, which affects how the modeled populations respond to variations in growth. In particular, the i-Tree climate zone affects how property value benefits are modeled for large trees. In the Northeast climate zone (which includes Detroit and Philadelphia), property value benefits and total benefits increase linearly with tree size, while in the Lower Midwest climate zone (which includes Indianapolis), property value benefits start to decrease after the tree reaches a DBH of 30 cm (12 inches) and total annual benefits do not increase any more beyond a DBH of 60 cm (24 inches) (Figure 8). These differences are an artifact of the limited data collected in the different climate zones (*i.e.*, in one reference city per zone) and may or may not actually be representative of how street trees grow across the region.

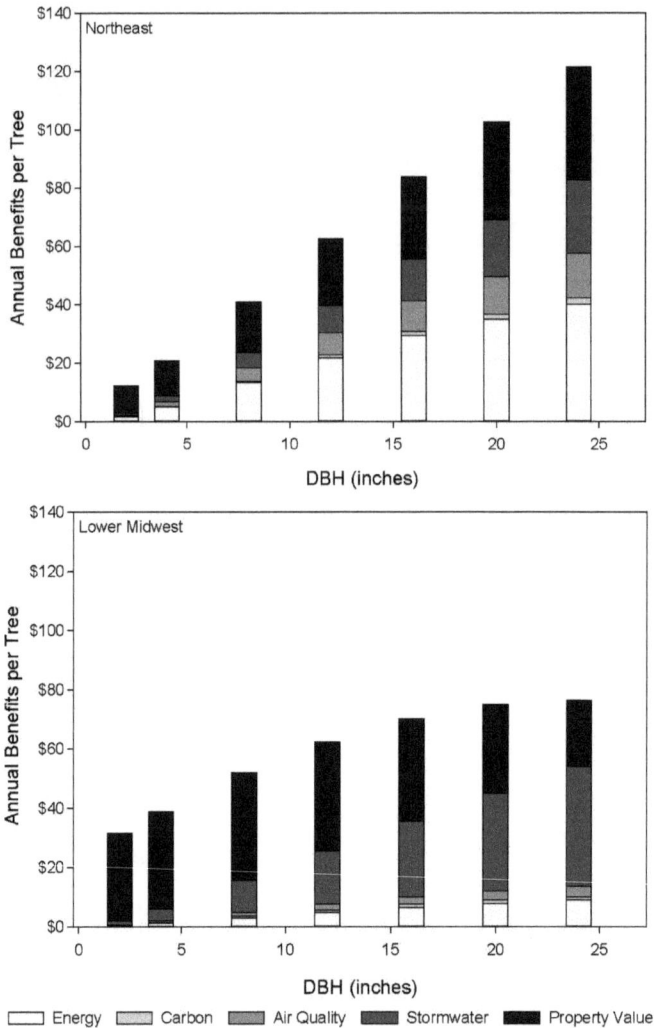

Figure 8. Annual benefits per tree by size for a hypothetical sugar maple (*Acer saccharum*) in the Northeast and Lower Midwest climate zones.

4.3. Effect of Survival vs. Growth Rate

Our results indicate that tree survival is more important than growth for providing future benefits. This is a particularly relevant finding given the lack of follow-up on tree success (survival and/or growth) in many tree-planting programs/campaigns. Street tree survival rates observed 3–5 years post-planting in our study are not particularly high compared to other results in the literature [30,36,37] (see Table 12 for comparison). As a population, the planted street trees in this study will not provide many more benefits 10 years in the future if establishment-phase annual survival rates continue. However, if the annual survival rate increases modestly to 96.4%, a reasonable expectation since most mortality is expected during the first years after transplanting, the populations will provide between $22,564 (Detroit) and $24,484 (Philadelphia) in total annual benefits in 10 years, even if the growth rates do not increase (Table 8).

Table 12. Annual survival rates from other *in situ* studies of urban tree survival.

Annual Survival Rate	Time Since Planting	Location	Study
87%	3–5 years	Philadelphia, PA	This study
93%	3–5 years	Detroit, MI and Indianapolis, IN	This study
94%	3–4 years	20 cities in Iowa	[36]
95.5%	2–10 years	Philadelphia, PA	[30] (street tree survey)
95.6%	2 years	New York City, NY	[37]
94.9%–96.5%	Various	Various	[30] (meta-analysis)

4.4. Maintenance Costs and Net Benefits of Recently Planted Street Trees

Whether the net benefits are positive after 13–15 years of tree growth depends on the cost calculation used; when annual benefits per tree are compared to the annual costs per tree from the i-Tree community tree guides, the net benefits are positive. However, this comparison does not include the costs for trees that were planted and then died—the per-tree costs would be higher if money spent on dead trees were included. In contrast, our comparison of cumulative benefits to the cumulative costs of planting and watering the trees shows that net benefits *per planted tree* are still negative after 13–15 years of growth for all tested scenarios (Table 11). Clearly, investing in the planting and watering of trees is a long-term investment. However, this back-of-the-envelope calculation implies that increasing survival and growth rates as much as possible decreases the payback period (or, amount of time until net benefits become positive and the investment in planting and maintenance is paid back as tree benefits) for a population of planted trees.

If even higher-than-measured growth rates cannot make up for tree mortality, efforts to maintain street trees that survive in the landscape are important. In order to get the best return on investment from planting street trees, follow-up maintenance and monitoring to ensure high survival rates are necessary. For instance, Boyce [38] observed that mortality rates were three times lower (that is, survival rates were higher) for trees with "stewards" (designated groups or individuals assigned to provide maintenance for a tree, including watering and tree pit care) than for trees without stewards. Gilman [39] compared the planting and maintenance costs of irrigated and not irrigated trees and found that mortality of non-irrigated trees resulted in higher costs per live tree 1 year after transplanting. This implies that a population of irrigated trees with low mortality rates should yield higher total *net* benefits than a population of non-irrigated trees with lower planting and maintenance costs but higher mortality rates (see Vogt, *et al.* [10] for a net benefits calculation using data from Gilman [39]).

4.5. Other Costs of Street Trees

The cost estimates from the i-Tree community tree guides [8,9] include planting, pruning, removal/disposal, infrastructure damage, irrigation, cleanup, liability/legal, and administrative costs. In addition, biogenic VOC and CO_2 emissions associated with trees are accounted for in the air quality and carbon benefit estimates, respectively [8,9]. Although these cost estimates do not include every possible costs associated with trees, we believe they represent those costs most typically associated with young, planted street trees.

4.6. Comparison to Similar Studies

McPherson and colleagues [40] modeled high and low survival scenarios 35 years into the future for Los Angeles's Million Trees LA tree-planting campaign, and found that a drop from "high" survival (1% annual mortality for the first 5 years and 0.5% annual mortality thereafter) to "low" survival (5% annual mortality for the first 5 years and 2% annual mortality thereafter) resulted in a 30% decrease in cumulative benefits after 35 years. A follow up study by McPherson [41] found that survival of selected street trees five years after planting more closely matched the low survival scenario from [40]. These results provide evidence that studies predicting the benefits of tree-planting programs (e.g., [40]) can easily overestimate the actual survival rates (and thus, future benefits) of planted trees.

In a similar study conducted in New York City to evaluate the MillionTreesNYC initiative, Morani and colleagues [42] predicted that if 100,000 trees per year were planted over the course of 10 years (thus reaching the goal of 1 million trees planted), the trees would remove 11,000 tons of air pollutants over the next 100 years if they maintain an annual mortality rate of 4% or less. However, the amount of air pollutants removed would drop to 3000 tons over the next 100 years if the annual mortality rate increased to 8% per year [42]. Other researchers have also discussed the relative importance of mortality rates over growth rates for determining future tree benefits. For example, Strohbach and colleagues [43] found that a modeled tree population sequesters 70% less carbon over its 50-year lifetime when annual mortality increases from 0.5% to 4%.

4.7. Limitations of This Study

There are two main sources of uncertainty in this analysis: the uncertainty in projecting survival and growth rates into the future, and the uncertainty in using i-Tree Streets to assign monetary values to the benefits provided by a population of street trees. We have tried to address the first source of uncertainty by modeling different survival and growth scenarios without making predictions as to which scenario is the most likely outcome; few studies have examined long-term growth or survival of city trees (but see [30,37,41,44–48]), and even fewer have examined species-specific growth or survival (e.g., [34]). The second source of uncertainty is more difficult to address. We have done our due diligence in investigating how i-Tree Streets estimates monetized benefits so we are aware of its sensitivities, but there are many aspects of the benefit estimates that are beyond our control. We acknowledge that the benefits experienced by the residents of these cities are likely different from the benefit estimates presented here, but these rough estimates are the best we can do with the tools available to us.

Other sources of uncertainty include our use of approximate caliper-at-planting rather than actual measurements and citywide growth rates rather than species-specific growth rates. These limitations are the result of the tradeoff between having a large sample size and having more details on each individual tree—we are willing to use approximate caliper-at-planting because it is what our nonprofit partners could provide for the number of trees we needed for this study. We chose not to use species-specific growth rates because it was unclear whether calculating growth rates by species or genus would provide more accurate growth rates, given the limitations in precision of caliper-at-planting data. Finally, we know there are problems with estimating benefits for large trees in i-Tree (see Figure 8), but this is less of a concern given that most of the trees in the 10-year projections were between 20 and 25 cm (8–10 inches) DBH.

Limitations of i-Tree

Perhaps the biggest limitations of this study are the limitations imposed by i-Tree. i-Tree calculates several important types of tree benefits: increased property values, air quality, carbon sequestration, energy savings, and stormwater mitigation. However, this leaves out many additional benefits that are less easily monetized, such as increased retail sales [49], reduced crime rates [3–5], and improved human health and well-being [6,7]. Furthermore, there are additional costs in time, money, and resources of managing trees that would need to be included in any true cost-benefit analysis or calculation of net benefits [12] (see Section 1.3 or [10] for a full discussion of costs associated with trees).

The i-Tree benefit estimates are not perfect; they involve many assumptions about how urban trees grow, the similarity of performance of different tree species, and the value urban trees will provide in different contexts [8,9]. Nowak and colleagues [50] named reliance on models to estimate functions of the urban forest, the potential estimation error of the biomass equations used, and the lack of data on ornamental and tropical tree species as some of the limitations of the UFORE model (used in i-Tree Eco, a sibling module of Streets in the i-Tree family). Dobbs and colleagues [51] also outlined some limitations of the UFORE model, including the limited data incorporated into the models, the bias towards species native to the northern U.S., and the difficulty of obtaining representative samples in the urban environment. The STRATUM model (used in i-Tree Streets) is subject to many of the same limitations. Given these limitations, the benefit estimates generated by i-Tree should be viewed as estimates only, and should not be interpreted as justification for investing in trees at the expense of investing in other types of urban infrastructure (*i.e.*, stormwater and sewer systems).

5. Conclusions

The main management implication of this study is that ensuring the survival of planted street trees is most important for providing future benefits; each dead tree represents a loss of $40–$50 in annual benefits for the city each year after the tree dies, while faster growth only increases the annual benefits per tree by $11.30 at the most (Table 10). This increase is not inconsequential, but also is not as dramatic as the increase in benefits gained from increasing annual survival. The differences in future total annual benefits between the establishment-phase survival rate scenarios and the no additional mortality scenarios (Figure 5) are striking, and indicate that tree-planting campaigns that do not track the survival of planted trees may be significantly overestimating the benefits provided by the trees if they assume 100% survival. If planting and establishment (*i.e.*, watering) costs are considered and net benefits are calculated (Table 11), high survival rates have the potential to decrease payback time for a population of planted trees (when average net benefits per planted tree become positive). The results of this study provide evidence for the importance of tracking the survival and growth of planted trees. If "success" is below expectations, planting and/or maintenance activities should be modified in the future to improve survival and/or growth rates as possible. Additionally, we have learned that high survival rates during the early establishment period (years 1–3) are critical to maximizing benefits; conversely, low initial survival rates can have a negative impact from which it is impossible to recover. High survival rates (greater than 96% annually) are necessary to ensure that the trees are able to provide the maximum amount of benefits possible over their lifetime.

Acknowledgments: A version of this research was presented at the 2015 International Society of Arboriculture Annual Conference and Trade Show, in the Arboriculture Research and Education Academy (AREA) student presentations; SEW received an AREA Travel Grant that enabled this presentation. The U.S. Forest Service (USFS) National Urban and Community Forestry Advisory Council (NUCFAC) and the USFS Northern Research Station provided major funding for the project from which tree survival and growth data used in this paper emerged. Partnerships with Alliance for Community Trees (ACTrees) and several ACTrees member organizations (Keep Indianapolis Beautiful, Pennsylvania Horticultural Society, The Greening of Detroit, Trees Atlanta and Forest ReLeaf of Missouri) made the aforementioned USFS grants possible. Additional funding was provided by: the Indiana University Office of Sustainability and the Garden Club of America, and additional support was provided by: The Vincent and Elinor Ostrom Workshop in Political Theory and Policy Analysis, the Center for the Study of Institutions, Population and Environmental Change, and the School of Public and Environmental Affairs, all at Indiana University Bloomington. Most importantly, we are grateful to the additional team members who participated in the larger research effort on which this paper was derived: Shannon Lea Watkins, Sarah K. Mincey, Rachael Bergmann, and Lynne Westphal. We are also grateful for the comments of four anonymous reviewers and the journal editor that significantly improved this paper.

Author Contributions: B.F. and J.V. conceived and designed the broader experiment on which this project was based; S.W. and B.F. conceived and designed the experiments specific to this project; S.W. performed the experiments and analyzed the data; S.W., J.V. and B.F. wrote the paper.

Conflicts of Interest: The authors declare no conflict of interest. The funding sponsors had no role in the design of the study; in the collection, analyses, or interpretation of data; in the writing of the manuscript, and in the decision to publish the results.

References

1. Xiao, Q.; McPherson, E.G.; Simpson, J.R.; Ustin, S.L. Rainfall interception by Sacramento's urban forest. *J. Arboric.* **1998**, *24*, 235–244.
2. Nowak, D.J.; Crane, D.E.; Stevens, J.C. Air pollution removal by urban trees and shrubs in the United States. *Urban For. Urban Green.* **2006**, *4*, 115–123. [CrossRef]
3. Donovan, G.H.; Prestemon, J.P. The effect of trees on crime in Portland, Oregon. *Environ. Behav.* **2012**, *44*, 3–30. [CrossRef]

4. Garvin, E.C.; Cannuscio, C.C.; Branas, C.C. Greening vacant lots to reduce violent crime: A randomised controlled trial. *Inj. Prev.* **2012**, *19*, 198–203. [CrossRef] [PubMed]

5. Troy, A.; Grove, M.J.; O'Neil-Dunne, J. The relationship between tree canopy and crime rates across an urban-rural gradient in the greater Baltimore region. *Landsc. Urban Plan.* **2012**, *106*, 262–270. [CrossRef]

6. South, E.C.; Kondo, M.C.; Cheney, R.A.; Branas, C.C. Neighborhood blight, stress, and health: A walking trial of urban greening and ambulatory heart rate. *Am. J. Public Health* **2015**, *105*, e1–e5. [CrossRef] [PubMed]

7. Bell, J.F.; Wilson, J.S.; Liu, G.C. Neighborhood greenness and 2-year changes in body mass index of children and youth. *Am. J. Prev. Med.* **2008**, *35*, 547–553. [CrossRef] [PubMed]

8. Peper, P.J.; McPherson, E.G.; Simpson, J.R.; Vargas, K.E.; Xiao, Q. *Lower Midwest Community Tree Guide: Benefits, Costs, and Strategic Planting*; US Department of Agriculture Forest Service, Pacific Southwest Research Station: Albany, CA, USA, 2009.

9. McPherson, E.G.; Simpson, J.R.; Peper, P.J.; Gardner, S.L.; Vargas, K.E.; Xiao, Q. *Northeast Community Tree Guide: Benefits, Costs, and Strategic Planting*; United States Department of Agriculture Forest Service, Pacific Southwest Research Station: Albany, CA, USA, 2007.

10. Vogt, J.M.; Hauer, R.J.; Fischer, B.C. The costs of maintaining and not maintaining trees: A review of the urban forestry and arboriculture literature. *Arboric. Urban For.* **2015**, *41*, 293–322.

11. Von Dohren, P.; Haase, D. Ecosystem disservices research: A review of the state of the art with a focus on cities. *Ecol. Indic.* **2015**, *52*, 490–497. [CrossRef]

12. Escobedo, F.J.; Kroeger, T.; Wagner, J.E. Urban forests and pollution mitigation: Analyzing ecosystem services and disservices. *Environ. Pollut.* **2011**, *159*, 2078–2087. [CrossRef] [PubMed]

13. Randrup, T.B.; McPherson, E.G.; Costello, L.R. Tree root intrusion in sewer systems: Review of extent and costs. *J. Infrastruct. Syst.* **2001**, *7*, 26–31. [CrossRef]

14. McPherson, E.G.; Peper, P.J. Infrastructure repair costs associated with street trees in 15 cities. In *Trees and Building Sites*; Watson, G., Neely, D., Eds.; International Society of Arboriculture: Savoy, IL, USA, 1995; pp. 49–64.

15. Posner, R.A. Cost-benefit analysis: Definition, justification, and comment on conference papers. *J. Leg. Stud.* **2000**, *29*, 1153–1177. [CrossRef]

16. i-Tree Streets. i-Tree Software Suite v6.0.7. n.d. Available online: http:/www.itreetools.org (accessed on 29 August 2014).

17. i-Tree Streets User's Manual v5.0. n.d. Available online: http://www.itreetools.org/resources/manuals/Streets_Manual_v5.pdf (accessed on 29 August 2014).

18. Bloomington Urban Forestry Research Group. About the Bloomington Urban Forestry Research Group (BUFRG) at CIPEC. Available online: https://www.indiana.edu/~cipec/research/bufrg_about.php (accessed on 10 March 2016).

19. Bloomington Urban Forestry Research Group. Urban Forestry in 5 U.S. Cities—Ecological & Social Outcomes of Neighborhood & Nonprofit Tree Planting: NUCFAC Grant. Available online: https://www.indiana.edu/~cipec/research/bufrgproj_nucfac.php (accessed on 10 March 2016).

20. Vogt, J.M.; Fischer, B.C. A protocol for citizen science monitoring of recently-planted urban trees. *Cities Environ.* **2014**, *7*, 4.

21. Vogt, J.M.; Mincey, S.K.; Fischer, B.C.; Patterson, M. *Planted Tree Re-Inventory Protocol*; Version 1.1; Bloomington Urban Forestry Research Group at the Center for the Study of Institutions, Population and Environmental Change, Indiana University: Bloomington, IN, USA, 2014.

22. StataCorp. *Stata Statistical Software: Release 13*; StataCorp LP: College Station, TX, USA, 2013.

23. Peper, P.J.; McPherson, E.G.; Mori, S.M. Equations for predicting diameter, height, crown width, and leaf area of San Joaquin Valley street trees. *J. Arboric.* **2001**, *27*, 306–317.

24. Peper, P.J.; McPherson, E.G.; Mori, S.M. Predictive equations for dimensions and leaf area of coastal Southern California street trees. *J. Arboric.* **2001**, *27*, 169–180.

25. Xiao, Q.; McPherson, E.G.; Ustin, S.L.; Grismer, M.E.; Simpson, J.R. Winter rainfall interception by two mature open-grown trees in Davis, California. *Hydrol. Process.* **2000**, *14*, 763–784. [CrossRef]

26. Anderson, L.M.; Cordell, H.K. Influence of trees on residential property values in Athens, Georgia (USA): A survey based on actual sales prices. *Landsc. Urban Plan.* **1988**, *15*, 153–164. [CrossRef]

27. DTE Energy. Your Natural Gas Pricing Options. Available online: https://www2.dteenergy.com/wps/wcm/connect/11919444-45e5-4a7c-afab-33d6db22b198/Gas+Rate+9-14_internet.pdf?MOD=AJPERES (accessed on 31 October 2014).

28. Citizens Energy Group. Gas Rate No. D2. Available online: http://www.citizensenergygroup.com/ratesriders/RateNo%20D2%20ResHeatDelivery%20-%20eff.%209-6-11.pdf (accessed on 31 October 2014).

29. Pennsylvania Public Utilities Commission. Natural Gas Shopping Tool. Available online: http://www.puc.pa.gov/consumer_info/natural_gas/natural_gas_shopping/gas_shopping_tool.aspx (accessed on 31 October 2014).

30. Roman, L.A.; Scatena, F.N. Street tree survival rates: Meta-analysis of previous studies and application to a field survey in Philadelphia, PA, USA. *Urban For. Urban Green.* **2011**, *10*, 269–274. [CrossRef]

31. Brand, D.G. The establishment of boreal and sub-boreal conifer plantations: An integrated analysis of environmental conditions and seedling growth. *For. Sci.* **1991**, *37*, 68–100.

32. Samyn, J.; de Vos, B. The assessment of mulch sheets to inhibit competitive vegetation in tree plantations in urban and natural environment. *Urban For. Urban Green.* **2002**, *1*, 25–37. [CrossRef]

33. Richards, N.A. Modeling survival and consequent replacement needs in a street tree population. *J. Arboric.* **1979**, *5*, 251–255.

34. Miller, R.H.; Miller, R.W. Survival of selected street tree taxa. *J. Arboric.* **1991**, *17*, 185–191.

35. Gilman, E.F.; Black, R.J.; Dehgan, B. Irrigation volume and frequency and tree size affect establishment rate. *J. Arboric.* **1998**, *24*, 1–9.

36. Thompson, J.R.; Nowak, D.J.; Crane, D.E.; Hunkins, J.A. Iowa, US, communities benefit from a tree-planting program: Characteristics of recently planted trees. *J. Arboric.* **2004**, *30*, 1–10.

37. Lu, J.W.T.; Svendsen, E.S.; Campbell, L.K.; Greenfeld, J.; Braden, J.; King, K.L.; Falxa-Raymond, N. Biological, social, and urban design factors affecting young street tree mortality in New York City. *Cities Environ.* **2010**, *3*, 1–15.

38. Boyce, S. It takes a stewardship village: Effect of volunteer tree stewardship on urban street tree mortality rates. *Cities Environ.* **2010**, *3*, 1–8.

39. Gilman, E.F. Effect of nursery production method, irrigation, and inoculation with mycorrhizae-forming fungi on establishment of *Quercus virginiana. J. Arboric.* **2001**, *27*, 30–39.

40. McPherson, E.G.; Simpson, J.R.; Xiao, Q.; Wu, C. *Los Angeles 1-Million Tree Canopy Cover Assessment*; US Department of Agriculture Forest Service, Pacific Southwest Research Station: Albany, CA, USA, 2008.

41. McPherson, E.G. Monitoring Million Trees LA: Tree performance during the early years and future benefits. *Arboric. Urban For.* **2014**, *40*, 285–300.

42. Morani, A.; Nowak, D.J.; Hirabayashi, S.; Calfapietra, C. How to select the best tree planting locations to enhance air pollution removal in the MillionTreesNYC initiative. *Environ. Pollut.* **2011**, *159*, 1040–1047. [CrossRef] [PubMed]

43. Strohbach, M.W.; Arnold, E.; Haase, D. The carbon footprint of urban green space—A life cycle approach. *Landsc. Urban Plan.* **2012**, *104*, 220–229. [CrossRef]

44. Sklar, F.; Ames, R.G. Staying alive: Street tree survival in the inner-city. *J. Urban Aff.* **1985**, *7*, 55–66. [CrossRef]

45. Roman, L.A.; Battles, J.J.; McBride, J.R. Determinants of establishment survival for residential trees in Sacramento County, CA. *Landsc. Urban Plan.* **2014**, *129*, 22–31. [CrossRef]

46. Roman, L.A.; Battles, J.J.; McBride, J.R. The balance of planting and mortality in a street tree population. *Urban Ecosyst.* **2013**, *17*, 387–404. [CrossRef]

47. Jack-Scott, E.; Piana, M.; Troxel, B.; Murphy-Dunning, C.; Ashton, M.S. Stewardship success: How community group dynamics affect urban street tree survival and growth. *Arboric. Urban For.* **2013**, *39*, 189–196.

48. Koeser, A.; Hauer, R.; Norris, K.; Krouse, R. Factors influencing long-term street tree survival in Milwaukee, WI, USA. *Urban For. Urban Green.* **2013**, *12*, 562–568. [CrossRef]

49. Wolf, K.L. Business district streetscapes, trees, and consumer response. *J. For.* **2005**, *103*, 396–400.

50. Nowak, D.J.; Crane, D.E.; Stevens, J.C.; Hoehn, R.E.; Walton, J.T.; Bond, J. A ground-based method of assessing urban forest structure and ecosystem services. *Arboric. Urban For.* **2008**, *34*, 347–358.

51. Dobbs, C.; Escobedo, F.J.; Zipperer, W.C. A framework for developing urban forest ecosystem services and goods indicators. *Landsc. Urban Plan.* **2011**, *99*, 196–206. [CrossRef]

Article

Managing Tree Diversity: A Comparison of Suburban Development in Two Canadian Cities †

Sophie A. Nitoslawski * and Peter N. Duinker

School for Resource and Environmental Studies, Dalhousie University, Halifax, NS B3H 4R2, Canada; peter.duinker@dal.ca
* Correspondence: sophie.nitoslawski@dal.ca; Tel.: +1-514-833-8108
† Originally presented as a conference paper at the 2nd International Conference on Urban Tree Diversity.

Academic Editors: Francisco Escobedo, Stephen John Livesley and Justin Morgenroth
Received: 2 April 2016; Accepted: 30 May 2016; Published: 31 May 2016

Abstract: Is (sub)urban forest diversity shaped by previous land use? This study was designed to quantitatively assess the impacts of subdivision development on urban tree-species composition in two Canadian cities: Halifax, Nova Scotia, and London, Ontario. The main goal was to determine whether cities with contrasting pre-urbanized or pre-settlement landscapes—woodlands in Halifax and agricultural fields in London—also revealed differences in urban tree diversity losses and/or gains due to urbanization. In each city, four residential neighbourhoods representing two age categories, older and newer (40–50 years, <15 years), were examined and trees on three land types were sampled: public (street), private (residential), and remnant (woodland). All public street trees within the chosen neighbourhoods were inventoried and approximately 10% of the residential property lots were sampled randomly. Plots were examined in remnant forests in or near each city, representing the original forest habitats prior to agricultural and/or urban landscape transformations. Diameter at breast height, species richness and evenness, and proportions of native and non-native trees were measured. In both cities, streetscapes in newer neighbourhoods exhibit greater species richness and evenness, and are characterized by substantially more native trees. Despite this trend, developers and home owners continue to intensively plant non-native species on newer and smaller property lots. Older neighbourhoods in Halifax containing remnant forest stands hold the greatest number of native trees on private property, alluding to the importance of residual forest buffers and patches in promoting naturalness in the private urban forest. These results suggest that identifying and quantifying flows of species between green spaces during and after development is valuable in order to effectively promote native species establishment and enhance overall urban forest diversity.

Keywords: tree diversity; subdivision development; suburb; native; non-native; biodiversity

---

1. Introduction

Biodiversity conservation and enhancement are key considerations in urban forest research and policy, given rapid rates of urbanization worldwide and growing concerns about biodiversity declines [1]. Many studies have explored the range of benefits and services that diverse urban forests provide, including greater ecosystem productivity, higher resilience to environmental perturbations, and increased wildlife habitat [2–4]. Consequently, diversity targets have been incorporated into the core objectives of urban forest management plans [5,6].

Despite the recognition that diversity is crucial to urban forest health and city sustainability, urban foresters and scholars continue to disagree on the importance of native tree species contributions to overall forest diversity; some argue that native tree species promote the establishment of other native organisms and increase the ecological integrity of urban ecosystems [1,3,7], while others question the prioritization of nativeness over non-native, non-invasive species that are more resilient to urban stresses and are vital for urban biodiversity enhancement [8,9]. Furthermore, biotic homogenization has been identified as a pressing challenge to biodiversity enhancement in cities, as widespread exotic species can outcompete and replace native ones in anthropogenic environments, potentially contributing to species extirpations and threatening local, unique ecosystems [2,10,11]. These arguments and concerns illustrate the importance of elucidating the factors shaping urban forest diversity as well as determining how diversity benefits can be strategically and effectively maximized.

Suburban development is of particular importance to urban biodiversity research, as studies have shown that both native- and non-native tree-species richness can peak in neighbourhoods developed at the periphery of cities, located between the countryside and the urban core [1,12]. This trend has been attributed to the high level of landscape and habitat heterogeneity (e.g., private gardens, streetscapes, public green spaces including parks) that can occur in suburban and peri-urban areas [13–15]. Research has revealed that the type of land cover (e.g., fields, grassland, forests, roads, ruderal habitat) can dictate suburban plant diversity, as habitat edges often promote greater species richness as well as the establishment of rare species [16,17]. Furthermore, relatively undisturbed forest patches have been shown to be more suitable for protecting native plant and tree species compared to areas that have undergone significant anthropogenic change [18]. Diversity on residential properties is also influenced by home owner attitudes towards species selection, which often focus on aesthetics and promote exotic and ornamental species plantings [2,4]. Patterns in landscape structure and species composition emphasize the need to consider the planning and design of subdivisions during and after development, especially if urban forest diversity is to be considered a priority.

The heterogeneity of suburban landscapes, including the multiple land types on which trees can grow, necessitates the involvement of various urban forest professionals and practitioners in urban forest management. Trees found in parks and along streets generally fall under the jurisdiction of the municipality, and are subject to policies and targets determined by urban foresters and planners pertaining to tree planting guidelines, species selection, and diversity goals [5,6]. The land types found in suburban areas represent differences in tree establishment and management, reinforcing the importance of considering other factors such as policy and administration along with biophysical characteristics and home owner preferences when exploring how urban forest diversity is shaped.

Approximately two out of three Canadians live in suburban areas, which are currently experiencing much higher population growth rates compared to inner-city neighbourhoods located closer to city centres [19]. The extensive suburbanization occurring across the country highlights the need to consider how suburban development influences urban forest diversity and potentially provides opportunities for biodiversity enhancement in urban areas. Although the effects of urban development on urban forest biodiversity have been examined in a single Canadian city [1,4], researchers have yet to compare forest diversity losses and/or gains associated with development and previous land use in different cities. The goal of this study was thus to explore whether contrasting development patterns in two Canadian cities also determine differences in suburban tree diversity.

2. Materials and Methods

2.1. Study Sites

The two cities chosen for this study are Halifax, Nova Scotia (44.6478°N, 63.5714°W), and London, Ontario (42.9837°N, 81.2497°W). Although both cities are found in the eastern region of the country, they are situated approximately 1500 km apart (Appendix Figure A1). Halifax is located in the Acadian forest region, a mixed broadleaf and temperate forest comprised of both broadleaf and coniferous species including red spruce (*Picea rubens*), yellow birch (*Betula alleghaniensis*), sugar maple (*Acer saccharum*), and balsam fir (*Abies balsamea*) [20,21]. Some relatively undisturbed hinterland forest stands are found in suburban and peri-urban neighbourhoods in Halifax [6]. London is located at the northern range of the Carolinian forest region, characterized by a wide variety of broadleaf tree species like hackberry (*Celtis occidentalis*), black walnut (*Juglans nigra*), hickory (*Carya* spp.), and tulip tree (*Liriodendron tulipfera*). London is surrounded primarily by agricultural fields and some scattered remnant woodland corridors [22].

In each city, four residential neighbourhoods representing two age categories were examined (Appendix Figures A2 and A3, Tables 1 and 2). Neighbourhood selection was based on the pre-urbanized landscape, which was determined by consulting with urban foresters and planners as well as historical aerial imagery (Figures 1 and 2). Other criteria for neighbourhood selection included size and general similarities in urban morphology related to street and sidewalk design, as well as presence of street trees and remnant woodland in Halifax. Approximately 10% of private property lots were randomly selected based on total residential property land area on ArcGIS© (version 10) using the parcel data layers from each city's spatial database (Figure 3). The residents of selected properties were given a letter describing the study and requesting permission to access their property. If permission was not granted, the home immediately next to the original selected property was chosen. All streets within the boundaries of each neighbourhood were examined. In Halifax, four remnant stands located adjacent to each residential neighbourhood were sampled using five plots of 10 m × 10 m (0.20 ha total remnant forest sampled). In London, two remnant woodland sites located outside city limits were sampled using three plots of 20 m × 20 m (0.24 ha total remnant forest sampled).

Table 1. Neighbourhoods examined in Halifax.

Neighbourhood	Decade of Development	Total Number of Lots	Number of Selected Lots	Average Lot Size ± S.D. (m^2)
Tam O'Shanter	1960s–1970s	452	41	870 ± 420
Birch Cove	1960s	250	26	890 ± 280
Millview	2000s	478	49	830 ± 300
Russell Lake West	2000s	300	30	710 ± 320

Figure 1. Aerial imagery showing the development of Stoney Creek (London, Ontario) from field to subdivision. Source: University of Western Ontario and Google Earth © DigitalGlobe.

Figure 2. Aerial imagery showing the development of Millview (Halifax, Nova Scotia) from woodland to subdivision. Source: Google Earth © DigitalGlobe.

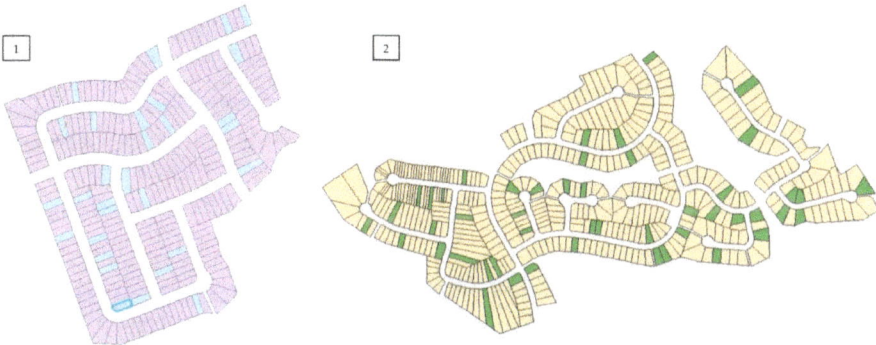

Figure 3. Neighbourhood extents and randomly selected private properties for Stoney Creek, London (**1**) and Millview, Halifax (**2**).

Table 2. Neighbourhoods examined in London.

Neighbourhood	Decade of Development	Total Number of Lots	Number of Selected Lots	Average Lot Size $\pm$ S.D. (m^2)
Masonville	1960s–1970s	304	33	1270 $\pm$ 300
Sherwood Forest	1960s	401	43	870 $\pm$ 250
Byron	2000s	315	32	560 $\pm$ 150
Stoney Creek	2000s	293	30	450 $\pm$ 80

2.2. Data Collection

All sites in Halifax were visited between June and July 2015, while all sites in London were visited between June and August 2015. The street trees in each neighbourhood were censused. All trees within the boundaries of randomly selected residential properties and within each plot in remnant woodland stands were measured. For each tree, the species was identified and diameter at breast height (DBH) was measured at 1.4 m from the base of the trunk using a diameter tape or calipers, depending on the size of the tree. Dead trees and shrubs (woody plants with height <5 m at maturity) were not measured.

Street trees were defined as trees either found on the road verge located between the street and the sidewalk, or, in the case where there was no sidewalk, located between the curb and private residential properties. Private property trees were defined as those found on landscaped and maintained areas of residential properties.

2.3. Data Analysis

Tree species identified were denoted as "native" or "non-native" based on whether the natural range of the species is found in the province; London is found in Ontario, which has 85 native species, while Halifax is located in Nova Scotia with 42 native species [21,23]. Naturalized species that were introduced after European colonization, like Norway maple (*Acer platanoides*), were considered non-native.

For each land type and neighbourhood, species richness was calculated by summing all identified tree species. Proportions of native and non-native species were calculated by dividing the count of native/non-native trees by the total count of trees measured.

Two indices were used to calculate species diversity [24]. The first is the Shannon-Weaver index, which takes into account both species richness and evenness:

$$H' = -\sum_i p_i \ln(p_i)$$

where p_i represents the count proportion of the *i*th species. Two-sample *t*-tests were carried out to determine whether results from this calculation varied significantly across neighbourhoods within and between cities.

The second diversity measure is the Simpson index of dominance, which describes the evenness of a community based on the probability that two randomly selected individuals belong to different species:

$$1 - D = \sum_i p_i^2$$

where p_i represents the count proportion of the *i*th species.

Four separate chi-squared analyses were carried out on SPSS© (version 22) using 2 × 2 contingency tables to identify the factors that influence proportions of native and non-native trees on residential properties. For these calculations, tree data was combined from the two neighbourhoods of equal age class within the same city. Two variables were tested: the pre-urbanized landscape and the decade of neighbourhood development (or neighbourhood age). The tests compared counts of native and non-native trees in old neighbourhoods with different pre-urbanized landscapes, in new neighbourhoods with different pre-urbanized landscapes, in both old and new neighbourhoods in Halifax, and in both old and new neighbourhoods in London.

3. Results

3.1. Site Descriptions

The size of the neighbourhoods ranged from approximately 25 to 50 hectares. Newer neighbourhoods were generally comprised of smaller residential property lots compared to older neighbourhoods. Properties in all neighbourhoods examined contained detached or semi-detached homes surrounded by lawn. Many home owners chose to maintain garden beds or other landscaped areas to grow shrubs, hedges, herbaceous plants, and trees. Both newer and older neighbourhoods in

Halifax retained some woodland area during development, including corridors located behind and between homes, and in some cases along pedestrian footpaths. Newer neighbourhoods in London were primarily surrounded by fields with some scattered forest corridors (Figure 4). Some areas of an older neighbourhood in London bordered a ravine with some woodland, while the other older neighbourhood housed a small patch of regenerated woodland.

(**a**) Russell Lake West, Halifax   (**b**) Byron, London

Figure 4. (**a**) An example of remnant woodland adjacent to a new subdivision development in Halifax; (**b**) An example of fields adjacent to a new subdivision development in London. Remnant stands in Halifax are generally found between property lines behind and beside housing developments as well as along pedestrian footpaths. In London, the scattered forest corridors that remain often exist further away from the development.

In Halifax, the remnant woodland sites were found adjacent to the subdivision, and were determined to be forested immediately prior to development, and remained so throughout the development process. Red maple (*Acer rubrum*) and red spruce dominated remnant sites in Halifax. Dominant trees were estimated to range between 50 and 100 years old. Other species sampled included yellow birch, striped maple (*Acer pensylvanicum*), and eastern hemlock (*Tsuga canadensis*). Two non-native species, Norway maple and European alder (*Alnus glutinosa*), were sampled in remnant plots located in close proximity to Birch Cove and Tam O'Shanter, respectively.

In London, remnant woodland was sampled at two sites located outside city limits. Dominant species included red maple, black cherry (*Prunus serotina*), and red oak (*Quercus rubra*); these trees were estimated to be between 100–200 years old. More uncommon species sampled included blue beech (*Carpinus caroliniana*), flowering dogwood (*Cornus florida*) and tulip tree.

3.2. Species Diversity

3.2.1. General Summary

In total, 82 tree species were identified in Halifax and 104 tree species were identified in London (Table 3). In both cities, the total species richness of private properties was greatest, followed by streetscapes and remnant woodland. Across all land types, London had a higher number of native species compared to Halifax. However, the selected neighbourhoods in Halifax housed a larger proportion of native species (33 native species sampled/42 total native species within the area of Nova Scotia = 79%) compared to London (52/85 = 61%).

Table 3. Combined species richness and raw counts of native and non-native species on each land type for both cities. Percentages represent proportion of native species sampled out of the total number of native species found in the province.

Land Type	Halifax			London		
	Native	Non-Native	Total	Native	Non-Native	Total
Remnant	16 (38%)	2	18	25 (29%)	0	25
Street	10 (24%)	23	33	21 (25%)	19	40
Private	28 (67%)	44	72	39 (46%)	45	84
Total number of species	33 (79%)	49	82	52 (61%)	52	104

3.2.2. Street Trees

Similar patterns in street-tree composition were observed in Halifax and London (Figures 5 and 6). Older neighbourhoods exhibited low species evenness and were dominated by non-native species, including Norway maple, littleleaf linden (*Tilia cordata*), crabapple (*Malus* spp.), and tree lilac (*Syringa reticulata*). In streetscapes, Norway maple accounted for 48% of all older trees in Halifax and 40% of all older trees in London, illustrating its popularity in street-tree planting in Canadian neighbourhoods developed 40–50 years ago.

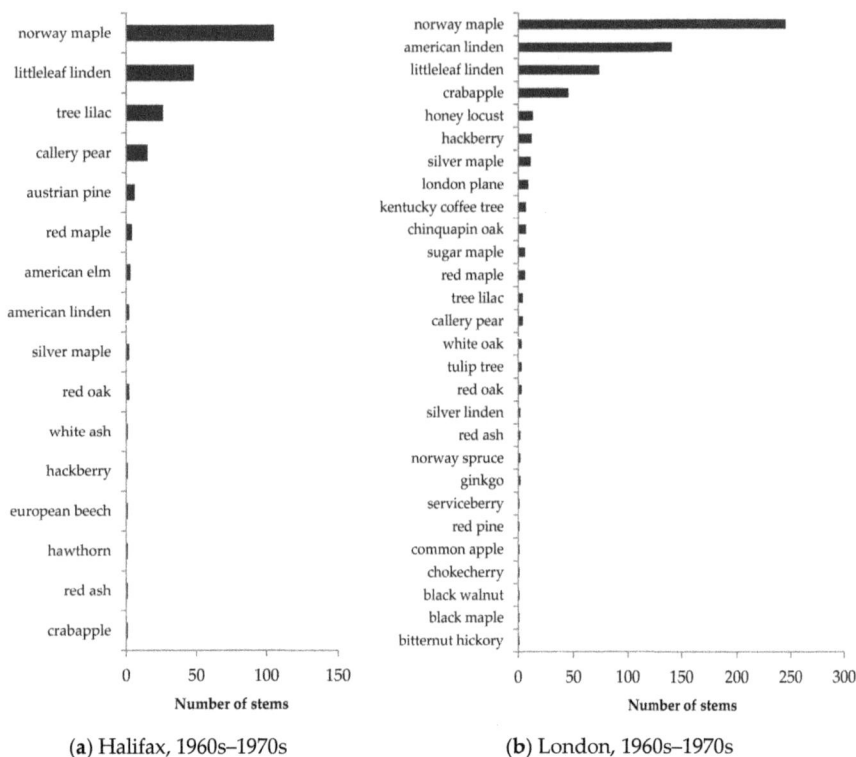

(a) Halifax, 1960s–1970s   (b) London, 1960s–1970s

Figure 5. Combined relative species abundance of street trees in neighbourhoods developed 40–50 years ago: (**a**) Relative species abundance of street trees in two neighbourhoods in Halifax; (**b**) Relative species abundance of street trees in two neighbourhoods in London.

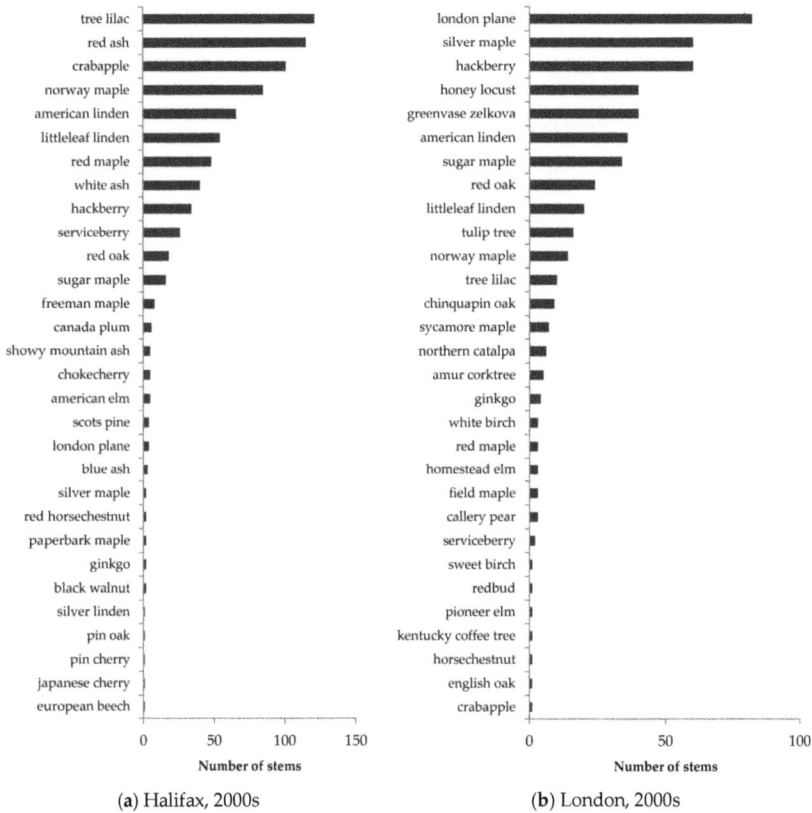

Figure 6. Combined relative species abundance of street trees in neighbourhoods developed <15 years ago: (**a**) Relative species abundance of street trees in two neighbourhoods in Halifax; (**b**) Relative species abundance of street trees in two neighbourhoods in London.

The results from the Shannon-Weaver diversity index calculation, which takes into account species richness and evenness, indicated that the diversity of streetscapes in older neighbourhoods in both cities is significantly different from streetscape diversity in newer neighbourhoods (Table 4). A two-sample *t*-test was used to compare differences in the Shannon index calculations (e.g., diversity) of street trees between older (mean = 1.30, SE = ±0.040) and newer (mean = 2.56, SE = ±0.105) neighbourhoods in Halifax ($t = 11.17$, $p = 0.008$, d.f. = 2). The same test found a significant difference between the diversity of street trees in older neighbourhoods (mean = 1.83, SE = ±0.040) and newer neighbourhoods (mean = 2.58, SE = ±0.055) in London ($t = 10.95$, $p = 0.008$, d.f. = 2). Older streetscapes in London were slightly more diverse than older streetscapes in Halifax ($t = 9.37$, $p = 0.011$, d.f. = 2), while a similar result was not found between newer streetscapes in both cities.

Street trees in newer neighbourhoods exhibited much greater species evenness, as illustrated by the more gradual slope in the species abundance distribution (Figure 6). The representation of Norway maple was also much smaller, accounting for approximately 11% of all newer street trees in Halifax and 3% of all newer street trees in London. Native stem count proportions were also higher in newer neighbourhoods compared to older ones, indicating that native species are more recently preferred for planting in city streets (Table 4).

Table 4. Diversity indices of street tree communities combined according to neighbourhood.

City	Decade of Development	Neighbourhood	Shannon Index (H') [1]	Simpson Index (1-D) [2]	Count Proportion of Native Trees
Halifax	1960s–1970s	Tam O'Shanter	1.34	0.65	0.11
	1960s	Birch Cove	1.26	0.57	0.03
	2000s	Millview	2.45	0.90	0.36
	2000s	Russell Lake West	2.66	0.91	0.37
London	1960s–1970s	Masonville	1.79	0.68	0.26
	1960s	Sherwood Forest	1.87	0.77	0.45
	2000s	Byron	2.52	0.89	0.57
	2000s	Stoney Creek	2.63	0.91	0.62

[1] Higher number indicates a more diverse community; [2] Higher number indicates a more even community.

3.2.3. Residential Property Trees

Contrasting patterns in the composition and diversity of residential property trees were observed between the two cities. In Halifax, older neighbourhoods had a significantly greater proportion of native trees compared to newer neighbourhoods (Figure 7). Older neighbourhoods in Halifax also displayed higher proportions of native trees compared to older neighbourhoods in London. In London, there were no notable difference between proportions of native and non-native trees across old and new neighbourhoods.

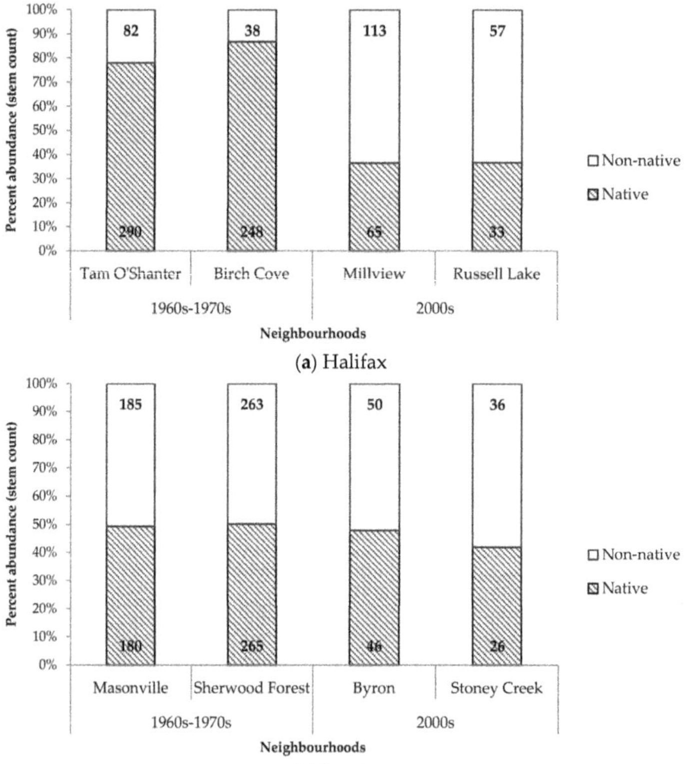

Figure 7. Counts and proportions of native and non-native trees for residential property trees in each neighbourhood: (**a**) Halifax; (**b**) London.

No statistical difference was observed between the diversity of property trees in older and newer neighbourhoods within each city (Table 5). However, a statistical difference did exist ($t = 5.77$, $p = 0.029$, d.f. = 2) between older neighbourhoods in Halifax (mean = 2.44, SE = ±0.155) and older neighbourhoods in London (mean = 3.46, SE = ±0.085). A similar relationship was not observed between newer neighbourhoods in both cities.

Table 5. Diversity indices of residential property tree communities combined according to neighbourhood.

City	Decade of Development	Neighbourhood	Shannon Index (H') [1]	Simpson Index (1-D) [2]	Count Proportion of Native Trees
Halifax	1960s–1970s	Tam O'Shanter	2.28	0.75	0.78
	1960s	Birch Cove	2.59	0.89	0.87
	2000s	Millview	3.09	0.91	0.37
	2000s	Russell Lake West	2.76	0.93	0.37
London	1960s–1970s	Masonville	3.37	0.95	0.50
	1960s	Sherwood Forest	3.54	0.95	0.50
	2000s	Byron	3.09	0.94	0.48
	2000s	Stoney Creek	3.14	0.95	0.42

[1] Higher number indicates a more diverse community; [2] Higher number indicates a more even community.

3.3. Size-Class Diversity

Size-class distributions were created to determine the composition of native and non-native trees according to DBH on residential properties, and to serve as a proxy for tree age. Although all size-class distributions were positively skewed, the distributions representing trees in older neighbourhoods differed considerably between the two cities.

In older neighbourhoods in Halifax, the distribution was more positively skewed compared to London. In Halifax, approximately 80% of small trees (DBH < 10cm) were native. In comparison, only 47% of trees of the same size class in older neighbourhoods in London were native. Furthermore, the number of native and non-native stems remained relatively equal and consistent across size classes in London (Figure 8). Conversely, in Halifax almost 60% of all native trees had a DBH below 20 cm.

In newer neighbourhoods, the size-class distribution of native and non-native trees was similar across both cities (Figure 9). Some older native trees were left standing in developments in both cities, while non-native trees dominated newly planted stems.

3.4. Chi-Squared Analysis

Four contingency tables were used to analyze the relationship between two factors (the pre-urbanized landscape and age of development) and counts of native and non-native trees on residential properties (Tables A1–A4). To review, the pre-urbanized landscape in Halifax is woodland, while the pre-urbanized landscape in London is field. A strong and statistically significant relationship was found between the type of pre-urbanized landscape and counts of native and non-native trees in neighbourhoods developed during the 1960s–1970s (Table 6). Halifax, developed onto woodland, had a significantly greater number of native trees compared to non-native trees in older neighbourhoods. London, developed onto field, had virtually equivalent numbers of native and non-native trees. The equivalent relationship in newer neighbourhoods was not found to be significant. Another important relationship existed in Halifax, but not in London, between decade of development and counts of native and non-native trees. Newer neighbourhoods in Halifax were dominated by non-native trees, while the opposite occurred in older neighbourhoods. In comparison, count proportions of native and non-native trees remained consistent across new and old neighbourhoods in London.

(a) Halifax, 1960s–1970s

(b) London, 1960s–1970s

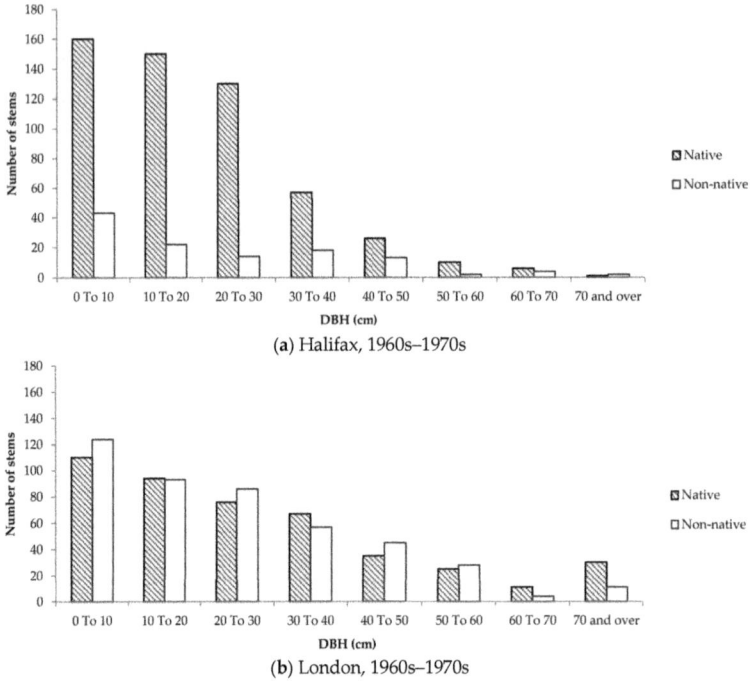

Figure 8. Size-class distribution of native and non-native trees on residential properties in neighbourhoods developed in the 1960s–1970s in: (**a**) Halifax; (**b**) London. Trees with a DBH = 10 cm, 20 cm, *etc.* are included in the lower histogram bin.

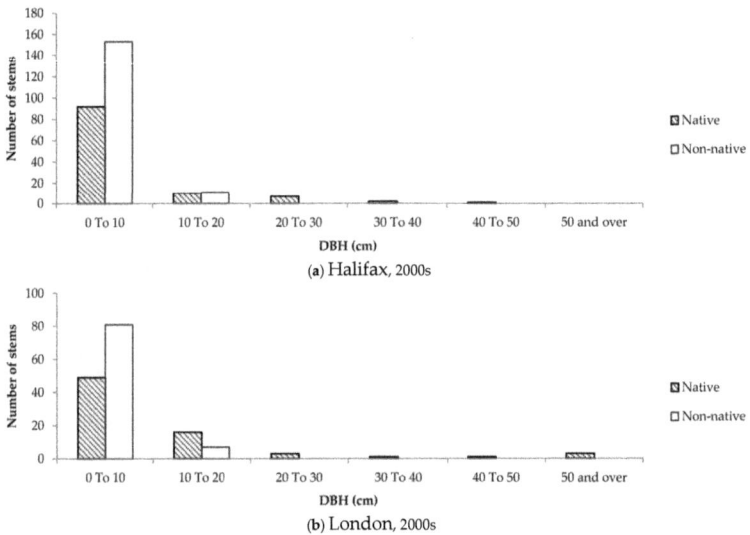

(a) Halifax, 2000s

(b) London, 2000s

Figure 9. Size-class distribution of native and non-native trees on residential properties in neighbourhoods developed in the 2000s in: (**a**) Halifax; (**b**) London. Trees with a DBH = 10 cm, 20 cm, *etc.* are included in the lower histogram bin.

Table 6. Results and statistical significance of chi-squared tests.

Tests	Degrees of Freedom (d.f.)	X^2	*p*-Value
O [1]	1	166.42	<0.005
N [2]	1	3.36	0.067
H [3]	1	180.85	<0.005
L [4]	1	0.98	0.323

[1] Relationship between the pre-urbanized landscape (Halifax and London) and counts of native and non-native trees in old neighbourhoods; [2] Relationship between the pre-urbanized landscape (Halifax and London) and counts of native and non-native trees in new neighbourhoods; [3] Relationship between neighbourhood age and counts of native and non-native trees in Halifax; [4] Relationship between neighbourhood age and counts of native and non-native trees in London.

4. Discussion

4.1. Development and Tree-Species Composition

The various land types that are found in suburban areas reflect differences in tree establishment, planting practices, maintenance, and ownership, and are thus influenced by different drivers of tree-species composition. The sites sampled in this study ranged from near-natural remnant forests to the urbanized streetscape. In both cities, streets in older neighbourhoods were dominated by only a few species, including the ubiquitous Norway maple. In comparison, streets in newer neighbourhoods had greater species richness and evenness as well as a better representation of native species, which have been shown to support insect and bird diversity more effectively than non-native species [3,25].

Remnant stands represented the most "natural" forest habitat found in suburban areas, and contained few (if any) non-native trees. Nevertheless, plots in Halifax supported two non-native tree species, both of which are invasive. Given the close proximity of the remnant sites to the adjacent residential area, it is probable that these species were planted on private property some time after development occurred, and managed to disperse into the remnant stands.

Many residential properties were intensively landscaped, and housed common ornamental non-native species like Norway maple, blue spruce (*Picea pungens*), and Japanese maple (*Acer palmatum*). Properties in newer neighbourhoods had fewer trees, of which a majority were non-native species. Conversely, properties in older neighbourhoods in Halifax held a large amount of native trees and resembled the remnant woodland sites more closely than properties in newer developments in terms of species composition. The size-class distributions suggest that some trees were kept during development in neighbourhoods in both cities; it is likely that large (and old) non-native trees found in London were planted on farms or along fence rows. These patterns illustrate the importance of examining changes in species distributions across remnant forests and adjacent green spaces, particularly as some non-native species can threaten native habitats and potentially displace native species [2,11].

The following sections elaborate on similarities and differences observed in the tree-species composition in streetscapes and on private properties in both cities.

4.2. Policy and Administration of Street Trees

Street trees are generally planted by a developer or contractor, and are maintained by the municipality. Despite the contrasting development patterns (e.g., the pre-urbanized landscape) of the selected neighbourhoods, the patterns exhibited in street-tree communities observed between the two cities are very similar. Given street-tree planting practices, it is certain that political and administrative factors play a large role in determining the types of trees found in city streetscapes.

The drastic difference between street-tree composition in newer *versus* older neighbourhoods in both cities could reflect changes in urban forest management and policy that have occurred in the last few decades. The infiltration of diseases and insects like Dutch elm disease and the emerald ash borer resulted in mass tree mortality rates in Canada in the last century, and has heightened awareness of the importance of diversifying planting stock [26–28]. Many urban forest management plans created more recently have developed targets to enhance native and overall species diversity, illustrating concerns about invasion events and urban forest canopy loss [5]. Higher evenness and native species representation in newer neighbourhoods in both cities may reflect a growing diversification trend among urban forest practitioners across Canada.

The differences observed in street-tree diversity between older and newer neighbourhoods could also be due in part to changes in nursery stock availability over time. Large-scale projects, like street-tree planting in new subdivision developments, often acquire trees from regional wholesalers instead of local nurseries [29]. At a given time, species selection may be limited due to location, resources, or customer demand; some tree species requested by the developer or urban forester may be substituted or outsourced, potentially resulting in more unfavourable species choices and reducing diversity enhancements [30]. However, some research in the United States has shown that nurseries are offering more species choices today compared to decades ago [31]. More recent demand for a wider range of species may therefore be influencing nurseries and wholesalers to diversify their stocks, in turn promoting higher species richness and evenness as well as native species representation in newer subdivision developments.

Various practitioners are both directly and indirectly involved in the process of street-tree species selection and planting, including municipal officials, arborists, urban planners, foresters, developers, and contractors. These groups reflect differences in expertise, values, priorities, and professional paradigms [29,32]; these in turn can dictate the species composition of street trees. For example, the popularity of tree lilac and London plane (*Platanus* × *acerifolia*) in newer neighbourhoods could be due to the preference of a particular urban forester or planner, based on factors such as tolerance to urban stresses, maintenance, growth rate, shade value, and cost.

4.3. The Influence of Remnant Woodland on Private Properties

Although little research has been carried out on the subject, it has been shown that development patterns, including the landscape present prior to urbanization, can influence the composition of urban tree species. One study found that residential neighbourhoods developed from forested areas had higher tree species richness and a greater proportion of native species compared to regions developed from prairie land [33]. Researchers have also suggested that the presence of remnant woodland located within or adjacent to residential areas can promote native species establishment and representation within the urban landscape [34,35].

The results of this study seem to illustrate these processes to some extent. The significant difference between both cities in native and non-native stem counts in older neighbourhoods suggests that native trees in Halifax may be dispersing from woodland areas retained during and after development onto adjacent residential properties. Furthermore, the much higher number of native trees found on older residential properties compared to newer ones in Halifax potentially demonstrates a lag effect; the tree-species composition of newer neighbourhoods may be influenced primarily by home owner planting preferences, while the species composition of older neighbourhoods may be shaped both by home owner planting decisions and the natural establishment of some trees over time as the development aged.

The size-class distributions for Halifax show that the majority of native trees are smaller, suggesting that many stems appeared on residential properties a while after development occurred. This phenomenon was not observed to the same degree in London, which could be due to the lack of remnant woodland in and around the selected neighbourhoods. The size-class distributions of native and non-native trees in London are very consistent across new and old neighbourhoods, therefore it is more likely that the species composition on private properties is influenced primarily by the planting attitudes and preferences expressed by home owners. Furthermore, the strikingly similar size-class distributions of trees in newer neighbourhoods in Halifax and London imply that the influence of remnant woodland is neither felt during nor immediately after development, but rather over time as trees are gradually given the chance to disperse into adjacent green space.

Despite the fact that private properties in Halifax had a better representation of native species, properties in London were generally more diverse, both in terms of species richness and evenness. This pattern could be attributed to differences in tree establishment; the tree-species composition of gardens in London mostly reflects individual planting preferences, and many home owners prefer non-native ornamental species for horticultural purposes, which contribute to overall species diversity [2,13,36]. In a similar vein, this trend could also explain why the diversity measures characterizing newer neighbourhoods in Halifax and London are more similar compared to the older neighbourhoods, where naturalization is likely occurring in Halifax, but not in London, due to native species dispersal.

Depending on diversity targets and priorities, neighbourhoods in cities like Halifax may benefit from retaining and protecting forest buffers and patches during and after development. Not only are remnant areas important for promoting native tree-species establishment, but other organisms may benefit from retaining woodland during development, illustrating the crucial role that urban forests play in biodiversity conservation [7].

4.4. Home Owner Attitudes and Opportunities for Diversity Enhancement

In both cities, residential properties were found to be most diverse compared to other land types, confirming the results from other studies on land tenure and tree-species composition [2,4,37]. Home owner attitudes and preferences likely explain the diverse range of planting decisions; many social and economic factors taking place at the property- and neighbourhood-level can influence residential species diversity, such as income [38], education [39], ethnicity and nationality [40], and planting fads [41]. These patterns emphasize the potential for private yards and gardens to contribute to urban forest biodiversity [34,42]. It is thus important to consider how landscape conversion and development history could influence private property trees, particularly in the context of biodiversity management and enhancement.

In the context of native species diversity, it is worthwhile to explore home owner attitudes towards native species establishment, and whether residents knowingly allow native trees to establish on their property. Research has shown that although urban dwellers may in theory support the planting of native species, they might not be knowledgeable enough to identify native trees and recognize their ecological importance [34]. Although results from this study have shown that native trees are probably dispersing onto private properties, the exact establishment mechanisms and potential decisions on the part of home owners have not been identified.

Subdivisions developed in cities like London, where most forested area has already been converted into agricultural fields, may benefit from targeted education initiatives and planting projects geared towards increasing native species representation and overall species richness, depending on the diversity goals of the municipality in question. Educating residents about the benefits of both native and non-native species may result in more informed planting choices, and in turn could influence the availability of local nursery stock [29]. Education initiatives geared towards nursery workers and wholesalers could also influence resource availability and encourage the dissemination of information about tree diversity and its benefits to customers [43]. London in particular has a relatively broad palette of native species to choose from, and encouraging a wider range of species selection on residential properties could give rise to diversity benefits at the city level.

Researchers have stressed the importance of increasing green space connectivity in urban and suburban areas for the purpose of increasing resilience to environmental change, providing wildlife habitat, and enhancing biodiversity [44]. Regardless of pre-urbanized landscape, suburban areas may effectively enhance diversity if the design of residential properties and neighbourhoods is considered and incorporated into biodiversity initiatives or urban forest management plans implemented at the city level [42]. It is worth noting that subdivision design should consider the possible negative impacts of residential encroachment onto remnant woodland, including loss and/or destruction of forest habitat, waste disposal and littering, and introductions of exotic species [45]. Neighbourhood-level urban forest strategies could educate residents about remnant forest protection (particularly in riparian areas) and identify specific biodiversity issues and opportunities at the community level, encouraging partnerships between municipalities, planners, developers, researchers, and residents [6,27,42]. Citizen engagement can be particularly useful in fostering public stewardship of the urban forest and in achieving municipal tree diversity goals [6].

4.5. Limitations, Challenges, and Future Research

The advantages of this study design relate to the large sample size of trees measured on different land types, allowing for robust analyses and conclusions about species composition at the neighbourhood level. However, it is difficult to generalize the results of this study to other residential areas or other cities, as only four neighbourhoods in each city were examined. Furthermore, this study did not examine parks, a land type that has been extensively researched and plays an important role in urban forest diversity enhancement and management [46–48].

Characterizing urban forest diversity is rendered more complex with the introduction of different land types (e.g., streets, private properties, remnant stands) that each require particular sampling protocols in order to effectively capture trends related to land use. Given the sampling design, the areas investigated for each land type in each neighbourhood are necessarily different. Species accumulation curves are included in the supplementary material, but should be interpreted with caution given the fact that urban species composition does not always reflect natural forest establishment and succession patterns that conform to traditional species-area relationships [49].

Although this study focused on the pre-urbanized landscape as a major driver of urban forest diversity, there are many other biophysical, socioeconomic, and political factors that influence urban tree diversity. In residential areas, lot size, which was not controlled for in this case, has been shown to correlate with urban tree diversity [50,51]. However, the results from this study showed a negligible correlation between property lot size and both number of stems and species richness (Appendix Figures A4 and A5). Exploring how other drivers pertaining to urban morphology and neighbourhood design could promote urban forest diversity and green space connectivity would also be an important endeavour, particularly as suburban areas continue to expand in Canadian metropolitan areas.

Future research should explore home owner decisions about species selection for planting and the natural establishment of native trees on residential properties from remnant woodland. It would be worthwhile to determine whether residents purposefully encourage native species establishment on private property, and what kind of motivations are involved. Furthermore, creating and implementing education and planting initiatives geared towards home owners and community groups could shed light on the types of strategies that are most effective for biodiversity promotion in residential areas.

5. Conclusions

The results of this study emphasize the importance of considering both spatial and temporal contexts associated with urban forest creation and management. The land types found in suburban areas represent divergences in tree establishment, planting, and maintenance practices, and thus reflect various manifestations in tree-species composition. Urban forest policies and management traditions likely explain differences observed in street-tree species composition between older and newer neighbourhoods. The pre-urbanized landscape, on the other hand, appears to be an important driver of urban forest composition on residential properties, as native species can naturally establish onto adjacent green spaces from forest buffers and patches, and potentially alter tree diversity over time.

Studies like these can encourage urban forest practitioners to reflect on the ways in which cities are developed, and how green infrastructure is created and molded based on the preferences and priorities of multiple professionals and urban dwellers. Envisioning how urban areas can be built to enhance urban forest diversity also allows for the opportunity to contemplate the kind of diversity that we would like to see in our cities. Although tree communities in London were generally more diverse, Halifax had greater proportions of native trees on residential properties. Is one case better than the other? What elements of urban forest diversity should be prioritized when developing and implementing management plans? These are crucial considerations, especially given growing concerns about biodiversity loss and environmental disturbances like pests, diseases, and climate change, which threaten the success of trees in urban landscapes. Given these difficulties, managers and practitioners should work towards achieving the *right* kind of biodiversity, instead of simply aiming for *more* biodiversity.

Acknowledgments: This work was funded by the Canada Graduate Scholarships-Master's program and the Killam predoctoral scholarship fund from Dalhousie University. The authors would like to thank Peter Bush for his support and guidance, and the organizers of the 2nd International Tree Diversity Conference for their generous publication fee waiver. The authors would also like to thank Gordon Flowerdew for his statistical expertise and advice.

Author Contributions: Sophie Nitoslawski and Peter Duinker conceived and designed the study; Sophie Nitoslawski collected the data, analyzed the results and drafted the manuscript; Peter Duinker contributed to the data collection and analysis, and reviewed and revised the draft manuscript.

Conflicts of Interest: The authors declare no conflict of interest.

Appendix A

Table A1. Contingency table for test "O" (old neighbourhoods).

	Native Trees	Non-Native Trees
Woodland (Halifax)	538	120
Field (London)	445	448

Table A2. Contingency table for test "N" (new neighbourhoods).

	Native Trees	Non-Native Trees
Woodland (Halifax)	98	170
Field (London)	72	86

Table A3. Contingency table for test "H" (all neighbourhoods in Halifax).

	Native Trees	Non-Native Trees
Old	538	120
New	98	170

Table A4. Contingency table for test "L" (all neighbourhoods in London).

	Native Trees	Non-Native Trees
Old	445	448
New	72	86

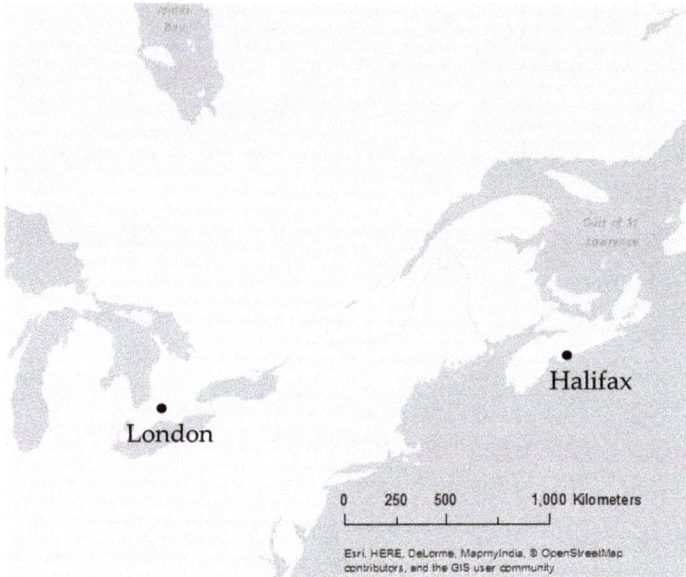

Figure A1. Halifax, Nova Scotia is located in Atlantic Canada and London, Ontario is located in Central Canada.

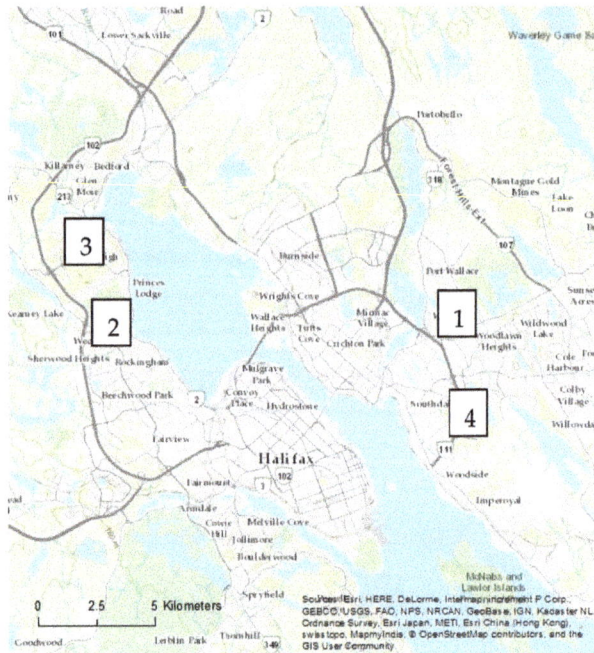

Figure A2. The neighbourhoods examined in Halifax: Tam O'Shanter (**1**); Birch Cove (**2**); Millview (**3**); and Russell Lake West (**4**).

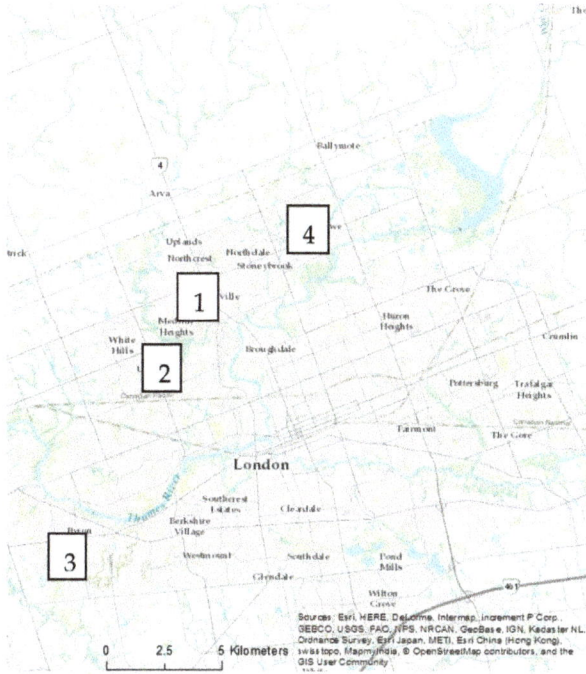

Figure A3. The neighbourhoods examined in London: Masonville (**1**); Sherwood Forest (**2**); Byron (**3**); and Stoney Creek (**4**).

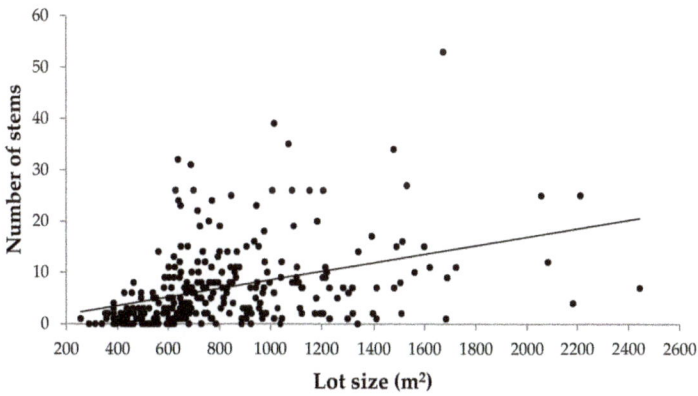

Figure A4. Correlation between private property lot size and number of stems found on each lot ($R^2 = 0.14$, $p < 0.05$, 284 lots sampled in total).

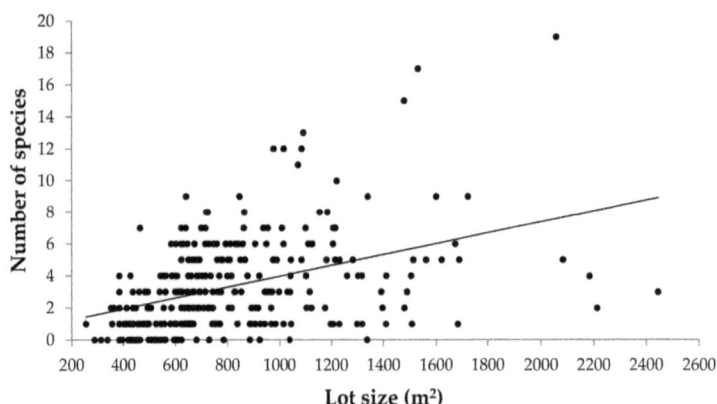

Figure A5. Correlation between private property lot size and number of species found on each lot ($R^2 = 0.17$, $p < 0.05$, 284 lots sampled in total).

References

1. Alvey, A.A. Promoting and preserving biodiversity in the urban forest. *Urban For. Urban Green.* **2006**, *5*, 195–201. [CrossRef]
2. Turner, K.; Lefler, L.; Freedman, B. Plant communities of selected urbanized areas of Halifax, Nova Scotia, Canada. *Landsc. Urban Plan.* **2005**, *71*, 191–206. [CrossRef]
3. Ilkin, K.; Knight, E.; Lindenmayer, C.B.; Fischer, J.; Manning, A.D. The influence of native *versus* exotic streetscape vegetation on the spatial distribution of birds in the suburbs and reserves. *Divers. Distrib.* **2013**, *19*, 294–306. [CrossRef]
4. Bourne, K.S.; Conway, T.M. The influence of land use type and municipal context on urban tree species diversity. *Urban Ecosyst.* **2014**, *17*, 329–348. [CrossRef]
5. Ordóñez, C.; Duinker, P.N. An analysis of urban forest management plans in Canada: Implications for urban forest management. *Landsc. Urban Plan.* **2013**, *116*, 36–47. [CrossRef]
6. Halifax Regional Municipality [HRM]. *Urban Forest Master Plan*, 2nd ed., 2013. Available online: http://www.halifax.ca/property/UFMP/documents/SecondEditionHRMUFMP.pdf (accessed on 8 March 2016).
7. Barth, B.J.; FitzGibbon, S.I.; Wilson, R.S. New urban developments that retain more remnant trees have greater bird diversity. *Landsc. Urban Plan.* **2015**, *136*, 122–129. [CrossRef]
8. Kendle, A.D.; Rose, J.E. The aliens have landed! What are the justifications for 'native only' policies in landscape plantings? *Landsc. Urban Plan.* **2000**, *47*, 19–31. [CrossRef]
9. Chalker-Scott, L. Nonnative, noninvasive woody species can enhance urban landscape biodiversity. *Arboric. Urban For.* **2015**, *41*, 173–186.
10. Olden, J.D.; Poff, N.L.R.; Douglas, M.R.; Douglas, M.E.; Fausch, K.D. Ecological and evolutionary consequences of biotic homogenization. *Trends Ecol. Evol.* **2004**, *19*, 18–24. [CrossRef] [PubMed]
11. McKinney, M.L. Urbanization as a major cause of biotic homogenization. *Biol. Conserv.* **2006**, *127*, 247–260. [CrossRef]
12. McKinney, M.L. Effects of urbanization on species richness: A review of plants and animals. *Urban Ecosyst.* **2008**, *11*, 161–176. [CrossRef]
13. McKinney, M.L. Urbanization, biodiversity, and conservation. *BioScience* **2002**, *52*, 883–890. [CrossRef]
14. Hansen, A.J.; Knight, R.L.; Marzluff, J.M.; Powell, S.; Brown, K.; Gude, P.H.; Jones, K. Effects of exurban development on biodiversity: Patterns, mechanisms, and research needs. *Ecol. Appl.* **2005**, *15*, 1893–1905. [CrossRef]
15. Deutschewitz, K.; Lausch, A.; Kühn, I.; Klotz, S. Native and alien plant species richness in relation to spatial heterogeneity on a regional scale in Germany. *Glob. Ecol. Biogeogr.* **2003**, *12*, 299–311. [CrossRef]
16. Čepelová, B.; Münzbergová, Z. Factors determining the plant species diversity and species composition in a suburban landscape. *Landsc. Urban Plan.* **2012**, *106*, 336–346. [CrossRef]

17. Godefroid, S.; Koedam, N. Distribution pattern of the flora in a peri-urban forest: An effect of the city-forest ecotone. *Landsc. Urban Plan.* **2003**, *65*, 169–185. [CrossRef]

18. Gong, C.; Chen, J.; Yu, S. Biotic homogenization and differentiation of the flora in artificial and near-natural habitats across urban green spaces. *Landsc. Urban Plan.* **2013**, *120*, 158–169. [CrossRef]

19. Gordon, D.L.A.; Janzen, M. Suburban nation? Estimating the size of Canada's suburban population. *J. Archit. Plan. Res.* **2013**, *30*, 197–220.

20. Loo, J.; Ives, N. The Acadian forest: Historical condition and human impacts. *For. Chron.* **2003**, *79*, 462–474. [CrossRef]

21. Farrar, J.L. *Trees in Canada*; Fitzhenry and Whiteside Limited: Markham, ON, Canada, 1995; p. 502.

22. The City of London. London Urban Forest Strategy, 2014. Available online: https://www.london.ca/residents/Environment/Trees-Forests/Documents/London%20Urban%20Forestry%20Strategy%20Final.pdf (accessed on 8 March 2016).

23. Kershaw, L. *Trees of Ontario*; Lone Pine Publishing: Edmonton, AB, Canada, 2001; p. 240.

24. Van Dyke, F. Biodiversity: Concept, Measurement, and Challenge. In *Conservation Biology: Foundations, Concepts, Applications*, 2nd ed.; Springer Science + Business Media BV: Dordrecht, The Netherlands, 2008; pp. 83–119.

25. Burghardt, K.; Tallamy, D.W.; Shriver, W.G. Impact of native plants on bird and butterfly biodiversity in suburban landscapes. *Conserv. Biol.* **2009**, *23*, 219–224. [CrossRef] [PubMed]

26. Raupp, M.J.; Buckelew Cumming, A.; Raupp, E.C. Street tree diversity in Eastern North America and its potential for tree loss to exotic borers. *Arboric. Urban For.* **2006**, *32*, 297–304.

27. Poland, T.M.; McCullough, D.G. Emerald ash borer: Invasion of the urban forest and threat to North America's ash resource. *J. For.* **2006**, *104*, 118–124.

28. Steenberg, J.W.N.; Duinker, P.N.; Charles, J.D. The neighbourhood approach to urban forest management: The case of Halifax, Canada. *Landsc. Urban Plan.* **2013**, *117*, 135–144. [CrossRef]

29. Conway, T.M.; Vander Vecht, J.V. Growing a diverse urban forest: Species selection decisions by practitioners planting and supplying trees. *Landsc. Urban Plan.* **2015**, *138*, 1–10. [CrossRef]

30. Sydnor, T.D.; Subburayalu, S.; Bumgardner, M. Contrasting Ohio nursery stock availability with community planting needs. *Arboric. Urban For.* **2010**, *36*, 47–54.

31. Pincetl, A.; Prabhu, S.S.; Gillespie, T.W.; Jenerette, G.D.; Pataki, D.E. The evolution of tree nursery offerings in Los Angeles County over the last 110 years. *Landsc. Urban Plan.* **2013**, *11*, 10–17. [CrossRef]

32. Kirkpatrick, J.B.; Davison, A.; Harwood, A. How tree professionals perceive trees and conflicts about trees in Australia's urban forest. *Landsc. Urban Plan.* **2013**, *119*, 124–130. [CrossRef]

33. Fahey, R.T.; Bowles, M.L.; McBride, J.L. Origins of the Chicago urban forest: Composition and structure in relation to presettlement vegetation and modern land use. *Arboric. Urban For.* **2012**, *38*, 181–193.

34. Doody, B.J.; Sullivan, J.J.; Meurk, C.D.; Stewart, G.H.; Perkins, H.C. Urban realities: The contribution of residential gardens to the conservation of urban forest remnants. *Biodivers. Conserv.* **2010**, *19*, 1385–1400. [CrossRef]

35. Ranta, P.; Viljanen, V. Vascular plants along an urban-rural gradient in the city of Tampere, Finland. *Urban Ecosyst.* **2011**, *14*, 361–376. [CrossRef]

36. Henderson, S.P.B.; Perkins, N.H.; Nelischer, M. Residential lawn alternatives: A study of their distribution, form and structure. *Landsc. Urban Plan.* **1998**, *42*, 135–145. [CrossRef]

37. Dobbs, C.; Kendal, D.; Nitschke, C. The effects of land tenure and land use on the urban forest structure and composition of Melbourne. *Urban For. Urban Green.* **2013**, *12*, 417–425. [CrossRef]

38. Hope, D.; Gries, C.; Zhu, W.; Fagans, W.F.; Redman, C.L.; Grimm, N.; Nelson, A.L.; Martin, C.; Kinzig, A. Socioeconomics drive urban plant diversity. *Proc. Natl. Acad. Sci. USA* **2003**, *100*, 8788–8792. [CrossRef] [PubMed]

39. Luck, G.W.; Smallbone, L.T.; O'Brien, R. Socio-economics and vegetation change in urban ecosystems: patterns in space and time. *Ecosystems* **2009**, *12*, 604–620. [CrossRef]

40. Fraser, E.D.G.; Kenney, W.A. Cultural background and landscape history as factors affecting perceptions of the urban forest. *J. Arboric.* **2000**, *26*, 106–113.

41. Nassauer, J.I.; Wang, Z.; Dayrell, E. What will the neighbours think? Cultural norms and ecological design. *Landsc. Urban Plan.* **2009**, *92*, 282–292. [CrossRef]

42. Goddard, M.A.; Dougill, A.J.; Benton, T.G. Scaling up from gardens: Biodiversity conservation in urban environments. *Trends Ecol. Evol.* **2010**, *25*, 90–98. [CrossRef] [PubMed]
43. Polakowski, N.R.; Lohr, V.I.; Cerny-Koenig, T. Survey of wholesale production nurseries indicates need for more education on the importance of plant species diversity. *Arboric. Urban For.* **2011**, *37*, 259–264.
44. Rudd, H.; Vala, J.; Schaefer, V. Importance of backyard habitat in a comprehensive biodiversity conservation strategy: A connectivity analysis of urban green spaces. *Restor. Ecol.* **2002**, *10*, 368–375. [CrossRef]
45. McWilliam, W.; Brown, R.; Eagles, P.; Seasons, M. Evaluation of planning policy for protecting green infrastructure from loss and degradation due to residential encroachment. *Land Use Policy* **2015**, *47*, 459–467. [CrossRef]
46. LaPaix, R.; Freedman, B. Vegetation structure and composition within urban parks of Halifax Regional Municipality, Nova Scotia, Canada. *Landsc. Urban Plan.* **2010**, *98*, 124–135. [CrossRef]
47. Godefroid, S.; Koedam, N. How important are large *vs.* small forest remnants for the conservation of the woodland flora in an urban context? *Glob. Ecol. Biogeogr.* **2003**, *12*, 287–298. [CrossRef]
48. Pennington, D.N.; Hansel, J.R.; Gorchov, D.L. Urbanization and riparian forest woody communities: Diversity, composition, and structure within a metropolitan landscape. *Biol. Conserv.* **2010**, *143*, 182–194. [CrossRef]
49. Zhao, J.; Ouyang, Z.; Xu, W.; Zheng, H.; Meng, X. Sampling adequacy estimation for plant species composition by accumulation surves—A case study of urban vegetation in Beijing, China. *Landsc. Urban Plan.* **2010**, *95*, 113–121. [CrossRef]
50. Tratalos, J.; Fuller, R.A.; Warren, P.H.; Davies, R.G. Urban form, biodiversity potential and ecosystem services. *Landsc. Urban Plan.* **2007**, *83*, 308–317. [CrossRef]
51. Kendal, D.; Williams, N.S.G.; Williams, K.J.H. Drivers of diversity and tree cover in gardens, parks, and streetscapes in an Australian city. *Urban For. Urban Green.* **2012**, *11*, 257–265. [CrossRef]

forests

Article

Regional Differences in Upland Forest to Developed (Urban) Land Cover Conversions in the Conterminous U.S., 1973–2011

Roger F. Auch [1,*], Mark A. Drummond [2], George Xian [1], Kristi L. Sayler [1], William Acevedo [1] and Janis L. Taylor [3]

[1] U.S. Geological Survey, Earth Resources Observations and Science (EROS) Center, 47914 252nd St., Sioux Falls, SD 57198, USA; xian@usgs.gov (G.X.); sayler@usgs.gov (K.L.S.); wacevedo@usgs.gov (W.A.)
[2] U.S. Geological Survey, Geosciences and Environmental Change Science Center, 2150 C Centre Ave., Fort Collins, CO 80526, USA; madrummond@usgs.gov
[3] Stinger Ghaffarian Technologies (SGT), contracter to the U.S. Geological Survey, Earth Resources Observations and Science (EROS) Center, 222 Big Ravine Drive, Whitefish, MT 48169, USA; janis.taylor.ctr@usgs.gov
* Correspondence: auch@usgs.gov; Tel.: +1-605-594-6086

Academic Editors: Francisco Escobedo, Stephen John Livesley and Justin Morgenroth
Received: 27 March 2016; Accepted: 15 June 2016; Published: 28 June 2016

Abstract: In this U.S. Geological Survey study of forest land cover across the conterminous U.S. (CONUS), specific proportions and rates of forest conversion to developed (urban) land were assessed on an ecoregional basis. The study period was divided into six time intervals between 1973 and 2011. Forest land cover was the source of 40% or more of the new urban land in 35 of the 84 ecoregions located within the CONUS. In 11 of these ecoregions this threshold exceeded in every time interval. When the percent of change, forest to urban, was compared to the percent of forest in each ecoregion, 58 ecoregions had a greater percent of change and, in six of those, change occurred in every time interval. Annual rates of forest to urban land cover change of 0.2% or higher occurred in 12 ecoregions at least once and in one ecoregion in all intervals. There were three ecoregions where the above conditions were met for nearly every time interval. Even though only a small number of the ecoregions were heavily impacted by forest loss to urban development within the CONUS, the ecosystem services provided by undeveloped forest land cover need to be quantified more completely to better inform future regional land management.

Keywords: Forest to urban developed land cover change; urbanization; conterminous U.S.; ecoregions; remote sensing

1. Introduction

Forests are substantial land cover sources for new urbanization both in the U.S. and globally [1–3]. Cumulatively, the increase in urban land and related types of development, such as roads and other exurban infrastructure, can cause a reduction of forest extent, fragmentation of wildlife habitat, and changes to hydrology and other regulating ecosystem services, such as carbon storage [4]. Due to geographic differences in human population and demographics, biophysical settings, and other factors, the impact of forest land cover conversion to new developed built-up land can be highly variable. Replacement of forest by urban development is also one of the most permanent changes to the environment [5] and may become even more important in regards to climate change effects on a growing number of people [6–8].

The growth of urban areas has been inescapable for decades, has tended to be sprawling, and is expected to continue to have substantial impact on land cover in the future [5,9]. However, mitigation is increasingly recognized as important, and there are new approaches to planning and managing urban ecological systems that could impact future trends, including consideration of urban forests and the sustainability of surrounding landscapes [10–13]. The role of forested land cover within, and surrounding, urban areas, and how best to mitigate the ongoing negative externalities of forest to urban developed land cover change is just one of the management pieces needed in understanding changed forest conditions in the near future [14].

Urbanization is a major driver of forest land cover change that needs renewed focus to analyze its widespread implications and potential impacts to human well-being [15]. A number of studies have conducted assessments of forest to urban developed land cover conversion either as their main emphasis or as part of the overall aspect of increased urbanization but these works tend to be scale limited by metropolitan area [16–19] or by region [20–23] or by temporal interval if done at a near national scale [24]. This research is the first to access near-national scale (CONUS) forest land cover to urban land change across a much longer time span (1973–2011) using similar remote sensing-derived datasets for six time-step intervals. Although near-national in overall scale, results are presented using a meso-scale ecoregional geographic framework that links similar land forms, vegetation, soils, and land use [25,26]. Using several proportional and rate conversion metrics, this work shows what ecoregions have been heavily impacted by forest to urban developed land cover conversion during the study period and where this type of land-use change has been much less of an issue.

2. Materials and Methods

2.1. Definitions of "Forest" and "Developed Land" Land Covers

This investigation does not explore urban tree cover or urban forestry; rather, we focus attention on conversion of forest land cover, as defined by two U.S. Geological Survey (USGS) datasets, to a land cover that has more anthropogenic characteristics than other types of features. Both the USGS Land Cover Trends project (LC Trends) and the National Land Cover Database (NLCD) definitions of forests are fairly simple; LC Trends defines forest land cover as 10% or more tree density and NLCD defines it as 20% or more for tree density as well as adding a height greater than five meters [27,28]. Urban developed land cover definitions for each data set are more complex but include that the land is either dominated by built and impervious features or a matrix of structures and vegetation or highly managed vegetation such as NLCD's "developed, open space" which is mostly lawn grasses [27,28]. Although such land cover definitions leave the impression of great precision where the semantics of what is "forest" and what is "urban" can be debated, most of the land change described in this research is of non-human occupied tree-dominated land (undeveloped forest land cover) being converted to residential subdivisions, commercial and industrial centers, road networks and right-of-ways, and other built features that are different land uses than what was found previously in the same location.

2.2. Materials

The land cover change data used in this research come from two different published (see additional citations in the 2.4 Limitations sub-section) USGS datasets that span two different time periods which, together, provide a nearly 40-year study period of forest land cover conversion to developed (urban and built-up) land. The first dataset is the USGS LC Trends project [1,29,30] that is based on sampled areas of U.S. Environmental Protection Agency (EPA) Level III ecoregions [31]. Each of the more than 2700 sample "blocks" had dates (circa 1973, 1980, 1986, 1992, and 2000) of modified Anderson 1 [32] land cover (e.g., forest, developed and built-up, agriculture, wetlands, and others) manually interpreted from imagery of various Landsat (Multi-Spectral (MSS), Thematic Mapper (TM), and Enhanced Thematic Mapper+ (ETM+) satellite sensors. Individual sample block land cover maps when compared between dates provide the change data that have estimates of land cover change that are statistically based at the ecoregion scale. The second USGS land cover change dataset is the

NLCD [33], which is a wall-to-wall mapping effort also derived using the data from several Landsat sensors (TM, ETM+) in an automated fashion to produce the Anderson II land cover [32] for the entire nation. Although the first NLCD (1992 iteration- [34]) was created using Landsat imagery from the early 1990s, we are using the iterations from circa 2001, 2006, and 2011 to complement and extend the land cover change record of the LC Trends project. The NLCD land cover data are at a 30 × 30 meter resolution (the innate resolution of the TM and ETM+ sensors) whereas the LC Trends sampled data are at a 60 × 60 meter resolution to enable comparison with MSS (data from 1973, 1980, and sometimes 1986) to the TM and ETM+ eras. At first glance, this may seem to be an issue but because individual maps from each of the datasets are not being directly compared to each other, only area estimates and percentages of land cover composition, the two different resolution sizes can work together.

2.3. Methods

We examined four land-cover change metrics that were easy to obtain from the datasets. These metrics included a threshold amount (⩾40%) of how much new urban developed land cover came from forest, the proportion of forest to developed land cover change that exceeded the proportion of forest land cover in the ecoregion, and a threshold rate of annual forest to urban developed land cover change for each time interval. Each of these metrics were used for each time interval. A final combination metric summed where the other metrics were met in most time intervals. Each of these metrics provides additional information about forest to urban developed land cover change.

The LC Trends project data already existed for estimating area of forest to urban developed land cover change for each ecoregion (See "LC Trends and NLCD" Excel in the supplemental material) for the first four time intervals (1973 through 2000). We were interested in determining the percent of forest to urban developed area in relation to the overall gain in urban developed land cover between dates, and did so by using the area estimates of forest to developed land cover change and dividing by the overall change in developed land cover (Equation (1), also see Supplemental Material).

$$\% \text{ new urban from forest} = \frac{amount \ (sq.km) \ of \ new \ urban \ from \ forest}{amount \ (sq.km) \ of \ new \ urban} \tag{1}$$

Equation (1) represents the percent of urban developed land cover from forest land cover.

A similar exercise was done with the NLCD land cover change data for the final two time intervals (2001 through 2011). However, because NLCD maps land cover at an Anderson Level II classification the changes from the three different types of forest (deciduous, evergreen, and mixed) to the four different types of urban developed land covers (developed-open space, developed-low intensity, developed-medium intensity, and developed-high intensity), had to be added up for each individual ecoregion (See "NLCD Classes to LC Trends Classes" in the supplemental material). The overall pixels of "forest to urban developed land cover change" (scaled up to Anderson I classifications here) were converted to square kilometers and then this area amount was divided by the area change in developed land between 2001 and 2006 and between 2006 and 2011 to derive the percentage of forest to urban developed land versus overall developed land cover change.

To determine the relationship between the percent of forest to urban development land cover change to the percent of forest land cover within the ecoregion a mean between each two dates of percent of forest land cover by ecoregion was calculated (Equation (2)). This allowed a single percentage for the land cover change data to be divided by a single percentage of forest land cover for each time interval (Equation (3)).The results of Equation (3) were then compared to the results from Equation (1).

$$\begin{aligned} average \ & amount \ (sq.km) \ of \ forest \\ &= \frac{amount \ (sq.km) of \ forest \ on \ first \ date + amount \ (sq.km) of \ forest \ on \ second \ date}{2} \end{aligned} \tag{2}$$

Equation (2) represents the average amount of forest land cover during a time interval.

$$\% \text{ of forest in ecoregion} = \frac{average\ amount\ (sq.km)\ of\ forest}{area\ (sq.km)\ of\ ecoregion} \tag{3}$$

Equation (3) represents the percent of forest land cover in an ecoregion per time interval.

For the LC Trends data, the percentage of forest as a proportion of each ecoregion's land cover composition was already provided (See "LC Trends and NLCD" Excel in the supplemental material). For the NLCD data, the total number of pixels classified as any forest type were added up and converted to km^2 for each ecoregion and each date and then divided by the total area in each ecoregion to create the percentage of forest. Then the mean of forest percentages of two dates was calculated as was done with the LC Trends data.

The annual rate of forest to urban developed land cover change for each ecoregion was calculated by taking the area of forest to developed land cover change divided by the area of forest land cover found in the first date of each time interval. This quotient was then divided by the number of years in each time interval (Equation 4).

$$A = ((amount\ [sq.km]\ of\ new\ urban\ from\ forest)/(amount\ [sq.km]\ of\ forest\ in\ first\ date))/$$
$$(number\ of\ years\ in\ time\ interval) \tag{4}$$

A = Annual rate of forest to urban developed land cover change.

Equation (4) represents the annual rate of forest to urban developed land cover change per ecoregion per time interval.

The same procedure was done with both the LC Trends and the NLCD data. National CONUS results from the LC Trends project [1] found that 1% annual overall land cover change for an ecoregion was considered a high rate. The threshold of 0.2% annual, or one fifth of what would be considered high in overall change, would translate to a 1% loss in forest land cover to urbanization every five years given that no replacement "to forest" source occurred and the conversion rate was sustained. The conversion to urban developed land cover is considered a near-permanent type of change in contrast to cyclic natural resource-based changes such as forestry, and so we used the threshold of 0.2% annual forest to urban developed land change as "high" for this type of change.

2.4. Limitations

One of the limitations in our results relates to scale. The regional scale of the investigation may mask the forest to urban developed land cover change dynamics of individual metropolitan areas by dampening the local intensity of change across a more extensive geographic area. This may be more of a factor in an ecoregion dominated by one large metropolitan area versus multiple urban centers. Another aspect of this limitation is that metropolitan areas are commonly spread across several ecoregions such as the Houston urban area, which occupies area in both the Western Gulf Coastal Plains and the South Central Plains (ecoregions #34 and #35, respectively, in Figure 1) or the New York metropolitan area spread across the Atlantic Coastal Pine Barrens, the Northeastern Coastal Zone, and the Northern Piedmont (ecoregions #84, #59, and #64, respectively, in Figure 1). There may be other spatial frameworks that can overcome this scale obstacle with the more recent wall-to-wall NLCD land change data but to include the longer 27-year record of the LC Trends sampled data, the Level III ecoregions provide the most appropriate estimates of change.

Large-area remote-sensing land cover mapping efforts always have a certain degree of error. The USGS LC Trends project and the USGS-led NLCD are no exception. Typically, remote-sensing land cover mapping projects use accuracy assessments to measure the uncertainty in their results. The LC Trends sample-based results give the uncertainties of the estimates in confidence intervals of how well the sampling captured specific types of change. Showing the sampling uncertainties in Table 1 does not mean that the LC Trends results are specifically better than NLCD numbers. The NLCD team uses accuracy assessments for their specific land cover classes at the national and large-region scale, although these accuracy assessments tend not to be completed the same time the land cover datasets are released to the public. At the current time, only the 2001–2006 land cover change data set has an accuracy assessment completed, but forest to urban developed land cover change was not separately assessed in this analysis. Rather, it was "bundled" with other land cover change classes into a "to developed" category. The NLCD "to developed" change category had an accuracy of 72% nationally for user's accuracy and regionally (EPA regions that are different than ecoregions) ranging from 58% to 81% [35]. NLCD users often use pixel-count change results for their specific areas of study because national and large-region accuracy assessments tend not to be spatially relevant for smaller regions. The LC Trends land cover change data did not have a formal accuracy assessment, but because the LC Trends research team used higher-resolution aerial photography (typically what is used in accuracy assessments) for at least two different dates as a way to augment the manual interpretation of the Landsat imagery, as well as team "block reviews" for each ecoregion, LC Trends change statistics are considered highly accurate [1,30].

EPA Level 3 Ecoregions

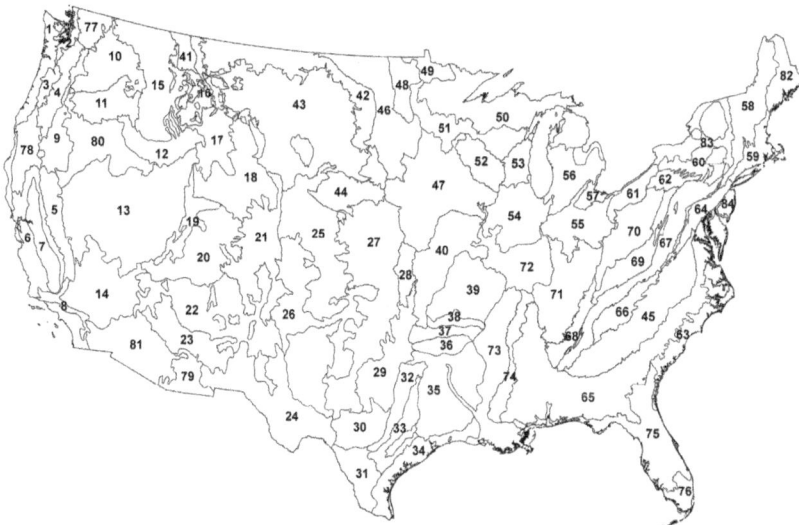

Figure 1. The numbers of the individual Level III EPA ecoregions (1999). Specific ecoregion numbers are called out in the text.

There is one time interval in which LC Trends and NLCD forest to urban developed land cover change overlap (1992 to 2000/01). Comparing these different datasets for the same time period is problematic, however, because NLCD 1992 [34] was created using methods different from those used for the subsequent NLCDs of 2001, 2006, and 2011 [33,36,37], and land cover change data found between the NLCD 1992 "Retrofit" [38] and NLCD 2001 [33] actually becomes a third dataset. Nonetheless, examining a map where this first NLCD change dataset compares area amounts for forest to urban developed land cover change and whether area amounts fall within LC Trends confidence intervals and where they do not is a worthwhile exercise. In a slight majority (46 out of 84) of the ecoregions, the NLCD 1992–2001 land cover change product did not have area amounts that were within LC Trends estimates confidence intervals (Figure 2). 39 LC Trends estimates were too low when compared to the NLCD change results and 7 LC Trends estimates were too high compared to NLCD (Figure 2). In some cases, the area difference between the two datasets was actually quite low. If a threshold of 10 km^2 or less was applied, 20 of the ecoregions where LC Trends were lower than NLCD would be eliminated and three ecoregions would be removed in cases where LC Trends estimates were higher than NLCD. Area amount discrepancies of over 100 km^2 between LC Trends and NLCD for the same time interval occurred in only two ecoregions, the Texas Blackland Prairies (ecoregion #32 in Figure 1) and the Western Allegheny Plateau (ecoregion #70 in Figure 1), 118 km^2 and 135 km^2, respectively.

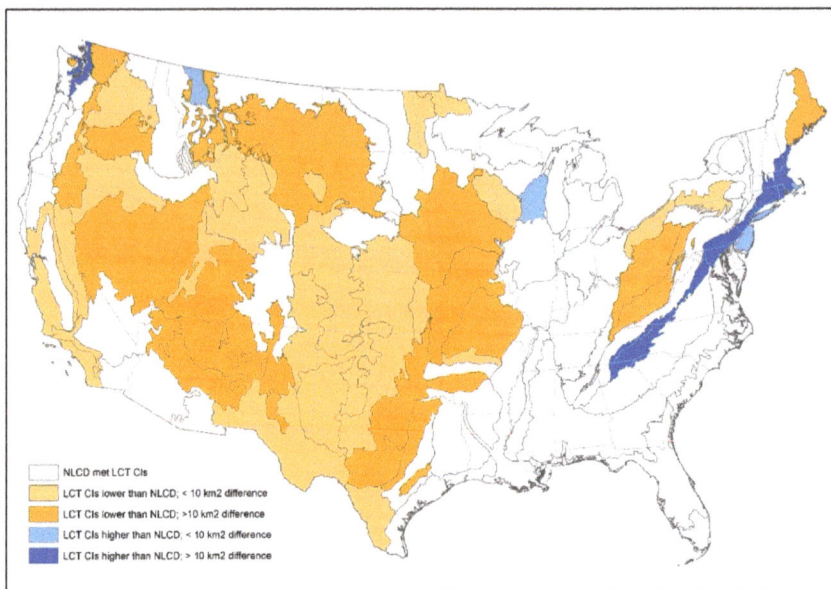

Figure 2. A comparison of forest to developed land cover area change between LC Trends (LCT) confidence intervals (CIs) estimates and the 1992 Retrofit NLCD- NLCD 2001 for the 1992–2000/2001 time interval.

Another potential limitation is the way in which the two different land cover change mapping efforts classified forested wetlands, especially in the Southeast and Gulf Coastal United States where these types of wetlands are prevalent. LC Trends placed forested wetlands into the broader "wetlands" land cover class, whereas NLCD mapped them as "woody wetlands" separate from emergent herbaceous wetlands. In neither case were these land cover classes included in our investigation. The issue here is not the difference between including forested wetlands within an Anderson I wetland classification or keeping them separate as an Anderson II class but that forested wetlands are notoriously hard to classify with high accuracy [39]. Some of the results from ecoregions from the above-listed larger regions may indicate classification confusion between what LC Trends called "upland" forest and what NLCD classified as "upland" forest compared to what was classified as "wetlands" and not included in the study. There may be other regional cases of differences between LC Trends and NLCD of how upland forest land cover is classified from other land covers.

Even though all of the above limitations may seem to call into question some of the results of this investigation, the combining of both the LC Trends and NLCD forest to urban developed land cover change provides the longest study period and the most geographically comprehensive inquiry into this type of land change within the United States. There are no other datasets comparable. The land cover change community has vetted numerous national- and regional-scale investigations using these two datasets [1,18,21,29,33–37,40–54] in spite of their imperfections. Instead of looking at the results of this investigation as precise measurements of change, they are better viewed as general observations of forest to urban developed land cover change at a CONUS regional scale.

Table 1. Ecoregions with forest as a substantial source of new development in any or all time intervals (all time intervals in bold). Ecoregions where at least 40% of new developed land cover came from forest.

Ecoregion	Forest to Urban, 1973–1980, km²	Forest to Urban as % of Total New Developed 1973–1980	Forest to Urban, 1980–1986, km²	Forest to Urban as % of Total New Developed 1980–1986	Forest to Urban, 1986–1992, km²	Forest to Urban as % of Total New Developed 1986–1992	Forest to Urban, 1992–2000, km²	Forest to Urban as % of Total New Developed 1992–2000	Forest to Urban, 2001–2006, km²	Forest to Urban as % of Total New Developed 2001–2006	Forest to Urban, 2006–2011, km²	Forest to Urban as % of Total New Developed 2006–2011
Coast Range	59 (±38)	75%	37 (±23)	58%	48 (±30)	72%	92 (±45)	72%	51	58%	1	47%
Puget Lowland	222 (±62)	85%	144 (±56)	73%	215 (±52)	71%	290 (±43)	66%			24	48%
Cascades	18 (±10)	92%	14 (±9)	63%	36 (±17)	78%	29 (±14)	85%				
Sierra Nevada					2 (±3)	100%						
Wasatch and Uinta Mountains	4 (±6)	93%	1 (±1)	68%								
Arizona/New Mexico Mountains	17 (±24)	61%										
Edwards Plateau			37 (±34)	66%	42 (±50)	49%	55 (±47)	54%	64	51%	51	47%
South Central Plains	167 (±83)	72%	374 (±239)	70%	103 (±38)	48%	367 (±297)	86%	156	52%	115	53%
Ouachita Mountains	11 (±7)	62%	12 (±9)	69%	15 (±15)	91%	17 (±12)	87%	17	62%	9	60%
Arkansas Valley					29 (±21)	41%					17	40%
Boston Mts.	3 (±3)	62%			4 (±3)	63%	4 (±3)	90%			5	52%
Ozark Highlands	112 (±121)	66%	42 (±33)	56%	61 (±42)	48%	55 (±38)	48%				
Canadian Rockies	3 (±3)	99%	4 (±3)	98%	2 (±2)	83%	5 (±3)	74%				
Piedmont	980 (±895)	79%	503 (±201)	58%	1569 (±859)	70%	2263 (±1374)	73%	837	59%	269	52%
Northern Minnesota Wetlands	5 (±5)	64%					1 (±1)	59%				
Northern Lakes and Plains	45 (±37)	97%	63 (±33)	85%	64 (±35)	89%	115 (±104)	94%				
Northeastern Highlands	85 (±53)	83%	67 (±52)	84%	161 (±116)	91%	194 (±140)	69%	34	62%	29	55%
Northeastern Coastal Zone	223 (±73)	78%	162 (±44)	75%	368 (±85)	71%	369 (±83)	75%	214	60%	137	57%
N. Appalachian Plateau and Uplands	24 (±15)	63%	11 (±11)	62%			19 (±12)	43%	4	43%		
Erie Drift Plains	67 (±36)	43%			137 (±77)	44%			53	45%		
North Central Appalachians	7 (±5)	75%	16 (±11)	83%	27 (±17)	82%	27 (±14)	90%	6	73%	12	68%

Table 1. Cont.

Ecoregion	Forest to Urban, 1973–1980, km²	Forest to Urban as % of Total New Developed 1973–1980	Forest to Urban, 1980–1986, km²	Forest to Urban as % of Total New Developed 1980–1986	Forest to Urban, 1986–1992, km²	Forest to Urban as % of Total New Developed 1986–1992	Forest to Urban, 1992–2000, km²	Forest to Urban as % of Total New Developed 1992–2000	Forest to Urban, 2001–2006, km²	Forest to Urban as % of Total New Developed 2001–2006	Forest to Urban, 2006–2011, km²	Forest to Urban as % of Total New Developed 2006–2011
Middle Atlantic Coastal Plain	444 (±270)	88%	498 (±336)	83%	493 (±305)	83%	306 (±178)	54%				
Southeastern Plains	483 (±325)	73%	578 (±367)	70%	578 (±330)	61%	1415 (±713)	69%				
Blue Ridge Mountains	112 (±59)	95%	95 (±68)	94%	66 (±53)	61%	191 (±71)	94%	38	67%	17	55%
Ridge and Valley	148 (±70)	60%	110 (±39)	41%	152 (±66)	47%	317 (±126)	46%	219	43%	15	56%
SW Appalachians	14 (±7)	77%	61 (±50)	72%	56 (±29)	64%	92 (±42)	70%	23	42%	5	51%
Central Appalachians	60 (±28)	59%	18 (±10)	65%	37 (±18)	40%	74 (±37)	61%				
Western Allegheny Plateau	47 (±26)	74%	30 (±11)	56%	76 (±46)	53%	79 (±26)	47%	87	57%	38	48%
Interior Plateau	105 (±73)	46%										
Mississippi Alluvial Plain	178 (±163)	50%	266 (±217)	47%			286 (±349)	41%				
North Cascades	1 (±2)	100%			3 (±3)	99%	28 (±27)	51%	3	47%		
Klamath Mountains												
Laurentian Plains and Hills	17 (±9)	91%	18 (±9)	78%	25 (±12)	85%	49 (±20)	81%	11	60%	6	58%
E Great Lakes and Hudson Lowlands	160 (±108)	64%	168 (±135)	62%			185 (±130)	42%				
Atlantic Coastal Pine Barrens	88 (±45)	45%	98 (±32)	41%					73	45%	45	45%

3. Results

A plurality of the ecoregions (35 out of 84) had conditions where at least 40% of their new developed land cover came from upland forest land cover at least one time during the study period. Geographically, these ecoregions tended to be clustered in the eastern U.S. outside of Florida (Figure 3). Other large regional clusters include the Pacific Northwest, the South-central U.S., and the Great Lakes North Woods as well as the "Texas Hill country" (Edwards Plateau, ecoregion #30 in Figure 1) and scattered ecoregions across the Inter-Mountain West, although most of the ones there were infrequent in occurrence. A number of these 35 ecoregions also had small area amounts of land being converted to developed land cover from forest (Table 1), making them appear more impressive on a map based on percentage of overall newly urban developed land than area affected.

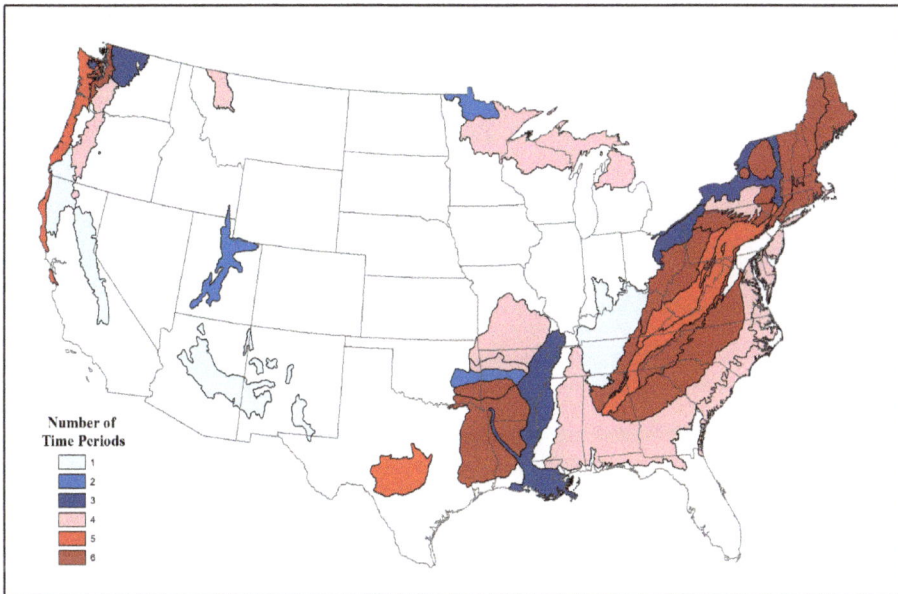

Figure 3. Ecoregions with forest as a substantial source of new development in any and all time intervals. Ecoregions where at least 40% of new developed land cover came from forest.

The number of ecoregions where forest was the source of at least 40% of the new urban development in every time period was more limited (11 out of 84). The number of clusters shrunk as well with only the Puget Lowland (ecoregion #2 in Figure 1) found in the Western U.S., a two-ecoregion cluster in the South-central U.S., three ecoregions in the Northeast, and five ecoregions scattered across the Appalachian Mountains and foothills (Figure 3). The Piedmont, Northeastern Coastal Zone, Puget Lowlands, and South Central Plains (ecoregions #45, #59, #2, and #35, respectively, in Figure 1) consistently had the most forest to urban developed land cover by area across time.

A majority of the ecoregions (58 out of 84) had at least one time interval where the proportion of forest to urban development in overall new urban land exceeded the proportion of forest land cover within the ecoregion. This is a useful metric because it can indicate where forested land is targeted more for conversion than other land covers within an ecoregion, and without replacement from another land cover forested land may face noticeable losses. The geographic pattern was more widespread and diffuse (Figure 4) than that seen in the forest as a substantial source of new urban developed land cover (Figure 3). However, in many ecoregions where this metric occurred, the threshold was met only occasionally (Table 2).

Table 2. Ecoregions exceeding their proportion of forest in forest to developed land cover change in any and all time intervals (all time intervals in bold). Ecoregions where the proportion of forest to developed exceeded the proportion of forest found within the ecoregion.

Ecoregion	% Forest to Urban, 1973–1980, LC Trends	% Forest to Ecoregion, 1973–1980, LC Trends	% Forest to Urban, 1980–1986, LC Trends	% Forest to Ecoregion, 1980–1986, LC Trends	% Forest to Urban, 1986–1992, LC Trends	% Forest to Ecoregion, 1986–1992, LC Trends	% Forest to Urban, 1992–2000, LC Trends	% Forest to Ecoregion, 1992–2000, LC Trends	% Forest to Urban, 2001–2006, NLCD	% Forest to Ecoregion, 2001–2006, NLCD	% Forest to Urban, 2006–2011, NLCD	% Forest to Ecoregion, 2006–2011, NLCD
Coast Range							72.0	71.7				
Puget Lowland	**85.5**	**55.5**	**73.0**	**53.0**	**70.7**	**50.0**	**66.2**	**47.6**	**58.4**	**42.9**	**48**	**41.2**
Cascades	99.0	81.8					85.3	81.7				
Sierra Nevada					100.0	72.8						
S. California Mountains			28.9	27.3								
MT Valley & Foothill Prairies					35.8	17.8						
Wyoming Basin											5	2.4
Wasatch and Uinta Mts.	92.8	61.7	68.0	61.6								
Arizona/New Mexico Mountains	61	58										
Chihuahuan Deserts			4.2	2.4								
Western High Plains											0.7	0.5
Southwestern Tablelands											5.7	2.8
Central Great Plains									3.8	2.5	3.2	2.5
Flint Hills			16.3	6.1					13.7	5.6	10	5.6
Central Oklahoma/Texas Plains							23.7	19.3				
Edwards Plateau			66.4	27.9	49.4	27.5	54.3	27.1	51.4	24.5	46.4	23.9
Southern Texas Plains	11.1	5.4							1.5	1.1	2.8	1.1
Texas Blackland Prairies									15.9	12.1	16.3	11.9
East Central Texas Plains					35.1	31.3	32.6	30.7	29.5	20.6	24.6	20.3
Western Gulf Coastal Plain	**21.2**	**12**	**12.7**	**11.9**	**21.5**	**11.9**	**28.3**	**11.7**	**12.9**	**5**	**11.3**	**4.8**
South Central Plains	72.4	62.7	69.9	60.6			85.9	59.3	51.8	47.2	52.8	45.8
Ouachita Mountains					90.5	76.9	86.8	78.5				
Boston Mts.							89.9	76.2				
Ozark Highlands	66.2	58.1										
Canadian Rockies	98.6	70.2	98.2	70.2	82.9	69.6	74.5	68.8			2.2	0.4
Nebraska Sandhills									1.0	0.4		

Table 2. Cont.

Ecoregion	% Forest to Urban, 1973–1980, LC Trends	% Forest to Ecoregion, 1973–1980, LC Trends	% Forest to Urban, 1980–1986, LC Trends	% Forest to Ecoregion, 1980–1986, LC Trends	% Forest to Urban, 1986–1992, LC Trends	% Forest to Ecoregion, 1986–1992, LC Trends	% Forest to Urban, 1992–2000, LC Trends	% Forest to Ecoregion, 1992–2000, LC Trends	% Forest to Urban, 2001–2006, NLCD	% Forest to Ecoregion, 2001–2006, NLCD	% Forest to Urban, 2006–2011, NLCD	% Forest to Ecoregion, 2006–2011, NLCD
Piedmont	78.6	59.4			70.4	57.2	72.6	55.8	59.4	57.3		
Northern Glaciated Plains			3.3	3.0					1.2	1.1		
Western Corn Belt Plains			5.4	3.3	4	3.3					6.2	4.4
Lake Agassiz Plain			6.3	5.6								
Northern Minnesota Wetlands	63.9	38.2					59.5	36.5	21.4	13.0		
Northern Lakes and Forests	97.3	64.1	84.5	63.2	88.6	62.4	94.4	61.9				
Southeastern Wisconsin Till Plains					15.5	11.9						
Central Corn Belt Plains	11	9.5	12.5	9.4					9.6	8.8	12.1	8.8
Eastern Corn Belt Plains					13.9	12.8			14.2	13.9		
S. Michigan/N. Indiana Drift Plains	33.4	24.4							23.0	20.1	20.3	20
Huron/Erie Lake Plains	17.5	12.8	13.0	12.7	21.5	12.7	18.8	12.6			11.6	8.9
Northeastern Highlands			83.9	83.1	91.3	81.9						
Northeastern Coastal Zone	77.9	50.2	75.3	49.5	71.2	48.7	75.3	47.5	60.4	45.6	56.6	44.9
N. Appalachian Plateau and Uplands	63.1	60.0	62.0	60.0								
Erie Drift Plains	43.0	37.5	38.0	37.4	44.0	37.2			45.3	37.5		
North Central Appalachians							90.4	86.7				
Middle Atlantic Coastal Plain	88.5	34.7	82.7	33.5	84.3	32.6	54.2	32	30.7	18.9	29.4	17.4
Northern Piedmont									35.9	30.4		
Southeastern Plains	73.3	52.6	70.4	51.9	60.6	51.8	69.3	51.9				
Blue Ridge Mountains	95.1	79.3	93.9	79			93.7	78.5				
Ridge and Valley	60.2	57.1										
Western Allegheny Plateau	74.3	64.3										
Interior Plateau	45.8	38.9										
Interior River Lowland									27.4	27	28.6	26.9

Table 2. Cont.

Ecoregion	% Forest to Urban, 1973–1980, LC Trends	% Forest to Ecoregion, 1973–1980, LC Trends	% Forest to Urban, 1980–1986, LC Trends	% Forest to Ecoregion, 1980–1986, LC Trends	% Forest to Urban, 1986–1992, LC Trends	% Forest to Ecoregion, 1986–1992, LC Trends	% Forest to Urban, 1992–2000, LC Trends	% Forest to Ecoregion, 1992–2000, LC Trends	% Forest to Urban, 2001–2006, NLCD	% Forest to Ecoregion, 2001–2006, NLCD	% Forest to Urban, 2006–2011, NLCD	% Forest to Ecoregion, 2006–2011, NLCD
Mississippi Alluvial Plain	49.7	10.3	46.5	9.9	28.4	9.6	41.3	9.6	11.4	4.5	14.8	4.5
Southern Coastal Plain	27.9	27.6	32.9	26.4	27.7	25.3	29.1	24.4				
Southern Florida Coastal Plain	35.6	2.8	21.2	2.7	14.9	2.6	9.9	2.6				
Northern Cascades	100.0	71.7			98.9	70.9	100.0	70.4				
Northern Basin and Range									3.0	1.7	2.1	1.6
Laurentian Plains and Hills	90.8	71.8	77.9	71.0	84.7	70.2	81.0	70.0				
E Great Lakes and Hudson Lowlands	62.4	39.3	62.5	39			42.2	38.9	39	34.2	34.4	34
Atlantic Coastal Pine Barrens	44.7	23.1	40.9	22.5	25.8	22	25.7	21.7	44.6	25.5	45	24.9

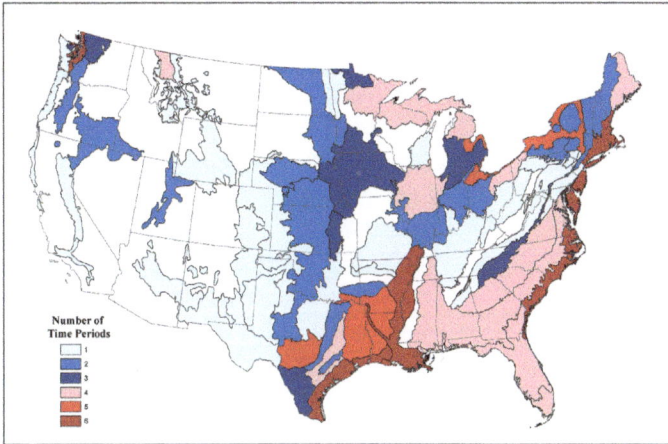

Figure 4. Ecoregions exceeding their proportion of forest in forest to urban developed land cover change in any and all time intervals. Ecoregions where the proportion of forest to developed land cover change exceeded the proportion of forest land cover found within the ecoregion.

The number of ecoregions where the proportion of forest to urban developed land cover change exceeded the proportion of forest within the ecoregion every interval was far fewer (6 out of 84,) than those exceeding it occasionally and only about half the ecoregions where forest was a substantial source of new urbanization every time interval. There was less geographic clustering of the six ecoregions that exceeded their proportion of forest every interval except the three along the eastern seaboard from southern Maine through Northern Florida (Figure 4), the Mississippi Alluvial Plain and Western Gulf Coastal Plain (ecoregions #73 and #34, respectively, in Figure 1), and the Puget Lowland (ecoregion #2 in Figure 1) in the Pacific Northwest. The six ecoregions that exceeded their forests' proportions when converting to urban development generally did so substantially.

Even though the annual rate of forest to developed land cover change was set at a fairly conservative number of 0.2%, only a minority of the ecoregions (12 out of 84) met or exceeded this rate at any time during the study period. Geographically, four clusters and one additional ecoregion are visible (Figure 5) although several of the clusters merge to create even larger contiguous regions. All of the ecoregions that front the Atlantic Ocean or Gulf of Mexico shoreline had a rate of 0.2% or greater annual change of upland forest converting to urban developed land cover at least once during the study period. Inland, the Northern Piedmont (ecoregion #64 in Figure 1) links highly urbanized areas of the Northeast coastal ecoregions and the Piedmont (ecoregion #45 in Figure 1) cities along the Fall Line and the foothills of the Appalachian Mountains, Gottmann's older "Megalopolis" of interspersed mosaics of urban, forest, and agricultural land covers [55] meeting up with Hart's and Morgan's emerging southern "Spersopolis" of low-density, but nearly continuous, residential housing along highways linking urban centers [56]. Another cluster is centered on the Erie Drift Plains and the Eastern Corn Belt Plains (ecoregions #61 and #55 respectively in Figure 1), whereas the Puget Lowland (ecoregion #2 in Figure 1) is the only ecoregion in the Western U.S.

The rate of "high" annual forest to urban developed land cover change ranged from three ecoregions reaching 0.2% at least during one time interval to the Southern Florida Coastal Plain (ecoregion #76 in Figure 1) reaching 0.61% annually during the 1986 to 1992 interval (Table 3). This ecoregion exceeded or nearly exceeded 0.5% annual change during the first three intervals of the LC Trends era, although with forest to urban developed land cover change declining to near zero during the NLCD intervals may bring into question the issue of how forest cover is classified as either "upland" or "wetland" between the two datasets. The Atlantic Coastal Pine Barrens (ecoregion

#84 in Figure 1), which includes the center of the New York metropolitan area, was the only ecoregion to reach or exceed the 0.2% annual rate during all the time intervals.

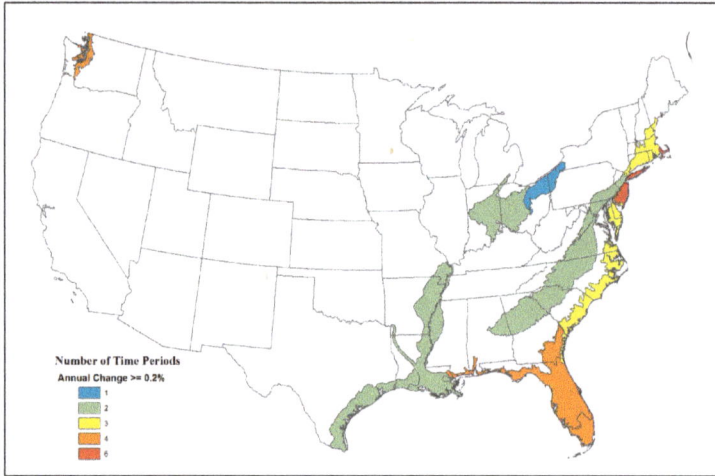

Figure 5. Ecoregions reaching or exceeding the rate of 0.2% annual forest to developed land cover change in any or all time intervals.

Table 3. Ecoregions reaching or exceeding the rate of 0.2% annual forest to developed land cover change in any or all time intervals (all time intervals in bold).

Eco	Annual Rate of Forest to Urban Change 73–80 LC Trends	Annual Rate of Forest to Urban Change 80–86 LC Trends	Annual Rate of Forest to Urban Change 86–92 LC Trends	Annual Rate of Forest to Urban Change 92–00 LC Trends	Annual Rate of Forest to Urban Change 01–06 NLCD	Annual Rate of Forest to Urban Change 06–11 NLCD
Puget Lowland	0.31%	0.25%	0.38%	0.42%		
Western Gulf Coastal Plain					0.26%	0.21%
Piedmont			0.27%	0.30%		
Eastern Corn Belt Plains			0.20%	0.24%		
Northeastern Coastal Zone			0.34%	0.26%	0.27%	
Erie Drift Plains			0.20%			
Middle Atlantic Coastal Plain	0.20%	0.27%	0.28%			
Northern Piedmont				0.25%	0.21%	
Mississippi Alluvial Plain		0.31%		0.26%		
Southern Coastal Plain	0.26%	0.26%	0.22%	0.27%		
Southern Florida Coastal Plain	0.56%	0.48%	0.61%	0.23%		
Atlantic Coastal Pine Barrens	0.28%	0.37%	0.24%	0.20%	0.35%	0.22%

The results of the composite metric shows that there are three ecoregions (Puget Lowland, Northeastern Coastal Zone, and the Atlantic Coastal Pine Barrens—ecoregions #2, #59, and #84, respectively, in Figure 1) that had 15 or above out of 18 "points" (Figure 6). Each of these are small ecoregions in size, heavily urbanized, and where continued urbanization has either been the leading or co-leading stories of land cover change during the study period.

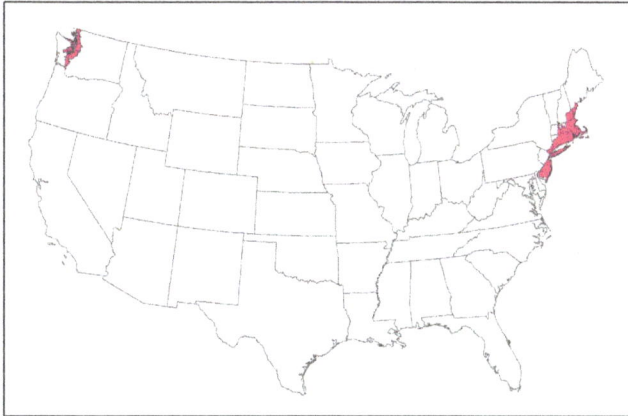

Figure 6. Ecoregions with a composite forest to urban developed land cover change score of 15 or greater. Puget Lowland and Atlantic Coastal Pine Barrens (ecoregions #2 and #84, respectively, in Figure 1) both had a score of 16, whereas the Northeastern Coastal Zone (ecoregion #59 in Figure 1) scored 15.

4. Discussion

Forest land cover across the U.S. is dynamic because of the geographic and temporal variability of many human and natural drivers including harvesting-replanting cycles (timber management), agricultural clearance or abandonment, natural disturbances, including wind throw, fire, and insects and disease, climate change and drought, as well as urbanization [1,44]. Monitoring and understanding these changes requires a long-term view. This analysis of the urban growth effects on regional forest land cover shows some of these long-term spatial dynamics.

Upland forest land cover at the ecoregion scale within a national context has not been heavily impacted by forest cover loss to urban development during the study period, and certainly not as cartographically displayed by Clement et al. [24] for the 2001–2006 interval. Small, already heavily urbanized ecoregions such as the Northeastern Coastal Zone and the Atlantic Coastal Pine Barrens of the northeast and the Puget Lowlands of the northwest U.S. may be the exceptions and may have been impacted the most. This does not mean that the loss in specific ecosystem services of former forested land, especially those services not found or found in greater amounts than in urban tree cover, in moderately affected ecoregions should be overlooked or discounted in importance. Land-cover modeling efforts for future dates, such as 2050, or even 2100, show sustained losses of forest land cover to urban development at both regional [57] and national scales [58]. Research into the quantification of ecosystem services provided by undeveloped forest land cover should continue to be encouraged. The growth or maintenance of urban forests may mitigate and moderate some of the loss of undeveloped forested lands in various ecosystems services, but do they truly replace their undeveloped counterparts in all aspects? Multi-scale land-use policies protecting more forest or slowing the rates of conversion may need to be augmented or even created, depending on location, to balance forest land cover ecosystem services with the opportunities and amenities found in urban regions. These multi-scale forest retention land-use policies may have special relevance because most people in the U.S. and, increasingly, around the world, live in cities for specific reasons. Increased forest land-cover preservation may clash with efforts to protect farmland and other natural or non-built-up land covers and land uses because urban areas continue to expand in size even with the efforts to increase density within existing developed land cover [24]. Americans have long pushed the boundaries of their cities and it is something not easily culturally undone [47,59,60]. The dilemma on how best to keep the most undeveloped land covers from being converted to highly urbanized conditions while cities expand in size will not be easily solved and will remain an issue into the future.

A way to improve the multi-scale regionalization of mapping forest to urban developed land cover conversion may be the use of Level IV ecoregions using available multi-date wall-to-wall land cover datasets. Drummond et al. [61] used this scale for the 2001–2006 era within two Level III ecoregions in the Southeast U.S. and showed urban growth at a finer scale without losing the next scale up in geographic size. Forested land preservation planning may be better articulated and discussed using the results from land change mapping using multi-scale ecoregions that commonly cross local and even state political jurisdictions. The impacts of land cover change from individual or multiple urban areas may be seen more clearly using Level IV ecoregions and wall-to-wall land cover data.

The inclusion of forested wetland land cover change to urban developed land may be a way to provide a more comprehensive overview of forested land conversion to urban areas especially in the Southeast coastal region of the U.S. where Xian et al. [54] reported that "woody wetlands" was a leading source of newly urbanized land cover. This has not been the case in other ecoregions, such as the Northeastern Coastal Zone, where wetlands conversion to urban developed land cover was a minor source of increased urbanization [62]. The inclusion of Anderson II "woody wetlands" with current and future wall-to-wall land cover mapping would negate the issue of whether forest is correctly classified as "upland" or "wetland" and provide a better indication of the total contribution of "forest" land cover as a source of new urban land.

5. Conclusions

This study was able to show which ecoregions in the CONUS that have been heavily impacted by the conversion of forest to urban developed land and those less affected. Forest land cover is an important component of the land conversion story of increased urbanization. In the past, forest often was a "leftover" part of the anthropogenic landscape or returned to forest after being used for other uses such as agriculture or mining. There is an increasing realization that forest land cover provides needed ecosystem services within and surrounding built-up areas. Increased human population and climate change impacts, both drive the need to better understand the overall, multi-scale geographic nature of such land cover change. Advances in remote sensing capabilities to produce more accurate and temporal dense land cover maps, along with the needed analysis and knowledge dissemination of what is learned from such information, will help us keep up with a dynamic world.

Supplementary Materials: The calculations and steps performed for Tables 1–3 can be found in the "LC Trends and NLCD" Excel. The steps in scaling up from the multiple NLCD Anderson II forest and developed classes to single Anderson I class each for forest and developed can be found in the "NLCD Classes to LC Trends Classes" Excel. These Excels are available online at www.mdpi.com/link.

Acknowledgments: The authors would like to thank the U.S. Geological Survey's Climate and Land Use Change, Climate and Land Use Research and Development Program and the U.S. Geological Survey's Land Change Science Program for support of this research. The authors would also like to thank James Vogelmann and Shuguang Liu, USGS Earth Resources and Observations Science Center and two anonymous reviewers for helpful comments and critiques that improved this paper.

Author Contributions: Roger Auch and George Xian conceived and designed the research, Roger Auch, Kristi Sayler, William Acevedo, and Janis Taylor analyzed the data; Roger Auch, Mark Drummond, and Janis Taylor wrote the paper.

Conflicts of Interest: The authors declare no conflict of interest. The founding sponsors had no role in the design of the study; in the collection, analyses, or interpretation of data; in the writing of the manuscript, and in the decision to publish the results".

Abbreviations

The following abbreviations are used in this manuscript:

MDPI	Multidisciplinary Digital Publishing Institute
DOAJ	Directory of open access journals
TLA	Three letter acronym
LD	linear dichroism

References

1. Sleeter, B.M.; Sohl, T.L.; Loveland, T.R.; Auch, R.F.; Acevedo, W.; Drummond, M.A.; Sayler, K.L.; Stehman, S.V. Land-cover change in the conterminous United States from 1973 to 2000. *Global Environ. Chang.* **2013**, *23*, 733–748. [CrossRef]

2. Seto, K.C.; Shepherd, J.M. Global urban land-use trends and climate impacts. *Curr. Opin. Environ. Sustain.* **2009**, *1*, 89–95. [CrossRef]

3. DeFries, R.S.; Rudel, T.; Uriart, M.; Hansen, M. Deforestation driven by urban population growth and agricultural trade in the twenty-first century. *Nat. Geosci.* **2010**, *3*, 178–181. [CrossRef]

4. Delphin, S.; Escobedo, F.J.; Abd-Elrahman, A.; Cropper, W.P. Urbanization as a land use change driver of forest ecosystem services. *Land Use Policy* **2016**, *54*, 188–199. [CrossRef]

5. Seto, K.C.; Fragkias, M.; Güneralp, B.; Reilly, M.K. A meta-analysis of global urban land expansion. *PLoS ONE* **2011**, *6*, e23777. [CrossRef] [PubMed]

6. Bounoua, L.; Zhang, P.; Mostovoy, G.; Thome, K.; Masek, J.; Imhoff, M.; Shepherd, M.; Quattrochi, D.; Santanello, J.; Silva, J. Impact of urbanization on US surface climate. *Environ. Res. Lett.* **2015**, *10*, 084010. [CrossRef]

7. Bagan, H.; Yamagata, Y. Land-cover change analysis in 50 global cities by using a combination of Landsat data and analysis of grid cells. *Environ. Res. Lett.* **2014**, *9*, 064015. [CrossRef]

8. Reinmann, A.B.; Hutyra, L.R.; Trlica, A.; Olofsson, P. Assessing the global warming potential of human settlement expansion in a mesic temperate landscape from 2005 to 2050. *Sci. Total. Environ.* **2016**, *545–546*, 512–524. [CrossRef] [PubMed]

9. Seto, K.C.; Güneralp, B.; Hutyra, L.R. Global forecasts of urban expansion to 2030 and direct impacts on biodiversity and carbon pools. *Proc. Natl. Acad. Sci. USA* **2012**, *109*, 16083–16088. [CrossRef] [PubMed]

10. Weber, T.; Sloan, A.; Wolf, J. Maryland's Green Infrastructure Assessment: Development of a comprehensive approach to land conservation. *Landsc. Urban Plan.* **2006**, *77*, 94–110. [CrossRef]

11. Gill, S.E.; Handley, J.F.; Ennos, A.R.; Pauleit, S.; Theuray, N.; Lindley, S.J. Characterising the urban environment of UK cities and towns: A template for landscape planning. *Landsc. Urban Plan.* **2008**, *87*, 210–222. [CrossRef]

12. Guo, X.; Li, W.; Da, L. Near-natural silviculture: Sustainable approach for urban renaturalization? Assessment based on 10 years recovering dynamics and eco-benefits in Shanghai. *J. Urban Plan. Dev.* **2015**, *141*. [CrossRef]

13. Salvati, L. Agro-forest landscape and the 'fringe' city: A multivariate assessment of land-use changes in a sprawling region and implications for planning. *Sci. Total. Environ.* **2014**, *490*, 715–723. [CrossRef] [PubMed]

14. Shifley, S.R.; Moser, W.K.; Nowak, D.J.; Miles, P.D.; Butler, B.J.; Aguilar, F.X.; Desantis, R.D.; Greenfield, E.J. Five anthropogenic factors that will radically alter forest conditions and management needs in the Northern United States. *For. Sci.* **2014**, *60*, 914–925. [CrossRef]

15. Millennium Ecosystem Assessment. *Ecosystems and Human Well-Being: A Framework for Assessment*; Island Press: Washington, DC, USA, 2005.

16. Ye, Y.; Zhang, J.E.; Chen, L.; Ouyang, Y.; Parajuli, P. Dynamics of ecosystem services values in response to landscape pattern changes from 1995 to 2005 in Guangzhou, Southern China. *Appl. Ecol. Environ. Res.* **2015**, *13*, 21–36.

17. Wu, Y.; Li, S.; Yu, S. Monitoring urban expansion and its effects on land use and land cover changes in Guangzhou city, China. *Environ. Monit. Assess.* **2016**, *188*, 1–15. [CrossRef] [PubMed]

18. Xian, G.; Crane, M. Assessments of urban growth in the Tampa Bay watershed using remote sensing data. *Remote. Sens. Environ.* **2005**, *97*, 203–215. [CrossRef]

19. Xian, G.; Crane, M.; McMahon, C. Quantifying multi-temporal urban development characteristics in Las Vegas from Landsat and ASTER data. *Photogramm. Eng. Remote. Sens.* **2008**, *74*, 473–481. [CrossRef]

20. Jeon, S.B.; Olofsson, P.; Woodcock, C.E. Land use change in New England: A reversal of the forest transition. *J. Land Use Sci.* **2014**, *9*, 105–130. [CrossRef]

21. Wu, Y.-J.; Thomas, V.; Oliver, R. Forest change dynamics across levels of urbanization in the eastern United States. *Southeast. Geogr.* **2014**, *54*, 406–420. [CrossRef]

22. Kennedy, R.E.; Yang, Z.; Braaten, J.; Copass, C.; Natalya, A.; Jordan, C.; Nelson, P. Attribution of disturbance change agent from Landsat time-series in support of habitat monitoring in the Puget Sound region, USA. *Remote. Sens. Environ.* **2015**, *166*, 271–285. [CrossRef]

23. Jiang, Y.; Fu, P.; Weng, Q. Assessing the impacts of urbanization-associated land use/land cover change on land surface temperature and surface moisture: A case study in the Midwestern United States. *Remote. Sens.* **2015**, *7*, 4880–4898. [CrossRef]

24. Clement, M.T.; Chi, G.; Ho, H.C. Urbanization and land-use change: A human ecology of deforestation across the United States, 2001–2006. *Sociol. Inq.* **2015**, *85*, 628–653. [CrossRef]

25. Omernik, J.M. Ecoregions of the conterminous United States. *Ann. Assoc. Am. Geogr.* **1987**, *77*, 118–125. [CrossRef]

26. Omernik, J.M.; Griffith, G.E. Ecoregions of the conterminous United States: Evolution of a hierarchical spatial framework. *Environ. Manag.* **2014**, *54*, 1249–1266. [CrossRef] [PubMed]

27. Sletter, B.M., Wilson, T.S., Acevedo, W. (Eds.) Appendix 3. In *Status and Trends of Land Change in the Western United States—1973 to 2000*; U.S. Geological Survey Professional Paper 1794-A; U.S. Geological Survey: Reston, VA, USA, 2012; p. 317.

28. U.S. Geological Survey. National Land Cover Database 2006 (NLCD2006) Product Legend. 2016. Available online: http://www.mrlc.gov/nlcd06_leg.php (accessed on 11 May 2016).

29. Loveland, T.R.; Sohl, T.L.; Stehman, S.V.; Gallant, A.L.; Sayler, K.L.; Napton, D.E. A strategy for estimating the rates of recent United States land-cover changes. *Photogramm. Eng. Remote. Sens.* **2002**, *68*, 1091–1099.

30. Auch, R.F.; Drummond, M.A.; Sayler, K.L.; Gallant, A.L.; Acevedo, W. An approach to assess land cover trends in the conterminous United States (1973–2000). In *Remote Sensing and Land Cover: Principles and Applications*; Giri, C., Ed.; Taylor and Francis CRC Press: Boca Raton, FL, USA, 2012; pp. 351–367.

31. U.S. Environmental Protection Agency. Level III Ecoregions of the Continental United States, 1999. National Health and Environmental Effects Research Laboratory, Scale 1:7,500,000 Map. 2015. Available online: ftp://ftp.epa.gov/wed/ecoregions/usgs/useco_March1999_v5.pdf (accessed on 26 March 2016).

32. Anderson, J.R.; Hardy, E.E.; Roach, J.T.; Witmer, R.E. *A Land use and Land Cover Classification System for Use with Remote Sensor Data*; U.S. Government Printing Office: Washington, DC, USA, 1976; p. 41.

33. Homer, C.G.; Dewitz, J.; Yang, L.; Jin, S.; Danielson, P.; Xian, G.; Coulston, J.; Herold, N.; Wickham, J.; Megown, K. Completion of the 2001 National Land Cover Database for the conterminous United States. *Photogramm. Eng. Remote. Sens.* **2007**, *73*, 337–341.

34. Vogelmann, J.E.; Howard, S.M.; Yang, L.; Larson, C.R.; Wylie, B.K.; van Driel, N. Completion of the 1990s National Land Cover Data Set for the conterminous United States from Landsat Thematic Mapper data and ancillary data sources. *Photogramm. Eng. Remote. Sens.* **2001**, *67*, 650–662.

35. Wickham, J.D.; Stehman, S.V.; Gass, L.; Dewitz, J.; Fry, J.A.; Wade, T.G. Accuracy assessment of NLCD 2006 land cover and impervious surface. *Remote. Sens. Environ.* **2013**, *130*, 294–304. [CrossRef]

36. Fry, J.; Xian, G.; Jin, S.; Dewitz, J.; Homer, C.; Yang, L.; Barnes, C.; Herold, N.; Wickham, J. Completion of the 2006 National Land Cover Database for the conterminous United States. *Photogramm. Eng. Remote. Sens.* **2011**, *77*, 858–864.

37. Homer, C.; Dewitz, J.; Yang, L.; Jin, S.; Danielson, P.; Xian, G.; Coulston, J.; Herold, N.; Wickham, J.; Megown, K. Completion of the 2011 National Land Cover Database for the conterminous United States—A decade of land cover change information. *Photogramm. Eng. Remote. Sens.* **2015**, *81*, 345–354.

38. Fry, J.; Coan, M.; Homer, C.; Meyer, D.; Wickham, J. Completion of the National Land Cover Database (NLCD) 1992–2001 Land Cover Change Retrofit Product. U.S. Geological Survey Open-File Report 2008-1379. Available online: http://pubs.usgs.gov/of/2008/1379/ (accessed on 17 May 2016).

39. Hollister, J.W.; Gonzalez, M.L.; Paul, J.P.; August, P.V.; Copeland, J.L. Assessing the accuracy of national land cover dataset area estimates at multiple spatial extents. *Photogramm. Eng. Remote. Sens.* **2004**, *70*, 405–414. [CrossRef]

40. Stehman, S.V.; Sohl, T.L.; Loveland, T.R. Statistical sampling to characterize recent United States land-cover change. *Remote. Sens. Environ.* **2003**, *86*, 517–529. [CrossRef]

41. Griffith, J.A.; Stehman, S.V.; Sohl, T.L.; Loveland, T.R. Detecting trends in landscape pattern metrics over a 20-year period using a sampling-based monitoring programme. *Int. J. Remote. Sens.* **2003**, *24*, 175–181. [CrossRef]

42. Stehman, S.V.; Sohl, T.L.; Loveland, T.R. Statistical sampling to characterize recent United States land-cover change. *Int. J. Remote. Sens.* **2005**, *26*, 4941–4957. [CrossRef]

43. Sleeter, B.M. Late 20th century land change in the Central California Valley Ecoregion. *Calif. Geogr.* **2008**, *48*, 27–60.

44. Drummond, M.A.; Loveland, T.R. Land-use pressure and a transition to forest-cover loss in the eastern United States. *BioScience* **2010**, *60*, 286–298. [CrossRef]

45. Napton, D.E.; Auch, R.F.; Headley, R.; Taylor, J.L. Land changes and their driving forces in the south eastern United States. *Reg. Environ. Chang.* **2010**, *10*, 37–53. [CrossRef]

46. Auch, R.F.; Sayler, K.L.; Napton, D.E.; Taylor, J.L.; Brooks, M.S. Ecoregional differences in late-20th-century land-use and land-cover change in the U.S. northern great plains. *Great Plains Res.* **2011**, *21*, 231–243.

47. Auch, R.F.; Napton, D.E.; Kambly, S.; Moreland, T.R., Jr.; Sayler, K.L. The driving forces of land change in the northern piedmont of the United States. *Geogr. Rev.* **2012**, *102*, 53–75. [CrossRef]

48. Soulard, C.E.; Sletter, B.M. Late twentieth century land-cover change in the basin and range ecoregions of the United States. *Reg. Environ. Chang.* **2012**, *12*, 813–823. [CrossRef]

49. Drummond, M.A.; Auch, R.F.; Karstensen, K.A.; Sayler, K.L.; Taylor, J.L.; Loveland, T.R. Land change variability and human-environment dynamics in the United States Great Plains. *Land Use Policy* **2012**, *29*, 710–723. [CrossRef]

50. Sohl, T.L.; Sohl, L.B. Land-sue change in the Atlantic coastal pine barrens ecoregion. *Geogr. Rev.* **2012**, *102*, 180–201. [CrossRef]

51. Soulard, C.E.; Wilson, T.S. Recent land-use/land-cover change in the Central California Valley. *J. Land Use Sci.* **2015**, *10*, 59–80. [CrossRef]

52. Auch, R.F.; Napton, D.E.; Sayler, K.L.; Drummond, M.A.; Kambly, S.; Sorenson, D.G. The Southern Piedmont's continued land-use evolution, 1973–2011. *Southeast. Geogr.* **2015**, *55*, 338–361. [CrossRef]

53. Wickham, J.D.; Riitters, K.H.; Wade, T.G.; Coan, M.; Homer, C. The effect of Appalachian mountaintop mining on interior forest. *Landsc. Ecol.* **2007**, *22*, 179–187. [CrossRef]

54. Xian, G.; Homer, C.; Bunde, B.; Danielson, P.; Dewitz, J.; Fry, J.; Pu, R. Quantifying urban land cover change between 2001 and 2006 in the Gulf of Mexico region. *Geocarto Int.* **2012**, *27*, 479–497. [CrossRef]

55. Gottmann, J. *Megalopolis: The Urbanized Northeastern Seaboard of the United States*; Twentieth Century Fund: New York, NY, USA, 1961; p. 820.

56. Hart, J.F.; Morgan, J.T. Spersopolis. *Southeast. Geogr.* **1995**, *35*, 103–117. [CrossRef]

57. Terando, A.J.; Costanza, J.; Belyea, C.; Dunn, R.R.; McKerrow, A.; Collazo, J.A. The southern megalopolis: Using the past to predict the future urban sprawl in the Southeast U.S. *PLoS ONE* **2014**, *9*, e102261. [CrossRef] [PubMed]

58. Sohl, T.L.; Sleeter, B.M.; Zhu, Z.; Sayler, K.L.; Bennett, S.; Bouchard, M.; Reker, R.; Hawbaker, T.; Wein, A.; Liu, S.; et al. A land-use and land-cover modeling strategy to support a national assessment of carbon stocks and fluxes. *Appl. Geogr.* **2012**, *23*, 111–124. [CrossRef]

59. von Hoffman, A.; Fleckner, J. *The Historical Origins and Causes of Urban Decentralization in the United States*; Joint Center for Housing Studies, Harvard University: Cambridge, MA, USA, 2001; p. 35.

60. Auch, R.F.; Acevedo, W.; Taylor, J.L. The historical development of the Nation's urban areas. *Rates, Trends, Causes, and Consequences of Urban Land-Use Change in the United States*; Acevedo, W., Taylor, J.L., Hester, D.J., Mladinich, C.S., Glavac, S., Eds.; U.S. Government Printing Office: Washington, DC, USA, 2006; pp. 1–12.

61. Drummond, M.A.; Stier, M.P.; Auch, R.F.; Taylor, A.; Griffith, G.E.; Hester, D.J.; Riegle, J.L.; Soulard, C.E.; Mcbeth, J.L. Assessing landscape change and processes of recurrence, replacement, and recovery in the Southeastern Coastal Plains, USA. *Environ. Manag.* **2015**, *56*, 1252–1271. [CrossRef] [PubMed]

62. Auch, R.F. Northeastern Coastal Zone. U.S. Geological Survey Land Cover Trends project web site. 2006. Available online: http://landcovertrends.usgs.gov/east/eco59Report.html (accessed on 26 March 2016).

forests

MDPI

Article

Urban Forest Indicators for Planning and Designing Future Forests

Sara Barron [1],*, Stephen R.J. Sheppard [1] and Patrick M. Condon [2]

[1]   Forest Resources Management, Faculty of Forestry, University of British Columbia, Vancouver, BC V6T 1Z4, Canada; stephen.sheppard@ubc.ca
[2]   School of Architecture and Landscape Architecture, Faculty of Applied Science, University of British Columbia, Vancouver, BC V6T 1Z4, Canada; pcondon@sala.ubc.ca
*   Correspondence: sara.fryer.barron@gmail.com; Tel.: +1-604-763-7728

Academic Editors: Francisco Escobedo, Stephen John Livesley and Justin Morgenroth
Received: 8 May 2016; Accepted: 9 September 2016; Published: 16 September 2016

Abstract: This paper describes a research project exploring future urban forests. This study uses a Delphi approach to develop a set of key indicators for healthy, resilient urban forests. Two groups of experts participated in the Delphi survey: International academics and local practitioners. The results of the Delphi indicate that "urban tree diversity" and "physical access to nature" are indicators of high importance. "Tree risk" and "energy conservation" were rated as indicators of relatively low importance. Results revealed some differences between academics and practitioners in terms of their rating of the indicators. The research shows that some indicators rated as high importance are not necessarily the ones measured or promoted by many municipal urban forestry programs. In particular, social indicators of human health and well-being were rated highly by participants, but not routinely measured by urban forestry programs.

Keywords: indicators; urban forest; Delphi; urban design; green infrastructure; climate change; ecosystem services

1. Introduction

Our world is rapidly urbanizing, while, at the same time, facing uncertain futures. If we want our cities to continue to be livable, dynamic places for the world's citizens, we must plan ahead for climate change. One important aspect of a city's functioning is in its urban forest. Urban forests provide a wide range of benefits, from ecosystem services [1–3], climate change adaptation [4,5] and climate change mitigation [6–8], to improvement in human health and well-being [9–11] Unfortunately, in many places in the world, urban forests are rapidly being lost to create residential spaces for the global migration to cities [12,13]. Urban forests are also threatened by changing climate, including spreading pests and diseases, changes to precipitation, and increased storm events.

In order to address these challenges, many cities are implementing tree-planting programs to increase the urban forest [14,15]. While these programs have excellent intentions, many lack an overall vision about the values to be achieved by the tree planting, and the urban forest design best fit to achieve these values. A dichotomy also exists between what cities strive for, and what they monitor as key indicators. For example, there is increasing awareness of the role of urban forests in enriching human health and well-being [9–11], but none of the indicator sets or management parameters surveyed included a measure of how trees can be planted to enhance their effects on human health.

The main aim of this study is to develop a decision support framework, centered on a set of key indicators, which can be used to build and test various scenarios of future urban forests. A decision support framework for urban forest design is proposed to help cities achieve their future forest vision.

A decision support framework translates "big picture" thinking at the vision level into values, goals, indicators, and targets, which will lead to specific strategies and actions to enable implementation [16]. A framework creates a transparent linkage between practical strategies and actions on the ground and the goals and objectives inspiring them.

The purpose of the Delphi research was to select a small, but broad set of key urban forest indicators to drive the design and measurement of future forest scenarios. Criteria and indicators are a commonly used tool in the profession of forestry [17–19] and urban design [20]. There have been a few studies looking at indicator sets for urban forestry [21–26]. None of the existing indicator sets captured the full range of urban forest benefits required for this study. This study is focused on urban forests in the context of climate change, and looks at issues of human health and well-being. Indicators addressing these issues were not in most indicator sets reviewed. Additionally, it is hoped that many communities implementing urban tree planting programs could use the indicators selected. This indicator set could provide communities with limited resources a simple but fairly comprehensive framework for planning future urban forests.

Indicators are a common tool used to measure progress towards an objective. For example, the use of criteria and indicator sets is well established in sustainable forest management plans [17–19] and literature exists on what constitutes a good indicator [17]. According to their literature review, Harshaw et al. [17] list the characteristics of a good indicator as:

- Relevant
- Credible
- Measurable
- Cost-effective
- Connected to [urban] forestry

These characteristics were used to evaluate the initial list of indicators for this project.

While indicators are primarily used to measure performance, Kellett [20] proposes "indicators are also very useful, arguably more useful, in community planning and design to direct alternatives and choices toward targets when opportunity is far greater to modify direction or approach" [20]. He differentiates between "enabling" indicators that can direct design from "performance" indicators that simply measure the results of design [20]. For example, tree canopy cover is an "enabling" indicator because it can be used to direct design decisions. A community could use a canopy cover target of thirty percent to guide design, while also using the indicator to measure the resulting forest to see if this cover was achieved. A performance indicator, on the other hand, can only be used once the design is in place. An example of a performance indicator could be "tree health". This indicator is important, but difficult to guide design decisions. For the purposes of this study, "enabling" indicators will be used that can both direct design choices for a series of alternative futures, as well as measure the results of these futures once they are modeled. Dobbs et al. argue that the urban forest structure determines its ability to provide ecosystem services [21]. Enabling indicators are those that determine the structure of the urban forest. No literature reviewed to date has proposed using indicators as a design input for visioning future forests.

There have been a few recent attempts to create indicator sets for urban forest resources [22–24]. Current indicator sets proposed for urban forestry have shortcomings. First, there is a lack of uptake for the indicators sets. Kenney et al. [22] noted that Clark et al.'s indicator set [23], while comprehensive, has had little uptake by municipalities in the decade since being proposed. If municipalities aren't using the indicators, this could show they aren't measuring the issues urban foresters are most interested in. While most cities have a set of management parameters, a study by Östberg et al. [24] found few parameters that were used by multiple cities. This lack of standardization between cities makes comparison of performance between cities difficult. For example, some cities may collect tree diversity data, while others may collect data about habitat areas. Many collect some data about tree cover, but even this measure is not easily compared between cities, as the definition of what constitutes a "city" is not standardized. Some cities include the suburban periphery, while others may focus on the downtown core. It is hoped that a shorter list of easily measurable indicators could provide a first step towards building common basis for cities to begin measuring and comparing urban forest assets.

2. Materials and Methods

This research used an email Delphi approach to solicit feedback from experts in the field of urban forestry. Participants were asked to rate indicators that could be used to design and measure urban forests, and then comment on a decision support framework for future urban forests. The goal for this study was to evaluate a set of key urban forest indicators.

2.1. Delphi Approach

The Delphi approach is a well-established research method that asks experts to lend their opinion to a structured problem in order to develop informed solutions [27,28]. The method provides structured feedback to participants after each round in order to facilitate an anonymous dialogue about the subject matter. It has been used once before to solicit expert feedback on parameters for urban forest inventories [24].

The Delphi method has known weaknesses [27]. These include: Its reliance on a careful selection of expert participants, the lengthy time required for a multiple round survey, participant attrition, ease of ability to be manipulated by the facilitator, an inability to handle discord easily, and limited interaction between participants. This study used a number of methods to help address these weaknesses. Academic participants were carefully screened based on relevant publication history. Local practitioners were selected based on their management positions within local governments. The study attempted to reduce time between rounds, but slow response rates were a problem. Participants were lost between the first and second rounds. This study used a combination of quantitative and qualitative feedback to help clearly communicate feedback received in order to increase transparency and reduce the possibility of facilitator manipulation.

Using the Delphi method, participants worked iteratively to select a small set of key indicators using both a likert scale for rating indicators and qualitative feedback. Qualitative feedback was encouraged, as suggested by Landeta [27] to get a better understanding of participant choices. The feedback was summarized and returned to experts in the second round of the Delphi. It was also used to modify the rationales for each indicator.

2.2. Email Delphi Method

This study used electronic mail to contact participants and distribute the survey. Before email, researchers undertaking a Delphi study used posted mail surveys to contact participants and collect feedback [28]. Email provides a faster method of communication than regular mail, but recent developments in social media and other online platforms could provide an even more user-friendly approach. Participant attrition was an issue in this study, and an exploration of other methods of delivering a Delphi survey is recommended.

2.3. Indicator Selection

Indicators were developed from relevant literature that represented a wide range of economic, environmental and social issues [21–26]. The indicators included both quantitative and qualitative issues (see Appendix Table A1). Indicators were reduced to those indicators that could function both as design drivers ("enabling indicators") and as measurement tools for the resulting design (Table 1). The urban forestry literature also underrepresents social values of urban forest measures [29]. The indicator sets reviewed contained few social indicators.

New indicators were created to capture social values of human health and well-being. Urban forests' contribution to human health and well-being has been the subject of much recent research [9–11], but the indicator sets reviewed did not sufficiently capture this dimension.

Dobbs et al. [21] did address human health and well-being, acknowledging "that quantifying the relationship between the urban forest and human well-being in terms of psychological and social values is critical in assessing ESG [Ecosystem Services and Goods]" [21] (p. 199). Their research focused on regulation indicators, such as air quality services of urban forests, in more detail than other dimensions of human health and well-being. This research builds on their indicator list to include two additional indicators that attempt to capture recreation and psychological benefits of urban forests.

Two indicators were introduced to test whether the experts agreed that these were important aspects of urban forestry. In order to fit within the study's requirement for indicators that could be used to direct design decisions, the indicators introduced had to be spatially explicit. The first new indicator: "physical access to nature", captures the equitable distribution of urban forest assets within a community. This indicator relates to the literature examining the human health benefits of additional physical activity linked to increased greenspace within a community [30–33]. Through these studies, access to nature has been tied to increased recreation, particularly walking, which is beneficial to both physical and mental health of residents.

The next indicator: "visual access to nature", attempts to capture the psychological health benefits of natural views from residences and workplaces. Views of nature have long been recognized as having positive benefits for human health and well-being [10,34]. Increasing visual access to nature for community residents could work towards lowering stress levels and improving mental wellness.

Indicators relating to the management of urban forests were well represented in existing indicators sets, but are not within the scope of this research project. This project recognizes that they are a critical contribution to the future of urban forests, but not within the scope of this research.

The reduced list was then sent to selected academic experts and local practitioners for the first round of the Delphi questioning. Participants were provided with identical questionnaires listing twelve indicators. Each indicator had a brief description, a rationale statement, and a likert rating scale. The likert scale asked participants to rate each indicator in terms of low importance (1) to high importance (5). Participants were also given space and encouraged to propose additional indicators they felt were missing from the list.

2.4. Expert Selection

The opinions of international and local expertise were sought. Linstone suggests a suitable minimum panel size of seven [28], which fit with the scope of this study. The targets were eight participants from each group.

2.4.1. Local Practitioners

The region of Metro Vancouver, in British Columbia, Canada, was chosen as the location for testing local practitioner preferences. Metro Vancouver faces all of the challenges outlined in the introduction of this paper, and it hosts a wide range of local governments, each with a unique approach to managing their urban forest [14]. The 13 local municipalities within the Metro Vancouver region with populations over 50,000 people were contacted. Emails were sent to either the identified urban forestry manager, parks manager, or to the general urban forest email if one was listed on the website. From the first email, names of appropriate local experts were suggested, and finally 18 local practitioners working in municipalities were asked to participate in the study. Ten of these expressed willingness to participate, and seven returned their first round survey. Only four local practitioners returned the second round survey. Two rounds of reminders were sent to encourage greater survey response.

2.4.2. Academics

Nineteen academics were identified and contacted for willingness to participate in the study. Academics were identified based on publication of relevant urban forestry research, including social science research. They were also selected based on geographical distribution. Academics from all continents were contacted, with the exception of Antarctica. Two academics suggested alternate names to stand in for them because of time constraints. Twelve academics expressed interest in participating in the study, and nine returned the first round survey. These nine represented eight countries from five continents. Six returned the second round survey. Again, two rounds of reminders were sent to encourage greater survey response.

3. Results

Few participants provided comments on the decision support framework, as the key focus of the exercise was to rate a list of indicators. Those that did comment questioned the terminology. Definitions of words such as: Goals, objectives, and values, have different implications within different fields of study. The discipline of Forestry often uses "criteria and indicators" to develop and measure scenarios. Social science research often discusses "values". Following the feedback, the decision support framework was modified slightly.

See Table 2 for decision support framework.

3.1. Delphi Round One

Table 1 shows the mean likert scale results of the first round with all participants combined. "Urban tree diversity" and "physical access to nature" were rated as the most important indicators. "Tree risk" and "energy conservation" were rated as the indicators with lowest importance. The order of the indicators within the survey was not changed, but seemed to have little effect on the ranking. The highest rated indicators were first and ninth within the survey, while the lowest ranked were fourth and sixth.

Table 1. Mean rankings and frequency of rating, all participants ($n = 15$).

Indicator	Rating	1	2	3	4	5
Urban Tree Diversity	4.9				2	13
Physical Access to Nature	4.9				2	13
Canopy Cover	4.3			1	8	6
Stormwater Control	4.2		1	3	3	8
Visual Access to Nature	4.1	1		1	7	6
Habitat Provision	4.1		1	1	9	4
Air Quality Improvement	4.0		2	1	7	5
Available Growing Space	3.8	3		1	4	7
Greenhouse Gas Sequestration and Storage	3.6	1	1	5	4	4
Energy Conservation	3.4	1	2	5	4	3
Property Value Benefits	3.3	1	1	9	1	3
Tree Risk	3.1	2	2	6	2	3

Generally, the mean ranking of indicators follows the consensus of the group. As seen in Table 1, indicators with high consensus (fewer different responses) are highest on the list. One notable exception to this rule was the "visual access to nature" indicator. One participant rated this 1: Low importance, one rated it 3, and the remaining participants rated it 4 or 5.

3.2. Comments

Participant comments are presented in the following section. They are presented by indicator in the order that all participants ranked the indicators. For each indicator, the rationale statement sent to participants is provided, as well as the revised statement following the Delphi, to give a better understanding of each indicator.

3.2.1. Urban Tree Diversity

Rationale statement provided in Round One: A diverse urban forest increases the ability of the forest to withstand change. Trees should be of diverse ages, species, genera, and families in order to ensure the forest can adapt to future climate change scenarios.

Revised rationale statement following Delphi: A diverse urban forest increases the ability of the forest to withstand change, which is of key importance to the long-term stability of the forest. Diversity is also influential on psycho-social outcomes of urban forests. Trees should be of diverse sizes, ages, species, genera, and families in order to ensure the forest can adapt to future climate change scenarios. Public opinion about desirable tree sizes/types/forms can help inform the creation of a diverse urban forest. This opinion will vary between communities.

In the participant comments, "urban tree diversity" was connected to climate change adaptation, ecosystem service provision, long term planning, decision making, psycho-social outcomes, cultural values, forest stability, and forest resilience. Participants commented that this was a useful indicator linked to many different aspects of urban forest planning. Good diversity is important to long term stability, resiliency, and is an important base for a range of ecosystem services. One participant noted that it "should be balanced with species selection for cohesive and consistent streetscapes" (Participant One). Another noted that companion plants should be included in the diversity discussion.

3.2.2. Physical Access to Nature

Rationale statement provided in Round One: Access to nature has been tied to increased recreation, particularly walking, which is beneficial to both physical and mental health of residents. It could also promote more sustainable commuting, as residents are more likely to walk, jog, or cycle to work along aesthetically pleasing routes. Ensuring equal access to nature for all residents within a community promotes greater equality.

Revised rationale statement following Delphi: Access to nature has been tied to increased recreation, particularly walking, which is beneficial to both physical and mental health of residents. "Play in nature" is very important for people to gain connection to nature and urban forests. It could also promote more sustainable commuting, as residents are more likely to walk, jog, or cycle to work along aesthetically pleasing routes. Ensuring equal access to nature for all residents within a community promotes greater equality. Urban forests should also provide a diversity of potential uses.

Participants commented that "physical access to nature" was an important indicator linking urban forests to human health. They noted that this indicator could start to address issues such as equitable distribution of green spaces and improved population health. Safety, structure, size, and accessibility of greenspace were noted as important considerations when using physical access as a measure of urban forest success. It was also noted that the type of community (dense urban vs. sprawling suburban) could change the way this indicator was measured.

3.2.3. Canopy Cover

Rationale statement provided in Round One: Canopy cover is a very common metric used to evaluate a city's urban forest. It is relatively easy to measure and communicate with the general public.

Revised rationale statement following Delphi: "Canopy cover" is a very common metric used to evaluate a city's urban forest. It is relatively easy to measure and communicate with the general public and a good starting point for quantifying a community's urban forest. "Canopy volume" estimates total leaf area of a tree's canopy, which provides more information about a tree's overall ecosystem service provision. Communities with sufficient means are encouraged to measure "canopy volume", as well as "canopy cover".

"Canopy cover" is an important indicator that is becoming commonly used. It was highly rated by both academics and practitioners. The ease of using and communicating this indicator was noted by participants as reasons to continue using "canopy cover" as an urban forest indicator.

Another was the ability to communicate long-term trends using "canopy cover". One participant noted the use of "canopy cover" in communicating the "borderless nature of the urban forest to the public and decision-makers" (Participant One). Some participants noted that with emerging three-dimensional technologies, such as LiDAR, canopy cover could eventually be measured in the third dimension as canopy volume, which would capture a more robust measurement of the urban forest in a community. One participant noted that "while canopy cover is an important measure, it does not tell as complete a story as canopy volume does in terms of the overall ecosystem services that trees provide" (Participant Two).

3.2.4. Stormwater Control

Rationale statement provided in Round One: Trees filter and infiltrate storm water, cleaning and moderating the amount of water running into engineered systems. If designed and planned with this function in mind, urban forests can provide both cost savings and enhanced environmental benefits for urban areas.

The rationale statement was not revised, as participants generally agreed with the statement provided.

"Stormwater control" was rated as an important indicator for measuring urban forests. Participants commented that this indicator was "widely recognized as one of the most important services provided by the urban forest" (Participant Three). Another mentioned that it was "an extremely powerful tool for raising support for green infrastructure" (Participant One). There was disagreement about the ease of measuring this indicator, though many participants noted that tools, such as iTree measure the stormwater control benefits of urban forests.

3.2.5. Visual Access to Nature

Rationale statement provided in Round One: There is growing evidence that access to nature, even when viewed through a window, is beneficial to well being. Increasing visual access to nature for community residents could work towards lowering stress levels and improving mental wellness.

The rationale statement was not revised, as participants generally agreed with the statement provided.

Participants commented that "visual access to nature" was an important indicator for connecting populations to nature. It was ranked highly by most participants, but one participant gave it a low importance rating on the scale. There was concern that the actual content of the view, and what constitutes "natural" would be difficult to define. There was consensus that natural should include trees, not just anything that is green, because trees "give us a vital third dimension (height) to our experience and interaction with them" (Participant Four). This indicator had the most comments and suggestions about possible measurement techniques.

3.2.6. Habitat Provision

Rationale statement provided in Round One: Urban forests can help protect biodiversity and provide habitat for urban flora and fauna. Different types of urban forest provide different amounts and quality of habitat. For example, a naturalized park would likely support more urban nature than a concrete planter holding a non-native species.

Revised rationale statement following Delphi: Urban forests can help protect biodiversity and provide habitat for urban flora and fauna. Quantity and quality of habitat varies in urban forests. For example, a large park with diverse trees and understory plantings would likely support more urban nature than a concrete planter holding a non-native species.

While some participants noted this was an important indicator, most were concerned about the difficulty of defining "habitat" and measuring success. One participant noted that, "it is a mistake to concentrate just on native species, particularly in the light of the pests and diseases issue and the changes that have already contributed to the urban heat island—changes that are only going to get worse. We need a resilient urban forest, comprising many species, as they have many roles to fulfill as well as 'nature'" (Participant Five). Other participants also linked this indicator to the "urban forest diversity" indicator.

3.2.7. Air Quality Improvement

Rationale statement provided in Round One: Research shows that the presence of trees is generally beneficial for the air quality and human health in an area. Trees absorb a variety of air pollutants in varying amounts, depending on a number of characteristics such as species, age, location, tree health, etc.

The rationale statement was not revised because this indicator was dropped from the list after the first round.

There was less consensus on the merits of measuring air quality improvement benefits of urban forests. Some participants noted that, "the air pollution reduction benefits of urban trees are under debate" (Participant Three). It was agreed that poor urban air quality is an important health concern, but that measuring the urban forests' contribution to this was difficult and would likely be relatively small. A few participants noted that pollen and VOC production would have to be taken into account when measuring a forest's net air quality improvement. It was also noted that this concept was not well understood or appreciated by the general public.

3.2.8. Available Growing Space

Rationale statement provided in Round One: The amount of available growing space indicates the potential of a community to increase and maintain their urban forest. Without space and suitable soil, the urban forest will be difficult to expand and manage.

Revised rationale statement following Delphi: The amount of available growing space indicates the ability of a community to increase and maintain their urban forest. Without sufficient soil volume and quality, the urban forest will be difficult to expand and manage. Focusing efforts on providing adequate space for trees, and planting the correct tree in the available space, will result in substantially reduced costs associated with maintenance (pruning) and infrastructure damage caused by trunk buttress flare and root expansion. The challenge is for communities to prioritize and plan for adequate growing space and soil volumes well in advance.

"Available growing space" was connected to tree health, tree canopy size, planning, and permeability. Participants commented that "space" should be measured in three dimensions, because adequate soil volume was of key importance to forest health. Along with property value benefits, this indicator generated the most comments, with every participant writing an opinion on the challenges in using this indicator. Many noted that it was an important indicator or priority during development of a new project or area, and that is was often overlooked. Participants also noted the need to consider public versus private landownership when looking at "available growing space".

3.2.9. Greenhouse Gas Sequestration and Storage

Rationale statement provided in Round One: Greenhouse gas sequestration and storage measures the amount of carbon dioxide absorbed and stored by trees and within the soil of urban forests. Carbon dioxide is the most abundant greenhouse gas derived from fossil fuels. The amount of carbon absorbed and stored reflects the contribution of urban forests to mitigating climate change. This indicator proposes bundling the two values to convey the full impact of trees on carbon mitigation.

The rationale statement was not revised, as participants generally agreed with the statement provided.

While climate change was acknowledged as an important concern, there was less consensus about the amount of carbon sequestered and stored by urban forests, and the value urban citizens place on this issue. Some participants suggested that other greenhouse gases, such as ozone and methane, be included. Others commented that inclusion of other components, such as urban soils, was important.

One participant suggested that "the only reason to make carbon calculations might be to choose species and spacings wisely, and to convey to the public that trees are exceptional agents of climate-change mitigation" (Participant Four). Another participant noted that the indicator "can be a part of a good case for advocacy for funding urban forest initiatives" (Participant Six).

3.2.10. Energy Conservation

Rationale statement provided in Round One: Energy conservation measures the contribution of the urban forest to reducing a community's energy use. This could be a reduction in building energy use through shading during hot summers.

The rationale statement was not revised because this indicator was dropped from the list after the first round.

Participants appreciated the energy conservation benefits of urban forests, but there was debate about how the scale of measuring this indicator might fit with the other urban forest indicators. Interventions and detailed measurements would make more sense on a site-specific scale, not at the scale of the entire urban forest. As one participant noted, "while I rate it low, it's possible that the public would really appreciate this as an indicator. It's one of the few indicators where the individual can feel they're making a measurable difference" (Participant Seven). A few different participants noted this personal benefit. It was also noted that the type of energy savings would differ globally, and that design opportunities would be regionally based.

3.2.11. Property Value Benefits

Rationale statement provided in Round One: Trees can contribute to an individual's economic well-being if they increase property value. Economic indicators such as this can be easy to communicate with residents.

The rationale statement was not revised because this indicator was dropped from the list after the first round.

Property value was included as an indicator for two reasons. The first is that it provided a clear economic indicator missing from the list. The second reason is to acknowledge the history of research surrounding urban forests and property value [1]. While difficult to rationalize as an enabling indicator, it could provide some guidance to designers about tree planting locations to promote more equitable distribution of urban forest resources.

"Property value benefits" was another indicator where scale was an issue noted by participants. At a finer scale, issues, such as tree maintenance, hazard trees, and other undesirable characteristics, would be important considerations. Many pointed to ethical concerns about an indicator that favors property owners, and single-family residences specifically. As one participant pointed out "higher prices mean that certain parts of the population are 'frozen out' and have less access to the urban forest" (Participant Three). Another pointed out that those living in high-rise towers would see little property value benefits in adjacent urban forests.

3.2.12. Tree Risk

Rationale statement provided in Round One: Often residents' concerns about urban forests stem from fears of potential damage to people, structures or utilities due to tree or limb fall during storms or from disease. Urban forests with lower risk due to healthy trees planted in appropriate locations might be more compatible with community members. Tree age by species should be considered when factoring risk for trees.

The rationale statement was not revised because this indicator was dropped from the list after the first round.

"Tree risk" was ranked as the least important indicator in the first round. While participants noted that tree risk can be a large part of the urban forestry discourse, and was important from a management perspective, they pointed out that perception of risk and actual risk were often at odds.

One participant pointed out that it was time consuming to measure, and "would be more trouble and cost than it would be worth" (Participant Eight). One participant argued that it is equally important to communicate the negative side of urban forestry, so the "tree risk" indicator helps achieve this.

3.3. Academic vs. Practitioner

The following section describes the differences between the academic and practitioner groups in round one of the Delphi.

The academic participants rated "urban tree diversity" and "access to nature" as top indicators (see Table 2). It is interesting to see social indicators connecting urban forests to their residents ranked as top indicators. These indicators were not specifically mentioned in previous urban forestry indicator sets. Practitioners placed less importance on "visual access to nature", but did rank "physical access to nature" as the highest priority, with every practitioner giving it a rating of five.

Table 2. Academic and practitioner ratings.

Indicator	Academic Rating	Practitioner Rating
Urban Tree Diversity	4.9	4.9
Physical Access to Nature	4.8	5.0
Visual Access to Nature	4.4	3.9
Canopy Cover	4.3	4.4
Stormwater Control	4.1	4.3
Habitat Provision	3.9	4.3
Available Growing Space	3.6	4.0
Energy Conservation	3.5	3.3
Air Quality Improvement	3.5	4.6
Tree Risk	3.1	3.1
Greenhouse Gas Sequestration and Storage	3.1	4.1
Property Value Benefits	2.8	3.9

Practitioners ranked "air quality improvement" as more important than academics. The academics that gave this indicator a low ranking cited recent research that found urban forests had a relatively minor impact on air quality improvement. This could demonstrate a lag time of results from published academic studies being shared amongst professionals working in the field. The survey had similar results for "greenhouse gas sequestration and storage", which was ranked seventh in overall importance by practitioners and eleventh by academics.

The major difference in comments between the groups was the regional focus of the practitioners. They connected each indicator with regional urban forestry issues. For example, many practitioners mentioned evergreen conifers as important storm water management trees in a region where most storm water falls during times when deciduous trees have lost their leaves.

3.4. Comments/Suggestions for Additional Indicators

Most suggestions for additional indicators included management and public perception/awareness indicators. This is a noted, though deliberate, shortcoming of this particular project, which is focused on indicators that can be used to drive design decisions for urban forests.

Suggested indicators included:

- Safety and Security
- Spirituality
- Sense of Place
- Products Derived from the Urban Forest
- Urban Forest Management
- Public Support for the Urban Forest

- Land Tenure
- Presence of Invasive Species
- Presence of Beneficial Pollinators
- Presence of Urban Wildlife

The first three indicators are specifically social issues that attempt to capture a dimension of urban forestry often missing in the literature. Although none of these were rated above the top indicators selected in the first round, they did receive mostly positive qualitative responses from participants during the second round. Each had a wide range of responses indicating a lack of consensus for these indicators. Each of these issues seemed more regionally appropriate, and could be used as a subset of indicators for specific locations.

3.5. Delphi Round Two

3.5.1. Academic

Academic participants were sent a new survey with the top seven indicators chosen by academics in round one and a list of indicators proposed by academic participants during round one. Revisions were made to the description and rationale statement for each indicator. Participants were asked to comment on these and to comment on metrics suggested for each indicator during round one.

The only change in indicator ranking from academic participants was the switching of canopy cover and visual access to nature (see Table 3). The rest of the indicators were ranked in the same order, so the survey reached consensus on their relative importance.

Table 3. Academic consensus after Round Two.

Indicator	Round Two Score	Round One Rank
Urban Tree Diversity	5.0	1
Physical Access to Nature	4.8	2
Canopy Cover	4.3	4
Visual Access to Nature	4.0	3
Stormwater Control	3.8	5
Habitat Provision	3.8	6
Available Growing Space	3.2	7

3.5.2. Practitioner

Practitioner participants were sent a new survey with the top seven indicators chosen in round one and a list of indicators proposed by practitioner participants during round one. Revisions were made to the description and rationale statement for each indicator. Participants were asked to comment on these and to comment on metrics suggested for each indicator during round one. There was a very low return rate on the practitioner responses; results are presented (see Table 4), but do not represent a group consensus.

Table 4. Practitioner results after Round Two.

Indicator	Round Two Score	Round One Rank
Physical Access to Nature	5.0	1
Urban Tree Diversity	5.0	2
Stormwater Control	5.0	5
Air Quality Improvement	4.5	3
Habitat Provision	4.0	6
Greenhouse Gas Sequestration & Storage	4.0	7
Available Growing Space	4.0	8
Canopy Cover	3.5	4

4. Discussion

Urban forests are valued for a range of reasons that are not often captured in a community's design, planning, and management of their forest. This research uses indicators within a decision support framework to create a more comprehensive approach to guide design and planning for future urban forests. As cities are increasingly acknowledging the important role of green infrastructure and natural systems in improving livability and viability, these indicators could be ubiquitous in planning approaches for greening global cities.

The study shows that some indicators ranked as high importance are not necessarily the ones measured or promoted by many urban forestry programs. As Kenney et al. [22] point out, simply using absolute canopy cover targets to guide urban forest management goals "does not provide a comprehensive assessment of urban forest stewardship in a community and does not account for an area's potential to support a forest canopy" [22] (p. 108). At the same time, an overly complex set of indicators is overwhelming for most communities to measure and communicate with the public. This study has created a basic set of indicators that captures a range of important urban forest values. It is hoped that this indicator set could provide a foundation for guiding planning decisions for global future urban forests.

The results indicate that social indicators, such as human health and well-being, are important considerations when planning urban forests. Recent research demonstrates that these values are not always included in urban forest valuation [29]. When included, social indicators were rated highly by participants, and made up a large portion of suggested additional indicators. More research into additional social indicators, both enabling and performance indicators, could yield measurement tools that better reflect values held about urban forests.

One participant pointed out that the indicators were "operating at different levels" and suggested following McPherson et al.'s conceptual approach of structure, function, and value [35], to categorize the indicators. Table 5 divides the indicators into structural and functional indicators.

Table 5. Structural versus functional indicators.

Structural Indicators	Functional Indicators
Urban Tree Diversity	Stormwater Management
Physical Access to Nature	Habitat Provision
Canopy Cover	Air Quality Improvement
Visual Access to Nature	Greenhouse Gas Sequestration and Storage
Available Growing Space	

The structural indicators can be conceptually divided into diversity, distribution, and density indicators. "Urban tree diversity" was the most highly rated indicator, which reflects the argument that urban forests of diverse species and ages "provide a wider range of benefits over the long term" [22] (p. 108). Distribution indicators, such as physical and visual access to nature, begin to describe where urban forest components should be located to provide benefits for all urban citizens. Finally, density indicators such as "canopy cover" and "available growing space" direct the amount of space dedicated to the urban forest resource. The structural indicators are easily categorized as "enabling" indicators; they direct urban forest design and planning in a spatially explicit way.

The functional indicators are not as easily spatialized, or conceived as "enabling" indicators. Fortunately, much recent research in urban forestry provides design directions to optimize these indicators [2–8,10]. We can use this research to make design choices that optimize storm water management (leaf area index, thresholds for impervious surface area), habitat provision (ideal patch sizes, connectivity corridors), air quality improvement (low VOC species, tree location near pollution sources), and greenhouse gas storage and sequestration (tree location to cool buildings, tree location to enhance physical activity). When combined (see Table 6), the structural and functional indicators, when used with targets set by a community, create a comprehensive set of instructions to guide planning and design of urban forests.

Table 6. Final indicators.

Selected Indicators
Urban Tree Diversity
Physical Access to Nature
Canopy Cover
Stormwater Control
Habitat Provision
Air Quality Improvement
Visual Access to Nature
Available Growing Space
Greenhouse Gas Sequestration and Storage

Future Research

Academic experts from a wide range of social, political, and ecological contexts came to consensus on a set of indicators that can be used to guide the design of future urban forests. Their results were quite similar to a group of practitioner experts from one region, indicating that these values are shared, not just within the academic community, but also with a wide range of professionals working in the field of urban forestry.

This study is the first part of a three phase research project that intends to test the physical arrangement of forest and urban form components in order to understand where there are synergies and where there are conflicts between these components in planning for future sustainable and resilient communities. The next phase of this project will test the ability of these indicators to guide design and planning decisions through the creation of a set of future forest scenarios for a community in the Metro Vancouver region. The scenarios will provide a method for evaluating trade-offs and conflicts between different urban forest structures, a need addressed by McPherson et al. [35]. Once the scenarios are developed, the same indicators will be used to measure the performance of each scenario. The indicators can then be used to compare the scenarios and assess the potential co-benefits or trade-offs between scenarios.

5. Conclusions

This study developed a framework, including a short list of key indicators, to guide planning and design of future urban forests. Experts and practitioners rated a set of indicators in terms of their importance for urban forest planning and design. As Dobbs et al. argue, indicators "are one approach that could be used to better understand the structure of an urban forest, the suite of ESG provided by urban forests, and their influence on human well-being using a simple, innovative and repeatable metric" [21] (p.1). This paper extends this idea to include indicators as inputs to guide design and planning of urban forest structure. The ranking of the indicators within this study reveals a range of values that are important to capture when planning and designing future urban forests. When planting trees, communities should think beyond basic canopy cover targets to focus on tree diversity, distribution, and other design requirements to maximize ecosystem services provided by urban forests, including social benefits such as human health and well-being.

Acknowledgments: The author would like to thank all participants for volunteering their time to answer the surveys. Funding for this project was provided through the University of British Columbia's Future Forest Fellowship.

Conflicts of Interest: The authors declare no conflicts of interest.

Appendix

Table A1. Existing Urban Forest Indicator Sets.

	i-Tree	Dobbs et al. 2011	Kenney et al. 2011	Clark et al. 1997	USDA Forest Health Indicators. (Woodall et al. 2011)	Parameters Östberg (2011)
Canopy Cover/volume	Number of Trees; Tree Density	Tree Canopy Cover; Tree Structure	Relative Canopy Cover	Canopy cover		Scientific name:tree specie & genera; Vitality
Diversity	Species Composition	Shannon diversity index	Species distribution; Age distribution	Species mix; Age distribution	Vegetation diversity	Coordinates; Hazard class
Tree Health/Risk	Tree Health; Pests risk analysis	Crown dieback; Damage to infrastructure	Condition of publicly owned trees		Crown condition; Ozone injury	Identification number; Presence of fruit bodies
Air Quality Improvement	Ozone, sulfur dioxide, nitrogen dioxide, carbon monoxide, particulate matter (<10 microns) removal; VOC emissions; Economic benefit based on effect of trees on air quality improvement	Air pollutant removal; Decrease in air quality; Pm 10 removal				Date of latest inventory
Allergenicity	Tree pollen allergenicity index	Allergenicity				Category of care; Conservation value
Greenhouse gas storage/sequestration	Carbon stored; Net carbon annually sequestered	C02 sequestration			Down woody material	Street or park trees; Age class
Energy conservation	Effects of trees on building energy use + consequent effects on co2 emissions	Temperature Reduction				Stem circumference at 1 m height at planting; Date of planting
Stormwater management	Canopy rainfall interception summarized by species or land use	Soil Infiltration				Name of disease or pest
Property value benefits	Property value increase					Reason for felling
Habitat provision		Ratio of native trees	Native vegetation	Native vegetation	Lichen communities	
Human health/well-being	Public health incidence reduction	Recreation cover				
Economic Value	Replacement value					
Soils/growing space	Soil Infiltration; Soil fertility; Soil bulk density; Soil nutrients; Heavy metals				Soils	
Other	Leaf area and distance to roads; Type of foliage; Curve Number; Fruit fall; Tree biomass	Species suitability	Species suitability			

267

Table 2. Decision Support Framework.

Vision	Goal	Objective	Proposed Indicators
Healthy and resilient urban forests that contribute to residents' well-being, climate mitigation, and ecosystem services.	A forest that adapts to predicted climate change	The urban forest is resilient to predicted changes in weather, water availability, and/or potential invasion by insects and diseases	Urban Forest Diversity; Available Growing Space; Canopy Cover; Tree Risk
	A forest that helps mitigate future climate change	The urban forest supports a community that releases fewer greenhouse gases through daily transportation and building energy uses, while storing and sequestering optimal amounts of carbon dioxide	Greenhouse Gas Sequestration and Storage; Energy Conservation; Air Quality Improvement
	A forest that contributes to local residents' well-being	The urban forest supports health and well-being by providing a variety of recreation opportunities, an aesthetically pleasing environment that lowers stress levels and maximizes filtration of air pollution	Visual Access to Nature; Physical Access to Nature; Property Value Benefits
	A forest that supports a resilient local ecosystem	The urban forest supports local flora and fauna through habitat and food provision, while infiltrating storm water to support health local streams and rivers	Storm water Control; Habitat Provision

References

1. Dobbs, C.; Kendal, D.; Nitschke, C.R. Multiple Ecosystem Services and Disservices of the Urban Forest Establishing Their Connections with Landscape Structure and Sociodemographics. *Ecol. Indic.* **2014**, *43*. [CrossRef]
2. Escobedo, F.J.; Kroeger, T.; Wagner, J.E. Urban Forests and Pollution Mitigation: Analyzing Ecosystem Services and Disservices. *Environ. Pollut.* **2011**, *159*, 2078–2087. [CrossRef] [PubMed]
3. Nowak, D.J.; Hoehn, R.E.; Bodine, A.R.; Greenfield, E.; Ellis, A.; Endreny, T.A.; Yang, Y.; Zhou, T.; Henry, R. *Assessing Urban Forest Effects and Values: Toronto's Urban Forest*; Resource Bulletin NRS-79; Department of Agriculture, Forest Service, Northern Research Station: Newtown Square, PA, USA, 2013; p. 59. Available online: http://www.nrs.fs.fed.us/pubs/43543 (accessed on 29 April 2016).
4. Gill, S.E.; Handley, J.F.; Ennos, A.R.; Pauleit, S. Adapting Cities for Climate Change: The Role of the Green Infrastructure. *Built Environ.* **2007**, *33*, 115–133. [CrossRef]
5. Hall, J.M.; Handley, J.F.; Ennos, A.R. The Potential of Tree Planting to Climate-proof High Density Residential Areas in Manchester, UK. *Landsc. Urban Plan.* **2012**, *104*, 410–417. [CrossRef]
6. Akbari, H. Shade Trees Reduce Building Energy Use and CO_2 Emissions from Power Plants. *Environ. Pollut.* **2002**, *116* (Suppl. 1), S119–S126. [CrossRef]
7. Nowak, D.J.; Crane, D.E. Carbon Storage and Sequestration by Urban Trees in the USA. *Environ. Pollut.* **2002**, *116*, 381–389. [CrossRef]
8. Yesilonis, I.D.; Pouyat, R.V. Carbon Stocks in Urban Forest Remnants: Atlanta and Baltimore as Case Studies. In *Carbon Sequestration in Urban Ecosystems*; Rattan, L., Bruce, A., Eds.; Springer: Dordrecht, Netherlands, 2012; pp. 103–120.
9. Donovan, G.H.; Butry, D.T.; Michael, Y.L.; Prestemon, J.P.; Liebhold, A.M.; Gatziolis, D.; Mao, M.Y. The Relationship between Trees and Human Health: Evidence from the Spread of the Emerald Ash Borer. *Am. J. Prev. Med.* **2013**, *44*, 139–145. [CrossRef] [PubMed]
10. Kaplan, R. The Nature of the View from Home Psychological Benefits. *Environ. Behav.* **2001**, *33*, 507–542. [CrossRef]
11. Hartig, T.; Richard, M.; Sjerp, V.; Howard, F. Nature and Health. *Annu. Rev. Public Health* **2014**, *35*, 207–228. [CrossRef] [PubMed]
12. Eigenbrod, F.; Bell, V.A.; Davies, H.N.; Heinemeyer, A.; Armsworth, P.R.; Gaston, K.J. The Impact of Projected Increases in Urbanization on Ecosystem Services. *Proc. R. Soc. B Biol. Sci.* **2011**, *278*, 3201–3208. [CrossRef] [PubMed]
13. Turner, W.R.; Nakamura, T.; Dinetti, M. Global Urbanization and the Separation of Humans from Nature. *BioScience* **2004**, *54*, 585–590. [CrossRef]
14. City of Vancouver. Developing Vancouver's Urban Forest Strategy, City of Vancouver. 2015. Available online: http://vancouver.ca/home-property-development/urba-forest-strategy.aspx (accessed on 22 April 2016).
15. MillionTrees NYC. Available online: http://www.milliontreesnyc.org/html/home/home.shtml (accessed on 22 April 2016).
16. Kellett, R.; Fryer, S.; Budke, I. Specification of Indicators and Selection Methodology for a Potential Community Demonstration Project. 2012. Available online: http://www.dcs.sala.ubc.ca/docs/cmhc_sustainability_indicators_final_report_sec.pdf (accessed on 22 April 2016).
17. Harshaw, H.W.; Sheppard, S.; Lewis, J.L. A Review and Synthesis of Social Indicators for Sustainable Forest Management. *J. Ecosyst. Manag.* **2007**, *8*, 17–37.
18. Gough, A.D.; Innes, J.L.; Allen, S.D. Development of common indicators of sustainable forest management. *Ecol. Indic.* **2008**, *8*, 425–430. [CrossRef]
19. Sheppard, S.R.; Meitner, M. Using multi-criteria analysis and visualisation for sustainable forest management planning with stakeholder groups. *For. Ecol. Manag.* **2005**, *207*, 171–187. [CrossRef]
20. Kellett, R. *Sustainability Indicators for Computer-Based Tools in Community Design, 1*; Canada Mortgage and Housing Corporation: Ottawa, ON, Canada, 2009.
21. Dobbs, C.; Escobedo, F.J.; Zipperer, W.C. A framework for developing urban forest ecosystem services and goods indicators. *Landsc. Urban Plan.* **2011**, *99*, 196–206. [CrossRef]
22. Kenney, W.A.; Van Wassenaer, P.J.; Satel, A.L. Criteria and indicators for strategic urban forest planning and management. *Arboric. Urban For.* **2011**, *37*, 108–117.

23. Clark, J.R.; Matheny, N.P.; Cross, G.; Wake, V. A model of urban forest sustainability. *J. Arboric.* **1997**, *23*, 17–30.
24. Östberg, J.; Delshammar, T.; Wiström, B.; Nielsen, A.B. Grading of Parameters for Urban Tree Inventories by City Officials, Arborists, and Academics Using the Delphi Method. *Environ. Manag.* **2013**, *51*, 694–708. [CrossRef] [PubMed]
25. Woodall, C.W.; Amacher, M.C.; Bechtold, W.A.; Coulston, J.W.; Jovan, S.; Perry, C.H.; Randolph, K.C.; Schulz, B.K.; Smith, G.C.; Tkacz, B.; et al. Status and Future of the Forest Health Indicators Program of the USA. *Environ. Monit. Assess.* **2011**, *177*, 419–436. [CrossRef] [PubMed]
26. Nowak, D.J.; Dwyer, J.F. Understanding the benefits and costs of urban forest ecosystems. In *Urban and Community Forestry in the Northeast*; Springer: Dordrecht, Netherlands, 2007; pp. 25–46.
27. Landeta, J. Current Validity of the Delphi Method in Social Sciences. *Technol. Forecast. Soc. Chang.* **2006**, *73*, 467–482. [CrossRef]
28. Linstone, H.A.; Turoff, M. (Eds.) *Delphi Method: Techniques and Applications*; Addison-Wesley: Boston, MA, USA, 1975.
29. Peckham, S.C.; Duinker, P.N.; Ordóñez, C. Urban Forest Values in Canada: Views of Citizens in Calgary and Halifax. *Urban For. Urban Green.* **2013**, *12*, 154–162. [CrossRef]
30. Takano, T.; Nakamura, K.; Watanabe, M. Urban Residential Environments and Senior Citizens' Longevity in Megacity Areas: The Importance of Walkable Green Spaces. *J. Epidemiol. Commun. Health* **2002**, *56*, 913–918. [CrossRef]
31. Bowler, D.E.; Buyung-Ali, L.M.; Knight, T.M.; Pullin, A.S. A Systematic Review of Evidence for the Added Benefits to Health of Exposure to Natural Environments. *BMC Public Health* **2010**, *10*, 456. [CrossRef] [PubMed]
32. Bell, J.F.; Wilson, J.S.; Liu, G.C. Neighborhood Greenness and 2-year Changes in Body Mass Index of Children and Youth. *Am. J. Prev. Med.* **2008**, *35*, 547–553. [CrossRef] [PubMed]
33. Lovasi, G.S.; Jacobson, J.S.; Quinn, J.W.; Neckerman, K.M.; Ashby-Thompson, M.N.; Rundle, A. Is the Environment near Home and School Associated with Physical Activity and Adiposity of Urban Preschool Children? *J. Urban Health* **2011**, *88*, 1143–1157. [CrossRef] [PubMed]
34. Ulrich, R.S. Human Responses to Vegetation and Landscapes. *Landsc. Urban Plan.* **1986**, *13*, 29–44. [CrossRef]
35. McPherson, E.G.; Nowak, D.; Heisler, G.; Grimmond, S.; Souch, C.; Grant, R.; Rowntree, R. Quantifying urban forest structure, function, and value: The Chicago Urban Forest Climate Project. *Urban Ecosyst.* **1997**, *1*, 49–61. [CrossRef]

MDPI AG

St. Alban-Anlage 66

4052 Basel, Switzerland

Tel. +41 61 683 77 34

Fax +41 61 302 89 18

http://www.mdpi.com

Forests Editorial Office

E-mail: forests@mdpi.com

http://www.mdpi.com/journal/forests